AF776840

CHINA – STRATIGRAPHY, PALEOGEOGRAPHY AND TECTONICS

China – Stratigraphy, Paleogeography and Tectonics

Arthur A. Meyerhoff
International Geological Consultant, Tulsa, Oklahoma, U.S.A.

Maurice Kamen-Kaye
Consulting Geologist, Cambridge, Massachusetts, U.S.A.

Chin Chen
Graduate Program of Oceanography and Limnology, Western Connecticut State University, Danbury, Connecticut, U.S.A.

and

Irfan Taner
Consulting Geologist, Tulsa, Oklahoma, U.S.A.

SPRINGER SCIENCE+BUSINESS MEDIA, B.V.

Library of Congress Cataloging-in-Publication Data

China--stratigraphy, paleogeography, and tectonics / Arthur A.
Meyerhoff ... [et al.].
p. cm.
Includes index.
ISBN 978-94-010-5678-6 ISBN 978-94-011-3770-6 (eBook)
DOI 10.1007/978-94-011-3770-6
1. Geology--China. 2. Geology, Stratigraphic. 3. Paleogeography--China. 4. Geology, Structural--China. I. Meyerhoff, A. A.
(Arthur Augustus), 1928-
QE294.C4834 1991
555.1--dc20 90-5340

ISBN 978-94-010-5678-6

Printed on acid-free paper.

All Rights Reserved
© 1991 by Springer Science+Business Media Dordrecht
Originally published by Kluwer Academic Publishers in 1991
Softcover reprint of the hardcover 1st edition 1991
No part of the material protected by this copyright notice may be reproduced or utilized in any form or by any means, electronic or mechanical including photocopying, recording or by any information storage and retrieval system, without written permission from the copyright owner.

Dedicated with gratitude to Professor Konrad B. Krauskopf, who taught the senior author the geology of China in the autumn of 1949.

Table of Contents

Preface

This volume concerns the geology of China, and it examines that concern by expositions of the stratigraphy, the paleogeography, and the tectonics of that remarkable country. In this sense, therefore, our aims and purposes are explicit in the title. The senior author and his colleagues, furthermore, do not have in mind any special or specific audience. This volume is quite simply for all geologists. By far the majority will be those whose native tongue is English, or those who understand English. Not to be overlooked, moreover, is the large number of Chinese geologists who not only read English but also who themselves write studies in English that appear in publications in both their homeland and abroad.

A constantly growing interest in the geology of China makes likely the appearance of several books on the subject in the remaining ten years of this century. Our own studies date back to 1949, when the senior author attended a course on the geology of China. From that time to this, he, and later he and his colleagues, have assumed, compiled and synthesized every significant presentation in geology and geophysics, including many written in Chinese. All four of us, additionally, have practised petroleum geology for many years, and all have published on a variety of subjects in that field. This is a type of experience that has brought about an appreciation of the vertical parameter in structure, and more, an understanding of the special meaning of the third dimension in geology.

In stratigraphy, we deal amply with China's regional patterns, from Archean gneiss more than three billion years old, through Proterozoic deposits, and onward in time through the Phanerozoic to the present. As few others are likely to do, two of our contributors have consulted twelve volumes of stratigraphy and peleontology written in Chinese in the 1950s, with the result that important data and concepts appear in our own compendium. All lithostratigraphic and chronostratigraphic systems are present at outcrop in China. While all such systems are important, the Proterozoic column possibly is unique in its continuous sedimentary development and in its reference section of global rank.

In paleogeography, this volume describes and illustrates first the broad distribution of Proterozoic deposits. Succeeding descriptions and illustrations trace the ebb and flow of shallow marine waters across China as Phanerozoic time of more than 600 million years elapses from the beginning of the Cambrian to the present.

In structure, this volume emphasizes the importance of paraplatforms, platforms, geosynclines, and great east-west zones of fracture in the Precambian, also the effects of these early structural elements on structure in the ensuing Phanerozoic. In the Phanerozoic itself, north-south stress developed in the pre-Phanerozoic continued through much of the Paleozoic. During the Mesozoic and the Cenozoic, on the contrary, the direction of stresses in many parts of the eastern half of China swung to NW-SE and finally to WNW-ESE. Both of these directions, and the systems of horsts and grabens that appeared with them, constitute significant factors in the accumulation of important deposits of oil and gas.

We believe that this study will prove to be distinguished by the alternatives that it offers to prevailing explanations put forward in paleogeography and structure. In paleogeography, explanations that depend on geodynamic separation of Gondwana on the south from Tethys on the north appear to us invalid on a full and careful study of the evidence. In structure, explanations that depend on geodynamic separation of large discrete blocks in some cases, on an geodynamic collisions in others, appear to us equally invalid. Cohesion, in our view, has characterized not only China itself but also large regions of greater Asia, through more than three billion years of geological passage.

May 20, 1990
Maurice Kamen-Kaye

Acknowledgements

The first of many acknowledgements is to our families for their support and understanding during the accomplishment of this book.

Numerous colleagues provided us with advice, suggestions, critiques, lengthy discussions of controversial points, and critical literature. Many of them gave us hours, days, and even weeks of their time to make this a better publication. Most notable among these are the late William B. Beatty, Huang Jiqing (C. K. Huang), K. Y. Lee, the late Peter Misch, the late Theodore Shabad, Curt Teichert, K. P. Wang, Jan-Olaf Willums, and Yang Qi. Important data and/or literature were supplied by numerous other colleagues, including M. Ismail Bhat, Arthur J. Boucot, Sankar Chatterjee, Stanley S. L. Chang, Ashok K. Dubey, Mrs. Nobu Ikeno Farrill, Fu Jiamo, Gerhard Fuchs, Augusto Gansser, Stephanie Green, John D. Haun, Edward Kinderman, Charles D. Masters, James T. Murray, Theodor Oresianu, James A. Peterson, Richard B. Powers, Takashi Saito and his colleagues with the Japan National Oil Company, M. N. Saxena, Brian J. Skinner, Gregory Ulmishek, Wang Longjiang, and L. A. Zawadýnsky. Prof. Zhang Qinwen read the manuscript and made several suggestions.

We thank Debra Richardson Burdeaux, Kathryn L. Meyerhoff, and Zoe Rasmussen for drafting. Typing was done by Donna Meyerhoff Hull, Nancy Runge, Ernestine R. Voyles, and Joyce Sober.

The fact that we have listed these persons' names does not in any way imply that they agree with our conclusions. The conclusions are ours and ours alone.

Chapter 1

Introduction

China occupies a major area of Eastern Asia and manifests a considerable variety of geographic and geomorphologic forms. Within its boundaries there flourished and developed some of the oldest civilizations in human history. In the spheres of technology and of utilization, the usage of metals dates back to the period of 2000 B.C. (Yang Zunyi et al., 1986). Technology continued to flourish in China through the centuries, and before 771 B.C. Chinese engineers had accomplished the feat of extracting hydrocarbons mechanically far in advance of other civilizations. Primitive tools were used to drill wells by hand (Li Kexiang, 1980). Confucius was among those who wrote of the drilling of deep wells for the recovery of salt brine, in the Sichuan basin of Southwest China about 600 B.C. (Owen, 1975): at depths of some 200 meters these wells were not only deep but also ultra-deep for their times. Strubell (1968) presented evidence that the drilling of a well specifically for gas took place also in the Sichuan basin at least as long ago as 211 B.C. Cheng Xirong (1986), in addition, reported that the tract 'Peaceful Miscellanea' of the North Song Dynasty written in 980 A.D. contains records of oil wells drilled in today's Yanchang County, Shaanxi Province (Eerduos basin).

Geological observations by Chinese scholars, in addition to the above achievements in economic geology, are recorded in Chinese literature as far back as at least 500 B.C. Needham (1959) summarized systematically all descriptions related to geology in the early civilizations. Of these summaries we give two examples cited by Grabau (1924).

1. 'Said Ma-Ka unto Wang-fan-pien: 'Since I followed you I have three times seen the Eastern Sea become a land of mulberry trees.' written by Ke-hung, 400 A.D.
2. 'In high mountains there are shells. They probably occur in the rocks which are the soils of older days, and the shells once lived in the water. The low places became high, and the soft mud turned into hard rock.' written by Chu-hsi, 1200 A.D.

Development of modern geology in China can be divided into three periods: (1) pre-1912; (2) 1912-1949; and (3) 1949 to present. During the first period, with no official institutions in China to carry out geological research and exploration, these were conducted mainly by foreign scientists. The first modern research in stratigraphy and paleontology in China began with the work of Pumpelly in the nineteenth century according to Grabau (1924). This was followed by the famous expedition of Ferdinand F. von Richthofen and his colleagues (1860-1872), an expedition which laid the foundation of Chinese geology. Five years later, L. von Loczy (1877-1880) and his team explored many regions not covered by von Richthofen's expedition and thus complement it. After these initial expeditions, several more were launched by scientists from various countries. Prominent among these are V. A. Obrutschew (1892-1909), Sven Hedin (1893-1908, 1929-1933), and B. Willis and E. Blackwelder (1903-1904).

During the second period (1912-1949), Chinese geological institutions were established. In 1912 two organizations were founded under the Ministry of Agriculture and Commerce: the first of these two was the section of geology that later became the Geological Survey of China, and the second was the Geological Institute, that was to educate geologists. Huang Jiqing (1986) briefly reviewed geological sciences in China for this period and the next (1949 to present) in his speech to the Geological Society of China's 60th-year celebration in 1982. Until 1949 the number of geologists in China did not pass a few hundreds (300 to 400), but their accomplishments were considerable.

Since 1949 (date of the foundation of the People's Republic of China), geological sciences have developed and improved both in quality and in quantity. The number of educational and research institutions as well as economic institutions (dealing with geology, oil, coal, mining companies) has increased dramatically to meet the demands of this most populous country in the world. In the early 1980s, about 60,000 geologists were active in geological sciences in China, a number that is likely to grow steadily through the decade of the 1990s.

Fig. I–1. Administrative-economic regions and provinces of People's Republic of China.

Physical Geography

In this section we describe and discuss briefly certain aspects of the physical features of China, using Zhao Songqiao (1986) as our main source of information.

Area and population

China, which covers 9.6 million square kilometers (6.5 percent of the earth's land surface) in eastern Asia, lies between longitudes 54° North and 18° North and between latitudes 73° East and 139° East. Its territory measures about 5,500 kilometers east-west. The land boundary that it shares with its neighbors measures about 22,800 kilometers in length. These neighbors are Korea to the northeast, Soviet Union to the northeast and northwest, People's Republic of Mongolia to the north, Afghanistan and Pakistan to the west, India, Nepal, Sikkim, and Bhutan to the southwest, and Burma, Laos, and Vietnam to the south (Figure I-1). In the east and the southeast, marginal seas (Yellow Sea, East China Sea, and South China Sea) separate China from Japan, Philippines, Malaysia, and Indonesia. The total length of coastline is about 18,000 kilometers.

China is divided into six administrative economic regions, and these are divided additionally into twenty-nine administrative units (Figure I-1, Table I-1; Xue Muqiao, ed., 1982). Some authors classify Qinghai Province and Xizang (Tibet) Autonomous Region together as a Qinghai-Xizang Plateau administrative economic region (e.g. Zhao Songqiao, 1986). In the ensuing text, where we refer to an official administrative division, we use capital letters (e.g. North China). For areas less exact than the official division we use small letters. Where we wish to designate an area not greatly different from the official one we use the shorter of two adjectives, e.g. 'north China;' for a more diffuse areas we use the longer adjective, e.g. 'northern China.'

China is the most populous country in the world, with a census of 1,008,880,000 as of July 1, 1982. Although about eighty percent of the population lives in rural areas, urban population is increasing drastically due mainly to emigration from such rural areas to the cities. Most of the population at present is concentrated in

Table I-1. Administrative-Economic divisions of China. These divisions are shown on Figure I-1 (from Xue Muqiao, 1982)

NORTH CHINA	CENTRAL-SOUTH CHINA
1. Beijing Municipality	16. Henan Province
2. Tianjin Municipality	17. Hubei Province
3. Hebei Province	18. Hunan Province
4. Shanxi Province	19. Guangdong Province
5a. Western Half of Nei Monggol Autonomous Region	20. Guangxi Zhuang Autonomous Region
NORTHEAST CHINA	SOUTHWEST CHINA
5b. Eastern half of Nei Monggol Autonomous Region	21. Sichuan Province
	22. Guizhou Province
	23. Yunnan Province
6. Liaoning Province	24. Xizang (Tibet) Autonomous Region
7. Jilin Province	
8. Heilongjiang Province	
EAST CHINA	NORTHWEST CHINA
9. Shanghai Municipality	25. Shaanxi Province
10. Jiangsu Province	26. Gansu Province
11. Zhejiang Province	27. Qinghai Province
12. Anhui Province	28. Ningxia Hui Autonomous Region
13. Jiangxi Province	
14. Fujian Province	29. Xinjiang Uygur Autonomous Region
15. Shandong Province	

North China, East China, and Central-South China.

Physiography

China's physical features vary in type and size. Mountains in China occupy about thirty-three percent of the total area, plateaux about twenty-five percent, hills about ten percent, basins about nineteen percent, and plains about twelve percent. Topographic relief in China decreases from west to east in stepwise fashion. Average elevation of the Qinghai-Xizang (Tibet) Plateau, 4,500 meters in the west, is followed by a region of plateaux and basins with average elevations of 1,000 to 2,000 meters across to the Da Hinggan-Taihang-Wu line of mountains (Figure I-2). The next step, between the Da Hinggan-Taihang-Wu line of mountains and the coast line, is represented by great plains whose elevation is below 500 meters. Interestingly enough, these steps of topography are reflected well on the map of Mohorovicic discontinuity, for which refer to our chapter on tectonics and structure.

The mountains in China can be grouped into four systems: (1) the east-west mountain system (also called latitudinal system), whose major elements are the Tian Shan-Yin Shan-Yan Shan, the Kunlun Shan-Qin Ling - Dabie Shan, the Tanggula Shan, the Gangdise Shan, the Himalaya, and the Nan Ling; (2) the north-south mountain system, which includes the Helan Shan, the Liupan Shan, the Longmen Shan and the Hengduan Shan; (3) the northeast-southwest and north-northeast-south-southwest mountain system, dominant only to the east of the north-south mountain system. The prominent elements of this system are the Da Hinggan Ling, Taihang Shan, Wushan, and Wuyi Shan; and (4) the northwest-southeast mountain system, examples of which are the Altay Mountains, the Qilian Shan, and the Xiao Hinggan Mountains.

Unusually large physiographic basins exist in central and northwestern China (Figure I-2), and these essentially are intermontane in character (e.g. Junggar, Tarim, Sichuan).

The Qinghai-Xizang (Tibet) Plateau lies in southwestern China, where it covers about 2.5 millions square kilometers (over twenty-five percent of the total land area of China). Elevations in the Qinghai-Xizang Plateau average 4,000 to 4,500 meters, reason enough to call this great area the 'roof of the world.' Three more plateaux lie in China, and three of them are aligned north-south in central China; the Nei Monggol Plateau and the Loess Plateau north of the Sichuan basin, and the Yunnan-Guizhou Plateau south of the Sichuan basin. The elevations of these plateaux range from 1,000 to 2,000 meters.

The plains and the lesser hills lie mainly in eastern China.

Climate

Because of its vast area and its complex topographic features, China experiences different and greatly contrasted types of climates. On the whole, however, the Qinghai-Xizang (Tibet) Plateau exerts a most powerful influence on climate in China.

China generally is divided into three climatic realms: (1) Eastern Monsoon realm, (2) Northwestern Arid realm, and (3) Qinghai-Xizang (Tibet) Frigid Plateau realm. The eastern monsoon realm naturally is subject to its monsoon winds, but also to decreases of temperature from south to north with sharp differences in winter, and with plentiful precipitation. The eastern monsoon realm is subdivided further into zones such as temperate, humid, and subhumid, warm-temperate humid and subhumid, subtropical humid, and tropical humid regions from north to south.

The northwest arid realm spreads from the Da Hinggan Ling westward to the Junggar and the Tarim basins, and it is characterized by moderate solar radiation, sharply different temperature between summer and winter, and by low precipitation. It is subdivided into two regions; temperate grassland, and warm-temperate and temperate desert.

The Qinghai-Xizang (Tibet) frigid plateau realm experiences low temperature (mean temperature below 0°C), high solar radiation, strong winds, and precipitation that ranges from moderate in the south to low in the north. The high peaks in this realm are snow-capped, and permafrost is widely distributed.

Hydrology

China contains about 50,000 rivers with drainage area greater than 100 square kilometers, and of these about

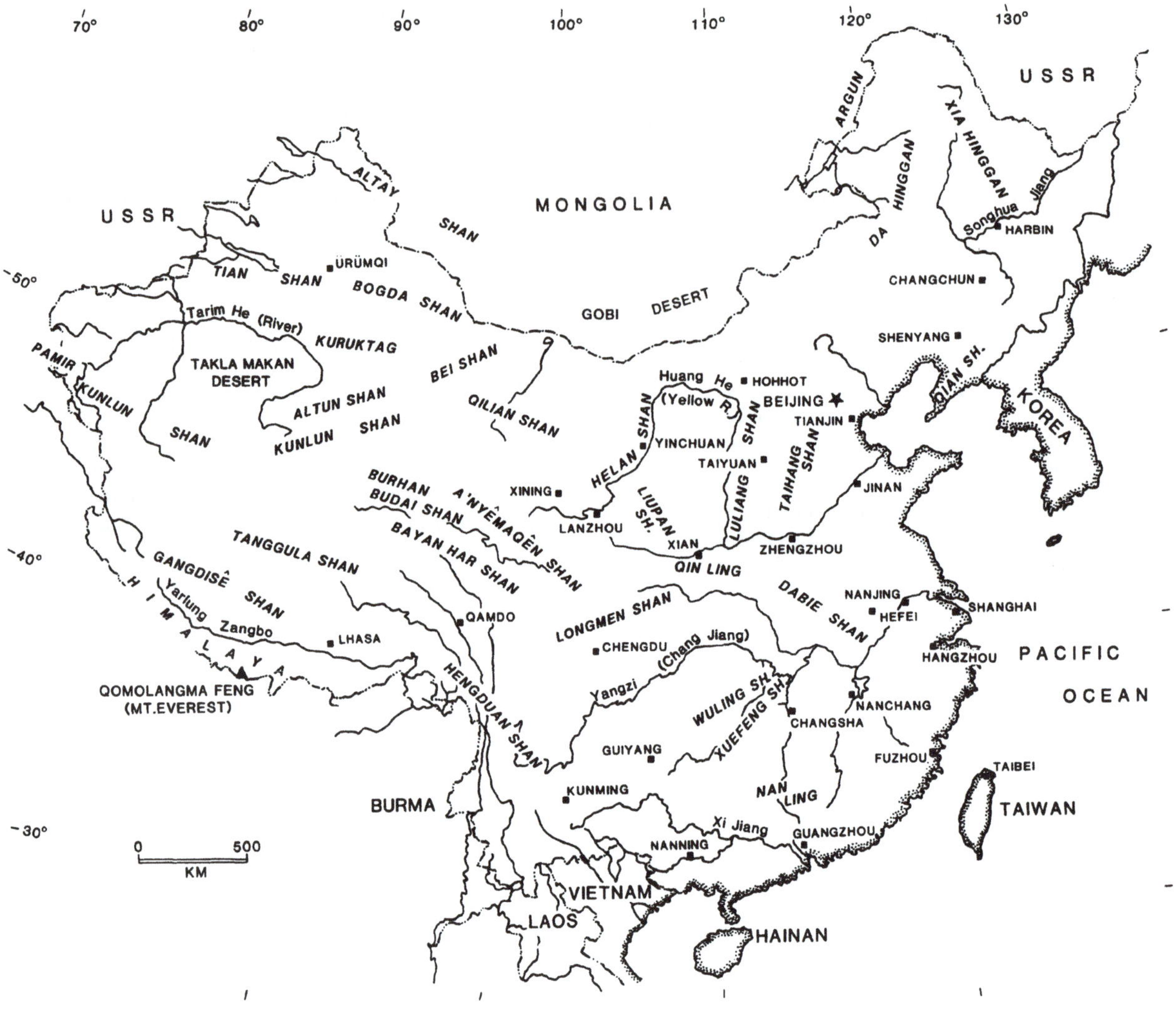

Fig. I-2. Major physical features of China.

1,500 possess a drainage area greater than 1,000 square kilometers. The Chang Jiang (Yangzi River), the largest, receives waters from a drainage area of 1.8 million square kilometers, and it flows along a total length of 6,300 kilometers.

Rivers in China divide into two large systems: inland and oceanic. The great divide between inland (internal) and oceanic (external) river systems in China runs essentially northeast-southwest. Beginning at the southern end of the Da Hinggan Mountains of northeastern China, this divide passes through the Yan Shan, Helan Shan, eastern Qilian Shan, Bayan Har, Nyainqentanglha and Gangdise mountains. The inland system, thirty-six percent of the total land area of China, lies mainly in Northwest China and in the northwestern area of the Qinghai-Xizang Plateau, with only a few perennial rivers. All inland rivers flow into saline inland lakes, or die away in sandy deserts or salt marshes. The inland river system in China is subdivided into four drainage areas: (1) Nei Monggol, (2) Gansu-Xinjiang, (3) Qaidam Basin, and (4) northern Qinghai-Xizang.

The oceanic river system in China, about sixty-four percent of the total land area of China, is subdivided into three drainage basins: Pacific, Indian, and Arctic. The Pacific drainage basin, the most important, accounts for 88.9 percent of the oceanic system and 56.7 percent of the total land area of China. A majority of large rivers in China flow into the Pacific Ocean. Among these are the Heilong Jiang (Jiang = River), Chang Jiang, Huang He (Yellow River; He = River), Zhujiang, and Yuan Jiang (Red River).

The Indian drainage subsystem occupies 10.3 percent of the total oceanic system and 6.5 percent of the total land area of China. Rivers flowing into the Indian Ocean come from upper reaches in China, and these are the Nu Jiang (Salween), the Yarlung Zangbo (Brahmaputra River), and the Shiquan (Indus River).

The arctic subsystem in China contains only one large river, the Ertix River, in the northern Junggar basin of Northwest China. The Ertix River is tributary to the Ob River which itself flows into the Kara Sea of the Arctic Ocean.

China contains also a large number of lakes, 2,800 natural lakes each with an area greater than one square kilometer. Total area of these lakes exceeds 80,000 square kilometers.

Transliteration

We use two systems of romanization for the transliteration of Chinese words. Wade-Giles was the most widely used system for many years, but since the 1960s, and especially since the late 1970s, Pinyin, a more phonetic system, has replaced Wade-Giles. In this book we use Pinyin as much as possible, and we take geographic names in Pinyin from the *Atlas of the Provinces of the People's Republic of China*, that is, *Zhonghua Renmin Gonghuego Fen Sheng Dituji* (Beijing (Peking), Atlas Publications, 1977). Some traditional Wade-Giles names are so well known that we attach them in parenthesis, as in *Atlas* ... , cited in the previous sentence. Other examples of this usage are: Tianjin (Tientsin), or Xizang (Tibet). Some names could not be found on maps, and we used whatever romanized word was available. Where this was done the word is followed by (sic).

Locations cited in the text

We have successfully located almost all geographic localities. In some cases, we were obliged to guess, and our guesses are labelled with (?). Political boundaries changed twice during preparation of this book; we hope we have caught the most important changes, but undoubtedly a few have slipped through the net.

Chapter II

Stratigraphy and Paleontology

INTRODUCTION

Sources

Stratigraphy and paleontology in China enjoy the advantage of a solid foundation, particularly that created by the pioneer studies of A. W. Grabau (Zhao, 1961). The two volumes of Grabau's *Stratigraphy of China* appeared in 1923–1924 and 1928 respectively. Both volumes, which present much of China's 'classic' geological information, contain still invaluable references, and they form a lasting monument to the profound influences that Grabau himself exercised upon the geology of China. Also of great importance in the history of China's geology was the publication in 1939 of Li Siguang's (J. S. Lee) *Geology of China* in English.

In 1959 the Chinese National Committee of Stratigraphy convened a symposium on the stratigraphy of China. The twelve volumes that resulted from this symposium appeared in 1962, and these constitute the first significant contributions to the stratigraphy and the paleontology of China since Grabau's time. The author of each volume was a leading Chinese authority (or authorities) on a specific aspect of China's stratigraphy and paleontology. Together, these volumes provide a fund of knowledge that is basic to the geology of Asia, and they also provide a tribute to advances in geology within the People's Republic of China (PRC). Because the text of all twelve volumes is in Chinese, their material remains almost unknown outside of China, especially because abstracts and summaries in other languages have not circulated (Research Group, 1962).

Each volume covers one major age group, and beginning with the oldest division, covers the following topics: (1) the regions in which rocks of each group occur (with or without subdivision of the strata of each age in each region, depending on the volume of knowledge available); (2) a standard lithostratigraphic column for each region; (3) biostratigraphy and zones of index fossils for each series; (4) variations of lithofacies and environments of deposition; (5) classification of each geologic system, and criteria for selecting its upper and lower boundaries; (6) correlations of the lithostratigraphic columns in each region; (7) correlation of the standard columns in China with those in other parts of the world; (8) paleogeography, with or without maps; (9) economic mineral deposits; (10) tectonics and tectogenesis of strata (geosynclines and platforms), including orogenies; and (11) discussion of other significant subjects.

In recent years, two important series of books on biostratigraphy have appeared.

The first series consists of three books, one each by Hao Yichun et al. (1974), Ye et al. (1976), and Gao and Zhao (1976). These books deal mainly with microfossil zonation of the Cretaceous and with assemblages of microfossils in the freshwater Songliao basin of Heilongjiang and Jilin Provinces. Those by Hao Yichun et al. and by Ye et al. deal with fossil ostracods. That by Gao and Zhao deals principally with fossil pollen and fossil spores.

The second series consists of six volumes that deal mainly with lower Tertiary biostratigraphic assemblages of microfossils and nannofossils from coastal Bohai Bay in the eastern part of the North China basins. Hou et al. (1979) described the foraminifera. Wang et al. (1978) described the charophytes. Song et al. (1978a) described the pollen and the spores. Song et al. (1978b) described the dinoflagellates and the acritarchs. Yu et al. (1978) reviewed the gastropods.

Additions to the two series described above are: (1) *Paleontological Stratigraphy* (Fang et al., 1979); (2) *An Outline of the Geology of China* (Compilation Group of the Geological Map of China, Chinese Academy of Geological Sciences, 1976); and (3) *Paleontology in China*, 1979 (Teichert et al., 1981). Fang et al. contains good summations of each geologic era, together with columnar sections and figures for each geologic system, including the Sinian interval of the Proterozoic. The Compilation Group (1976) outlined the structural framework of China, its stratigraphic sequences, and its igneous episodes. An English version of its outline appeared in 1976. The volume by Teichert et al. is entirely in English, and deals with different faunas and floras in China, from Cambrian through Miocene.

Beginning in 1979 and 1980, several significant regional works appeared, among them Han and Yang's (1979, 1980) *Coal Geology of China*, Huang Jiqing et al. (1980) *The Geotectonic Evolution of China*, Wang and Liu (1980) *Historical Geology of China*, and Wang Shangwen's (1983) *Petroleum Geology of China*. In 1986, Yang Zunyi et al. *Geology of China* appeared, the first such book in the English language since J. S. Lee's (Li Siguang) classic work in 1939. These are but five of more than 30 major works that have appeared within the last ten years.

Symbols

In this book, symbols used for the ages of strata, and for groups and formations, are of two types. Letters represent the major divisions of the Precambrian and the major divisions of the Phanerozoic (Table II–1). Subordinate numbers represent subdivision. For example: D_1 refers to Early Devonian rocks; D_2 to Middle Devonian rocks; and D_3 to Late Devonian rocks.

Table II–1. Letter abbreviations for Pre-Cambrian stratigraphic Divisions and Phanerozoic Geologic Systems

System	Letter (s)
Archean (Archeozoic)	Ar
Proterozoic	Pt
Sinian	Z
Cambrian	C
Ordovician	O
Silurian	S
Devonian	D
Carboniferous	C
Permian	P
Triassic	T
Jurassic	J
Cretaceous	Cr
Tertiary	Tr
Paleogene	Pg[a]
Neogene	N
Quaternary	Q

[a] In some cross sections prepared by the Ministry of Petroleum Industry and other petroleum-oriented groups, the letter 'E' is used instead of 'Pg'.

PRE-SINIAN

General (Tables II–2 and II–3)

The traditional pre-Sinian complexes of China (Compilation Group, 1976; Fang et al., 1979; Wang and Qiao, 1984) include Archeozoic (Archean) and lower Proterozoic metamorphic rocks, igneous intrusions, volcanic rocks, and some sedimentary strata. These yield radiometric ages that range from almost 3,200 Ma to 1,800 or 1,700 Ma. Other pre-Sinian rocks include a thick middle and upper Proterozoic section (1,800 to 850 Ma), formerly placed in the Sinian by Grabau (1923–1924). The section younger than the interval 1,800 Ma to 850 Ma, known generally as the Sinian suberathem, has been informally named the 'Jidong sequence' by Huang Jiqing (1984). Wang and Qiao's (1984) English-language summary presents the major changes and revisions made in recent years, some of them for the purpose of correlating Chinese sections more easily with those in other parts of the world. A major treatise on the Precambrian of China appeared in 1985 (Chen Jinbiao, comp., 1985), valuable because a great deal of petroleum formed and accumulated in the Precambrian itself.

Table II–2. Chronostratigraphic subdivision of the Pre-Cambrian and the Pre-Cambrian tectonic stages (from Wang and Qiao, 1984)

ERA AND SUB-ERA	PERIOD	TECTONIC STAGES		OROGENIC MOVEMENTS	GEOLOGICAL EVENTS	SCHEME (PRECAMBRIAN SUBCOMMISSION 1982)
PALEOZOIC	CAMBRIAN	MEGASTAGE III	EARLY CALEDONIAN STAGE	XINGKAIAN	FORMATION OF MOST MEDIAN MASSIFS	CAMBRIAN
						570
						EDIACARAN
	600					
UPPER PROTEROZOIC	SINIAN			CHENGJIANGIAN		700
						CRYOGIAN
	800			JINNINGIAN	FORMATION OF YANGZI PLATFORM	
	QINGBAIKOUAN	MEGASTAGE II	JINNINGIAN STAGE			900
						RIFTIAN
	1,050					
MIDDLE PROTEROZOIC	JIXIANIAN			SIBAOAN		1,200
						UNDESCRIBED
	1,400					1,400
	CHANGCHENGIAN					EUKARYOTIAN
						1,600
						ROUGIAN
	1,850			LULIANGIAN (ZHONGTIAOAN)	FORMATION OF N. CHINA AND TARIM PROTOPLATFORMS	1,800
LOWER PROTEROZOIC	HUTUOAN		LULIANGIAN STAGE			OROGENIAN
						2,100
	2,300			WUTAIAN		
	WUTAIAN					FERRIAN
	2,600			FUPINGIAN	FORMATION OF CONTINENTAL NUCLEI	2,500
UPPER ARCHEAN	FUPINGIAN	MEGASTAGE I	FUPINGIAN STAGE			ARCHEAN
	3,100					
LOWER ARCHEAN	QIANXIAN					
	3,800					
← HADEAN						
	4,500					

From Wang and Qiao (1984)

Table II–3. Stratigraphic correlation chart of Proterozoic (pre-Sinian) in China (from Wang and Qiao, 1984)

REGIONS	(1) YANSHAN WUTAI	(2) ZHONGTIAO-SONGSHAN	(3) LIAODONG	(4) HUAINAN	(5) NEI MONGOL	(6) QUFUK TAGH	(7) BEISHAN-ALXA	(8) CENTRAL QILIAN	(9) SOUTH QAIDAM	(10) WEST HUBEI	(11) CENTRAL YUNNAN	(12) NORTH JIANGXI	(13) CENTRAL ANHUI	(14) NORTH GUANGXI	(15) CENTRAL HUNAN	(16) SOUTH ANHUI	(17) WEST ZHEJIANG
	NORTH CHINA					NORTHWEST CHINA				SOUTH CHINA							
	Cm_1 FUJUNSHAN FM.	Z_2 LUOQUAN FM.	Cm_1 JIANCHANG FM.	Z JIAYUAN FM.	Cm	Z_1 BAIYISI FM.	Z	Cm_2	D_3	Z_1 LIANTUO FM.	Z_1 CHENGJIANG FM.	Z_1 DONGMEN FM.	Z_1 ZHOUGANG FM.	Z_1 CHANGAN FM.	Z_1 JIANGKOU GROUP	Z_1 XIUNING FM.	Z_1 ZHITANG FM.
800 — Pt_3 QINGBAIKOUAN	QINGERYU FM. LONGSHAN FM XIAMALING FM.	LUOYUKOU FM. SANJIAOTANG FM. CUIZHUANG FM.	XIHE GROUP: NANFEN FM., DIAOYUTAI FM. YONGNING GROUP	SHOUXIAN FM. LIULAOBEI FM. WUSHAN FM.		PARGANGTAG GROUP	DAHUOLUO-SHAN GROUP			MACAOYUAN FM.	LIUBATANG FM.	LUOKEDONG FM. ? XIUSHUI GROUP	ZHANGBALING GROUP	DANZHOU GROUP: GONGDONG FM., HETONG FM., BAIZHU FM.	PANXI GROUP: WUQIANGXJ FM., MADIYI FM.	PULING GROUP DENGIJA FM.	SHANGSHU GROUP HONGCHICUN FM. LUOJIAMEN FM.
1,050 — Pt_2 JIXIANIAN	TIELING FM. HONGSHUIZHUANG FM. WUMISHAN FM. YANGZHUANG FM.	BEITAIJIAN FM. BAICAOPIONG FM.				AIRJIGAN GROUP	PINGTOUSHAN GROUP	HUASHISHAN GROUP: BEIMENJIA FM.	BINGGOU GROUP	SHENNONGJIA GROUP: UPPER SUBGR.	KUNYANG GROUP: UPPER SUBGR.	JIULING GROUP: UPPER SUBGR.	BEIJIANGJUN FM.	SIBAO GROUP: YUXI FM., WENTONG FM.	LENGJIAXI GROUP	SHANGXI GROUP: NIUWUZ FM., MUKENG FM.	SHUANGXIWU GROUP
1,400 — CHANGCHENGIAN	GAOYUZHUANG FM. DAHONGYU FM. — 1,650 — TUANSHANZI FM. CHUANLINGGOU FM. CHANGZHOUGOU FM.	YUNMENGSHAN FM. XIONGER GROUP			SHINAGAN FM.	YANGJIBULAK GROUP	BAIHU GROUP	KESUER FM.		LOWER SUBGR.	LOWER SUBGR.	LOWER SUBGR.		JIUXIAO FM. BAIYANTANG FM.		BANQIAO ? FM. ZHANGQIAN ? FM.	
1,850 — Pt_1 HUTUOAN	HUTUO GROUP: GUOJIAZHAI SUBGR., DONGYE SUBGR., DOUCUN SUBGR.	DANSHANSHI GROUP ZHONGTIAO GROUP	LIAOCHE GROUP: QIXIAN FM., DASHIQIAO FM., GAOJIAYU FM. ?	FENGYANG GROUP: SONGJI FM., QINGSHI SHAN FM., YINJIAJIAN FM.	MAJIADIAN GROUP ERDAOWA GROUP	XINGDITAG GROUP ?	LONGSHOU-SHAN (DUNHUANG) GROUP	HUANGYUAN GROUP: QINGSHIPO FM., MOSHIGOU FM.	JINSHUIKOU GROUP		DAHONG-SHAN GROUP		? SUSONG GROUP				
2,300 — WUTAIAN	WUTAI GROUP: PUSHANG FM., TAIHUAI FM., SHIZUI FM.	JIANGXIAN GROUP	LIERYU FM. LANGZISHAN FM.	WUHE GROUP	SANHEMING GROUP	DAKLAKBU-LAK GROUP		DONGCHAGOU FM. LIUJIATAI FM.		SANDOUPING GROUP							
2,500 — Ar	Ar LONGQUANGUAN GROUP		ANSHAN GROUP		Ar WULASHAN GROUP								DABIE GROUP				

From Wang and Qiao (1984)

Terminologies used for the Precambrian are shown on Tables II-2 and II-3.

Distribution

Archeozoic and pre-Sinian Proterozoic complexes occur mainly in three parts of China - the northeastern part, the southeastern, and western China (Research Center, 1962; Figure II-1). The western outcrops are separated from those of the northeast and the southeast by a line of five north-south mountain ranges in Ningxia Huizu Autonomous Region and in the provinces Gansu, Shanxi, Sichuan, and Yunnan. In north-to-south order, these ranges are the Helan Shan, Liupan Shan, Longmen Shan, Hengduan Shan, and Ailoa Shan (Figure I-2). Northeastern and the southeastern China generally are separated by a series of east-west ranges, approximately on strike. These extend from the Kunlun Shan and Qilian Shan on the west through parts of the provinces of Shaanxi, Henan, Hubei, Anhui, and Zhejiang. In order, from west to east, these are the Huaiyang Shan, Qin Ling, Dabie Shan, and Huang Shan (Figure I-2).

The most extensive exposures lie in the northeast.

Fig. II-1. Distribution map of pre-Sinian outcrops of East China.

Exposures in the west thus are more limited in comparison. The least extensive exposures occur in the southeast (Figure II–1). Outcrops of the northern tier of China belong mainly to the Tian Shan tectonic belt on the west and to the Kunlun Shan-Qin Ling tectonic trend from west to center, as well as to the intervening folded uplift.

Wang and Qiao (1984) recognized the same three areas as the Research Group (1962), but subdivided them further into 'domains,' according to the stratigraphy. Thus domain I (Figure III–1) is the northern (Siberia-Mongolia) continental domain (Ia = Junggar-Altay superregion, and Ib = the Hinggan-Nei Monggol superregion); II is the North China continental domain (IIa = Northwest China superregion; IIb = North China superregion); III is the South China continental domain (Yangzi superregion); IV is the southern (Gondwana) continental domain (Xizang or Tibet superregion); and V is the Southeast China continental margin domain (Southeast superregion). The Precambrian basement below these is shown in Figure III–1 together with the major structural elements. We shall note the essential contiguity and continuity of China by the late Proterozoic.

Table II–2 is a chronostratigraphic correlation chart that also shows the tectonic stages of the Chinese Precambrian. Table II–3 presents more detail, region by region, of the Archeozoic (Archean), lower Proterozoic, middle Proterozoic, and most of the upper Proterozoic (pre-Sinian) of China.

Definition of Pre-Sinian

Baron von Richthofen (1882, 1912) invoked China (= Sino) in naming the Sinian System, and assigned to it all unfossiliferous sedimentary rocks between the underlying, metamorphosed, and intruded Wutai Group and the next higher major unconformity in the section – generally above the Ordovician. Thus, as defined originally, the Sinian was a Precambrian to early Paleozoic unit.

Willis and others (1911 – 1913) restricted the name Sinian to Cambro-Ordovician formations, but Grabau (1920) assigned all unmetamorphosed, or only mildly metamorphosed section above the Wutai and below the Cambrian to the Sinian. Although Grabau assigned the section that he called Sinian to the Paleozoic, Grabau's usage continued until the 1950s, after which detailed mapping and successive refinements resulted in a gradual reduction of the Sinian, first from the base of the Changchengian to the base of the Cambrian, and later to the post-Qingbaikouan section (Table II–2; Lee, 1939; Chang Da, 1959 (Chang Ta, 1963); Research Group, 1962; Liu Hung-yun et al., 1973; Compilation Group, 1976; Fang et al., 1979; Wang and Qiao, 1984).

The Sinian has been called 'Eocambrian,' but today most geologists would assign the Sinian to the Proterozoic, and they would correlate it with the Vendian System of the U.S.S.R. Rocks dated 1,800 to 1,700 Ma to 850 Ma that formerly comprised the basal unit of Grabau's Sinian have been called the Jidong sequence by Huang Jiqing (1980; see Table III–1).

Type Section

Like their Western and Japanese counterparts, Chinese geologists customarily establish type sections for each unit and succession of units within a region. Unlike geologists in Western countries, however, Chinese geologists have established type sections not only for formations and groups, but also for whole systems. The type sections for the systems consist of composite sections for one geologic province and obviously cannot be extrapolated into other geologic provinces (e.g. a platformal section cannot be extrapolated into a geosynclinal section). We present some of the type sections of China, especially those that are important to petroleum exploration. We omit illustrations of all other type sections.

Chinese geologists have established no type section for the oldest rocks in China, the lower Archeozoic. Archeozoic rocks, however, have been recognized for more than 100 years, and they were described and named by von Richthofen (1882, 1912). Wherever found, they consist of strongly metamorphosed and severely deformed gneisses (including hypersthene gneiss) and granulites. Their upper age limit is about 3,100 Ma (dated at 3,145 Ma in eastern Liaoning Province: Compilation Group, 1976), and an unconformity separates the lower Archeozoic from the upper (Table II–2).

The upper Archeozoic Fuping Group (Table II–2) and its equivalents overlie the lower Archeozoic, and ranges in age from approximately 3,100 to 2,600 Ma. The Anshan Group is a partial equivalent. The Fuping consists of gneiss, amphibolite, and granulite. In some localities, light-red or pink granitic gneiss and thick sections of amphibole-chlorite schist predominate (Academia Sinica, 1958; Compilation Group, 1962; Fang et al., 1979). An unconformity everywhere separates the Fuping from younger rocks. The type section for northern China is in the Wutai Shan and the Taihang Shan of northeastern Shanxi Province and western Hebei Province respectively. Figure II–2 is a generalized summary of the type section, the location of which appears on Figure II–1.

In North China and Northeast China, consolidation of the basement took place at the close of the early Proterozoic (Sino-Korean, or North China platform). Hence the basement consists here of Archeozoic and lower Proterozoic rocks. Its type section, also shown on Figure II–2, lies just west of the Wutai Shan on the Shanxi Plateau.

The Wutai is well exposed on the Shanxi Plateau, in the Wutai Shan, and in the Taihang Shan (Wang and Qiao, 1984). It consists of 5,000 to 6,000 m of granulite, amphibolite, various kinds of schist, and taconite (bed-

Age (b.y.)	Geologic time		Column	Major lithology	Thickness (m)	Tectonic movement and magmatic activities	Degree of metamorphism	Mineral deposits
	Sinian			Limestones			None	Building materials
1.85	Early Proterozoic	Hutuo Group (Pt1)		Upper: Low metamorphosed dolomitic conglomerates, feldspar, and quartz ss., slate. Middle: Dolomites, limestones with phyllites, mafic volcanic rocks with stromatolites. Lower: Dolomitic marbles with stromatolites. Bottom: metamorphosed gold-bearing cgl. & ss.	5,800 to 8,000	Luliang Movement. Intermediate to mafic volcanic activities and pegmatitic intrusions	Light metamorphism, local granitization	Copper, uranium, gold, building materials
2.3		Wutai Group (Pt1)		All types granulites, amphibolites, hornblende schists, plagioclase gneisses, ferruginous quartzites	8,200 to 16,500	Wutai Movement. Intermediate mafic to mafic volcanic rocks, granitic and pegmatitic intrusions	Medium metamorphism, weak migmatization	"Anshan-type" iron ore, copper, pyrite, beryl, micas
2.6	Late Archeozoic	Fuping Group (Ar2)		All types of gneisses, amphibolites, granulites, marbles, quartzites, and few magnetic iron quartzite	7,000 to 9,000	Anshan Movement. Large basic volcanic rocks, ultramafic to mafic intrusions and granites	High metamorphism, widely distributed, and strong migmatization and granitization	Aluminum, zinc, "Anshan-type" iron ore, graphite, phosphorus, corundum, chromite

Fig. II–2. Composite stratigraphic column of pre-Sinian in the area of Wutai Shan-Taihang Shan, North China (from Fang et al., 1979).

ded iron formation = BIF type). On the Shanxi Plateau, the unit originally included only a sequence of spilite and ophiolite in greenstone metamorphic facies. The equivalent section is widely exposed in the great uplifted blocks of northern China from the Kunlun Shan and Tian Shan on the west, to the Yin Shan and the Qin Ling on the east. In addition to spilite and ophiolite, the metamorphic sequence includes thick units of originally clayey rocks associated with the ophiolite (now metaophiolite). The present rocks consist of the medium- to low-grade metamorphic varieties; hence, migmatite and granitized rock are scarce. Locally, magnesian marble and ferruginous quartzite are strongly developed, and these form the Anshan-type, or banded, iron ores of China, best known in Liaoning Province of Northeast China (Chang Da, 1959; Compilation Group, 1976; Fang et al., 1979). These banded ores occur mainly in the middle and upper parts of the Wutai.

The Hutuo Group of the same region, generally believed to postdate the Wutài, also belongs to the early Proterozoic and is divided into three units: a lower terrigenous-clastic unit with marble; a middle argillaceous unit with basal metabasite and stromatolite-bearing rocks in the middle and upper parts; and, unconformable on the lower units, an upward-coarsening (reversed) molasse unit that formed after an early pulse of the Luliang orogeny (Figure II–2 and Table II–3). These units were first recognized by Fang et al. (1979), who noted that the lower unit contains uraniferous and auriferous conglomeratic beds; the stromatolites of the middle unit contain magnesian carbonate. Intrusions of granite and pegmatite are common in the lower and middle units. An angular unconformity, dated approximately 1,850 Ma, lies at the top, and is extensive in China (Table II–3).

Some Chinese geologists recently have challenged the traditional interpretation (adopted in this book) shown on Figure II–2, and have suggested that the Hutuo and Wutai are partly or wholly equivalent. We mention this dispute only to emphasize the many uncertainties that exist in old long-standing interpretations.

In southern China, an equivalent of the Hutuo forms

part of the basement. This equivalent consists of plagiogneiss and amphibolite below the Sinian in the Yangzi Gorges; in central Yunnan Province, a correlative unit is a sequence of alternate beds of metavolcanic and metasedimentary rocks. The equivalent unit in Northwest China is believed to be widespread, as it is in the Qinghai-Xizang Plateau (Wang and Qiao, 1984; Chang and Pan, 1984; Zhang et al., 1984). In the Qinghai-Xizang Plateau, we believe that an important and large stable Precambrian block is present, contrary to the conclusions of Huang Jiqing (1978) and Huang Jiqing et al. (1980).

The middle and upper Proterozoic consists, from base to top, of the Changcheng, Jixian, and Qingbaikou Groups which Wang and Qiao (1984) show as chronolithologic units or stages (Table II-3). These three units are well developed in Jixian County, Hebei Province, where the type section of North China is located (Figure II-3). The middle and upper Proterozoic on the Sino-Korean (North China) platform reaches a thickness of 10,000 m and occurs in two basins.

The Changcheng contains five formations (Figure II-3). The lower, limited in extent, appears fluviatile, littoral, and/or restricted marine; the second and fourth formations contain alkali volcanics (Chuanglinggou and Dahongyu) and transgressive carbonates (Gaoyuzhuang). The Jixian consists of a thick continuous sequence of carbonate rocks, some showing turbulent and slump structures. The Qingbaikou, formed on a stable platform, includes three formations. Additional details appear on Table II-3 and Figure II-3 (Compilation Group, 1976; Fang et al., 1979; Wang and Qiao, 1984).

In southern China, the middle and upper Proterozoic sections, preserved in three different areas, are represented by different sedimentary facies. One of these is the Shennongjia Group (Table II-3). Its type section

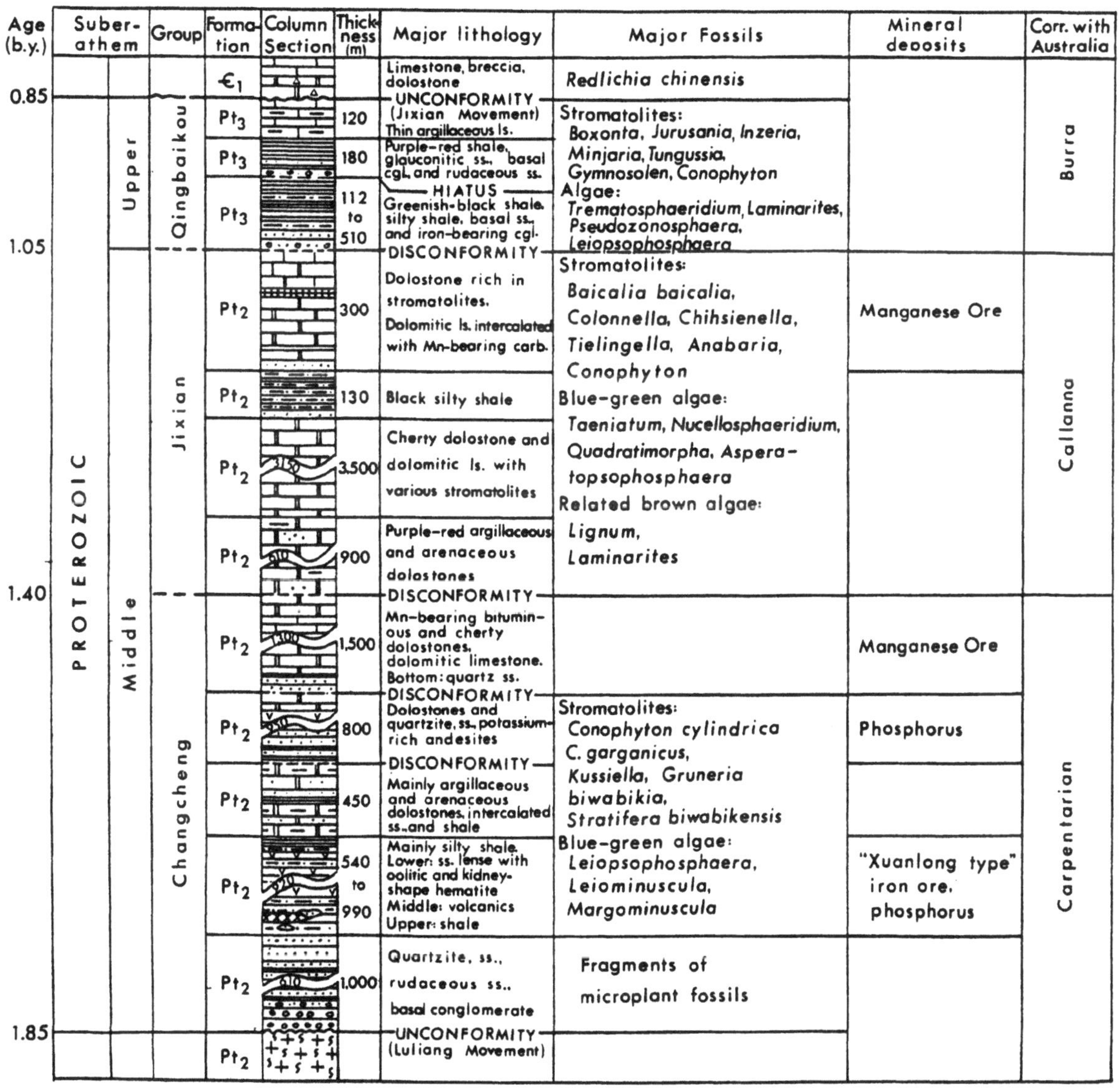

Fig. II-3. Stratigraphic column of Middle-Upper Proterozoic in Jixian County, Hebei Province, North China (modified from Fang et al., 1979 and Compilation Group, 1976).

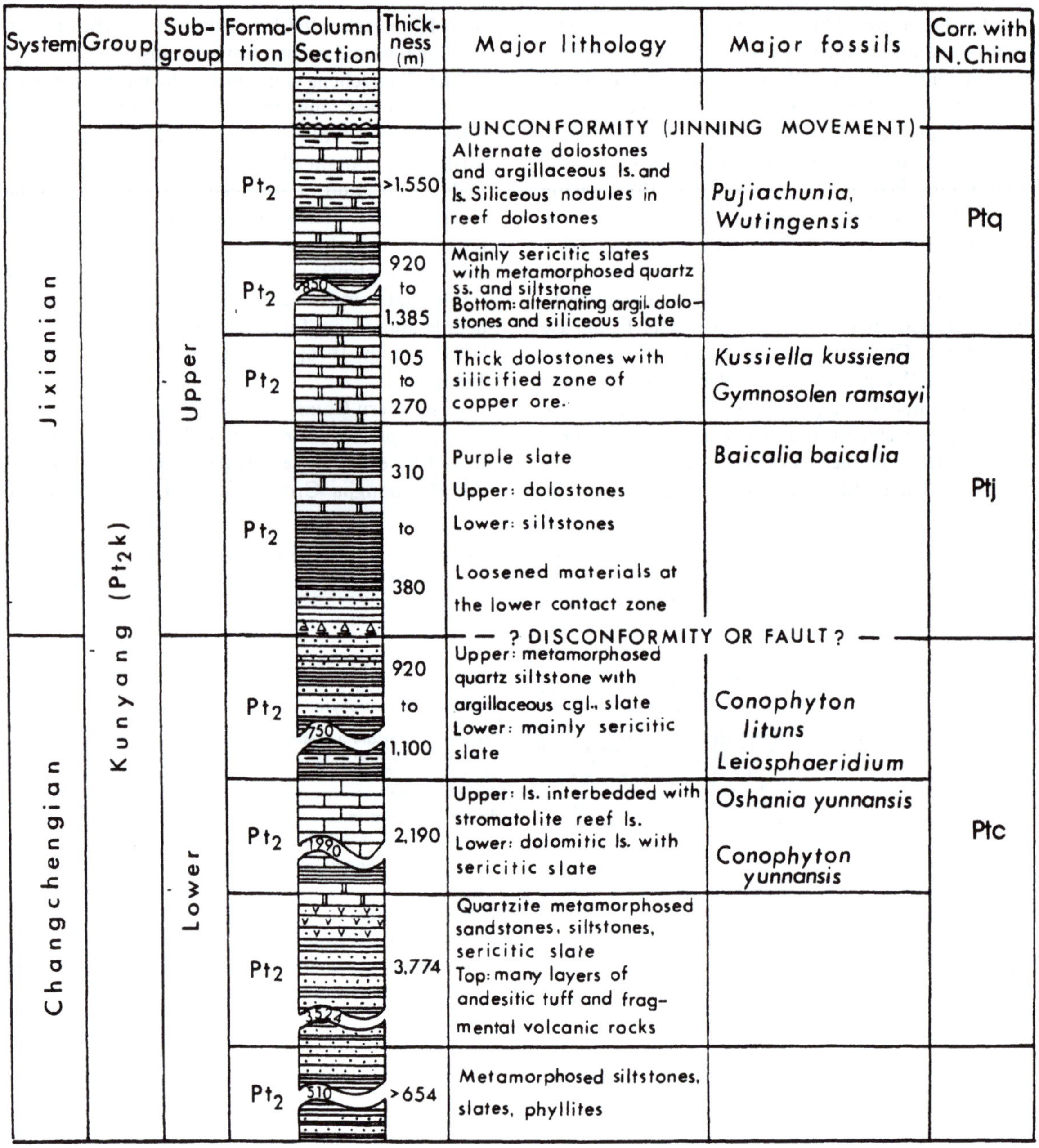

Fig. II-4. Stratigraphic column of Kunyang Group, Middle Proterozoic in eastern Yunnan Province, Southwest China (modified from Fang et al., 1979 and Compilation Group, 1976).

(in the northern interior, on the uplift of the same name) consists mainly of carbonate with terrigenous clastic intercalations of Changchengian and Jixianian ages. At the top is a molassic unit (equivalent to the Qingbaikouan).

A second assemblage of facies, called the Kunyang Group, crops out in eastern and southeastern Yunnan Province and in western Guizhou Province. The Kunyang Group (Figure II-4) consists of phyllite, slate, quartzite, schist, and red sandstone, which are intercalated with limestone lenses. Its thickness ranges from 2,000 to 10,000 m.

A third assemblage of facies, with various names in Table II-3, is well exposed for a distance of about 1,000 km along the Jiangnan uplift in the southeastern part of the Yangzi platform. The sequence of this assemblage extends from the border area between Anhui and Zhejiang Provinces, through northern Jiangxi Province and northern and western Hunan Province, and terminates in northern Guangxi Zhuang Autonomous Region and southeastern Guizhou Province. The exposed sections show well-developed zones of parallel facies. These are typically eugeosynclinal low-grade metamorphosed volcano-sedimentary sequences, with an aggregate thickness of more than 20,000 m (Guo et al., 1984; Huang Jiqing, 1984; Lee, 1984; Wang and Qiao, 1984; Zhang et al., 1984).

Middle and late Proterozoic sections crop out extensively in the Tian Shan, Kunlun Shan, Qilian Shan, and Alxa massif in Northwest China. The Quruktagh section, north of the Tarim basin, is well studied, and consists of immature terrigenous clastics (Changchengian) in the lower part, mainly carbonate (Jixianian) in the middle, and littoral terrigenous clastics (Qingbaikouan) at the top (Wang and Qiao, 1984).

On the Qinghai-Xizang (Tibet) Plateau, the Precambrian is poorly known, the only exception being the region of the High Himalayas adjacent to Nepal, India and Bhutan. Numerous large areas (on both the Lhasa block and the Qiangtang block) are known, however, where Pre-Cambrian is overlain unconformably by Late Cambrian to Ordovician and younger marine platformal strata (Chang and Pan, 1986).

A generalized paleogeographic map of the Middle and Upper Proterozoic, from Wang and Liu (1980) is shown as Figure II-5. The large blank areas of the Qinghai-Xizang Plateau contain scarcely-studied Precambrian rocks.

Magmatic Activity

The intensity of magmatism and of crustal movements differed from place to place and at different times during the pre-Sinian. In general the magmatic intrusions in southern China are more extensive in Middle to Late Proterozoic rocks; they consist, moreover, of types associated with tectonic belts. In the northern areas of China, magmatism associated with tectonic belts is less prominent. In the northeastern areas of China, this type of association does not exist at all. The Compilation Group (1976) recognized six magmatic episodes belonging to the Pre-Sinian.

Early Fuping Intrusive Episode

Several Archeozoic silicic intrusives thought to belong to the Fuping episode (2600 Ma) have been dated as early or pre-Fuping (3100 Ma). The extent of these ancient rocks is still unknown.

Fuping Intrusive Episode

Intrusives of the Fuping orogeny include granites, pegmatites, mafic rocks, and ultramafic rocks; some of the granites are partly migmatized.

Wutai Intrusive Episode

This is the earliest of the Proterozoic magmatic events. Intrusions include granite, pegmatite, diorite, and some mafic rocks. These intrusives occur in most of China except for areas of the northeast and western Shandong Province.

Luliang Intrusive Episode

These intrusives include the pegmatite dikes of the Wutai Shan and Taihang Shan, the Lianshanguan granite, the Gongzhangling granites of the Liaodong Peninsula of Liaoning Province, and some granites of the Longshou Shan in Gansu Province. Numerous pegmatites are present in the Daqing Shan and the Yan Shan.

Sibao Intrusive Episode

This episode is characterized by both silicic and alkaline intrusives. In North China the silicic rocks include oligoclase rapakivi granites (wiborgites) near Beijing, and pegmatites of the Wutai Shan and the Luliang Shan, both in Shanxi Province. In Central-South China (northern Guangxi Zhuang Autonomous Region), the igneous rocks consist of ultramafic and mafic associations with diorite. These alkaline rocks intrude the lower part of the middle Proterozic, and are overlain unconformably by its upper part.

Jinning Intrusive Episode

This is part of the Jinning tectogenesis, with radiometric dates of about 1,000 Ma, about 900 Ma, and 800 ± 50 Ma. In this episode the magmatic intrusions lie unconformable beneath the Sinian (S.S.). In Central-South China, granitic and other intrusives characterize the Jinning episode. In North China they are less abundant. In Northeast China they are scarce. The Sinian unconformably overlies all of these silicic rocks.

In Central-South China, strong crustal movements produced rugged topography and downfaulted basins. In North China, where tectonics were moderate, a general regression of marine waters took place, accompanied by local erosion.

SINIAN

Chinese geologists assign the Sinian System, both *sensu lato* (s.l.) and *sensu stricto* (s.s.) to the Proterozoic. Sinian s.s. indicates Late Proterozoic (850 to 570 Ma), Sinian s.l. indicates middle and later Proterozoic (1,400 to 600 Ma; Compilation Group, 1976). As noted above, the attribution of Sinian has been restricted to rock sections from 850 to 800 Ma at the base to 600 to 570 Ma at the top. These Sinian s.s., sections comprise the highest and youngest system of the Proterozoic (Table II-2; Wang and Qiao, 1984; Xing Yusheng, 1984).

Sinian s.s. applies also to a thick sequence of unmetamorphosed to slightly metamorphosed carbonate and terrigenous-clastic strata, mainly in marine facies. Volcanic rocks and presumed glacial beds also are present. The type section of the Yangzi Gorges in northwestern Hubei Province (Figure II-6), with an extension westward into eastern Sichuan Province, was designated, measured, and described by J. S. Lee (Li Siguang) and

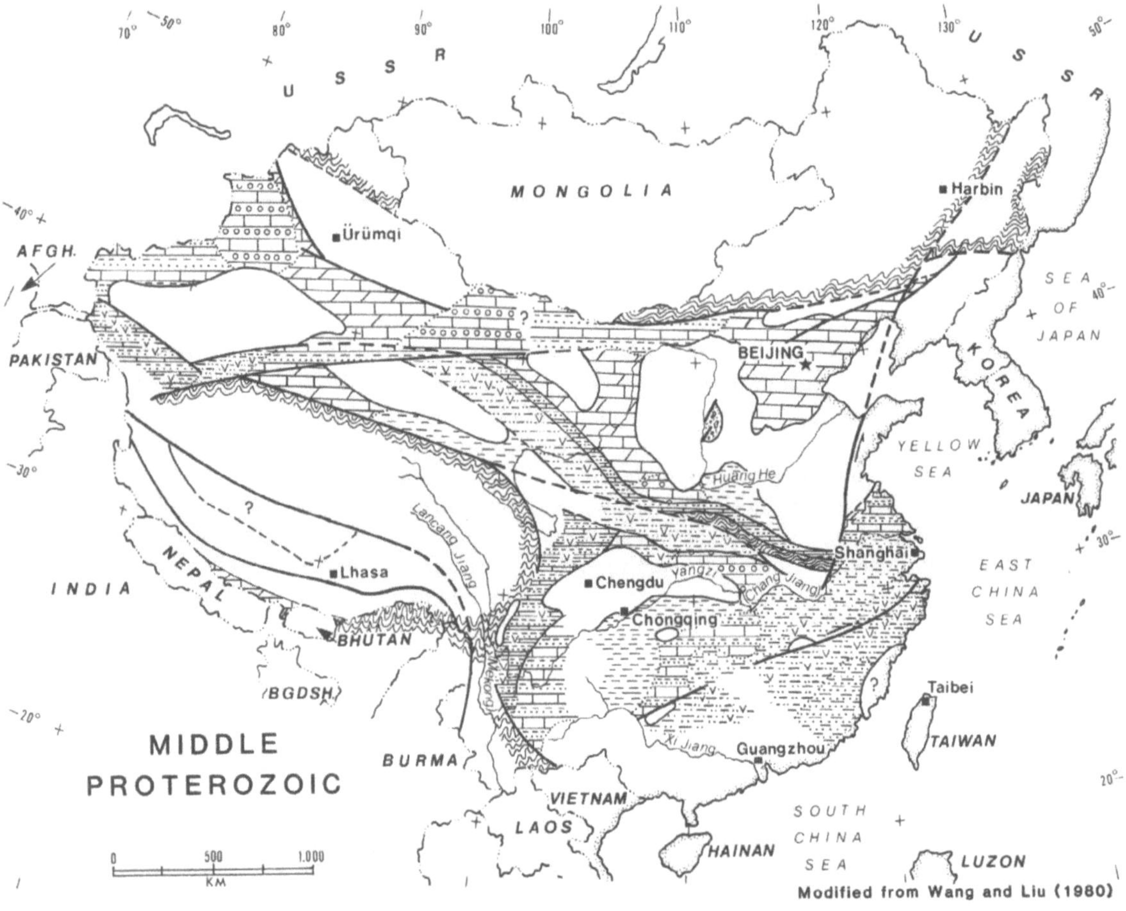

Fig. II–5. Paleogeographic and lithofacies map of Middle Proterozoic, China.

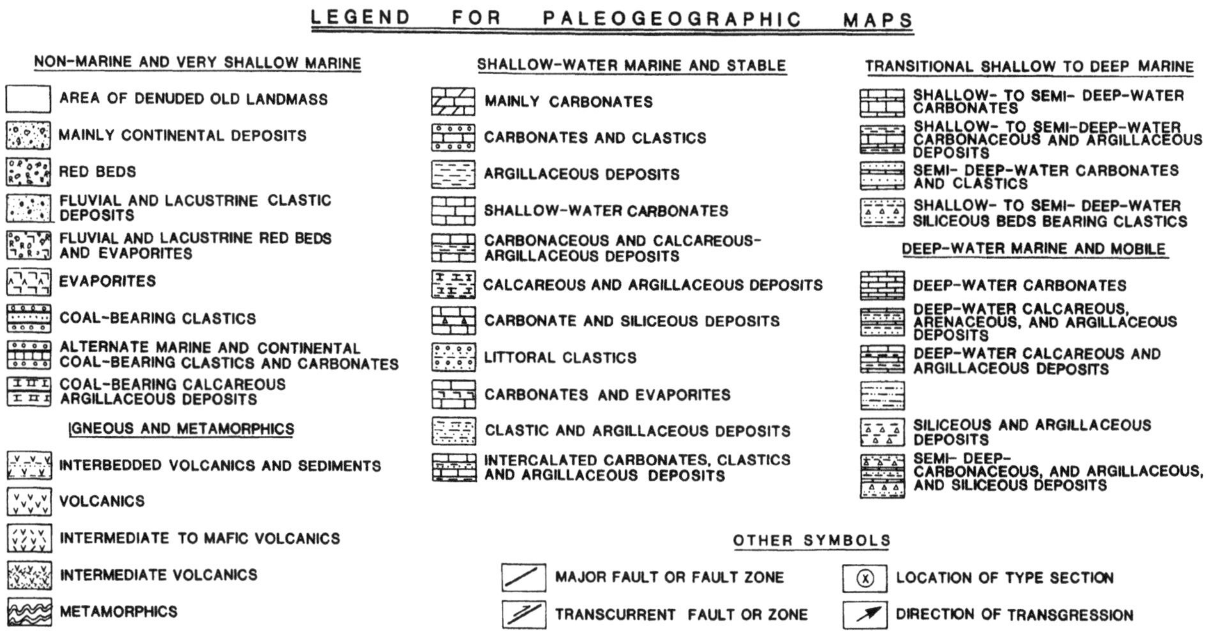

Chao (1924). Other 'classic' sections measured and described from North China by Kao et al. (1934) now are known to be older than the Yangzi Gorges section, and which belong to the Sinian s.l., which, in this work, we do not place in the Sinian. Instead, we follow Wang and Qiao (1984) and Xing Yusheng (1984).

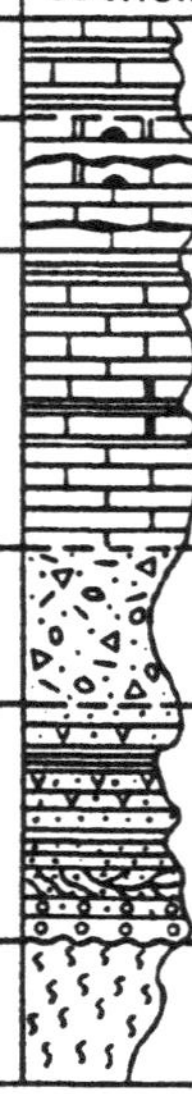

Suberathem	System	Series	Formation	Column section	Thickness (m)	Major lithology	Major fossils
Paleozoic	Cambrian	Lower	$€_1$			Shales with limestone; phosphorus at bottom	
						—DISCONFORMITY—	
Late Proterozoic	Sinian System (Zz)	Upper	Zz_4		60 to 300	Chert-bearing and nodular dolostones and dolomitic ls.	Hyolithes: *Circotheca* sp., *Pseudorthotheca* sp. Vermes: *Micronematites formosus*. *Saarina* sp. Spong spicules, Medusa: *Madigania annulata sprigg*, *Cyclomedusa davidi*. Algae: *Manicosiphonia*, *Nanamanicosiphonia*, *Varicamanicosiphonia*, *Vendotaenia*, *Praesolenopora*
			Zz_3		145 to 320	Mainly limestone, dolomitic limestone, with pyrite, black shale, phosphorus, glauconite	*Lophosphaeridium ichangense*, *Nostocomorpha prisca*, *Acanthomorphitae*, *Prismatomorphitae*
		Lower	Zz_2		0 to 100	Purplish-red and grayish-green glacial cgls., various size, no bedding, striations	
						—DISCONFORMITY—	
			Zz_1		50 to 260	Purplish-red feldspar and quartz sandstones with tuffaceous sandstone and shale. Basal cgl. and rudaceous sandstone.	*Trematosphaeridium holtedahlii*, *Leiopsophosphaera* sp., *Laminarites antigussimus*, *Trachysphaeridium rugosum*
						—UNCONFORMITY—	
Pre-Sinian						Gneiss and schist	

Fig. II–6. Continued.

Mineral deposits	Correlation: Australia		Correlation: USSR
	€		€
Phosphorus, gypsum, salt	Welpena	Adelaide	Vendian
Phosphorus	Umberatana (with glacial deposits)		
			Minyar (Riphean)
Iron and phosphorus ores			

Fig. II–6. Stratigraphic column of Sinian System in Sanjia, the gorge of Yangzi River (Chang Jiang), West Hubei Province. Formation names: Zz_1 = Liantuo, Zz_2 = Nantuo Tillite, Zz_3 = Doushantuo, and Zz_4 = Dengying (from Fang et al., 1979, Compilation Group, 1976, and Liu Hungyun, 1973).

Distribution and Facies

Sinian rocks crop out widely in northern areas of China and in southeastern China, as well as in the mountains of western China. In northern areas of China, the Sinian crops out mainly along the borders of the Sino-Korean (North China) platform (Figure II–7). In eastern Liaoning Province (Figure II–7), the section ranges in thickness from 2,200 to 4,350 m, and consists of littoral, neritic terrigenous-clastic, and carbonate lithofacies, many of whose units are rich in micro- and megascopic algae, stromatolites, and soft-bodied metazoans. Some units are rich in organic carbon and probably contained good source rocks for petroleum, as does the Proterozoic in many parts of China. The Sinian is divided into an Upper and Lower series (Table II–4). In eastern Liaoning Province, the Lower series includes two formations of grayish-white to white quartz sandstone, arkosic sandstone, glauconitic sandstone, and grayish-green to grayish-brown siltstone and shale. The Upper series consists mainly of gray, dark-gray, and purplish-red marl, stromatolitic limestone, dolomite, and some shale in the lower part and of yellowish-brown to gray sandstone, siltstone, shale, and limestone in the upper part. The Sinian is conformable on older rocks, but unconformable below the Lower Cambrian. In northern Anhui Province, the section is thinner; the Lower series includes a small amount of limestone; and the Upper series consists mainly of carbonates that contain some chert.

In southern to southeastern areas of China, the lithologic description of the Yangzi Gorges section

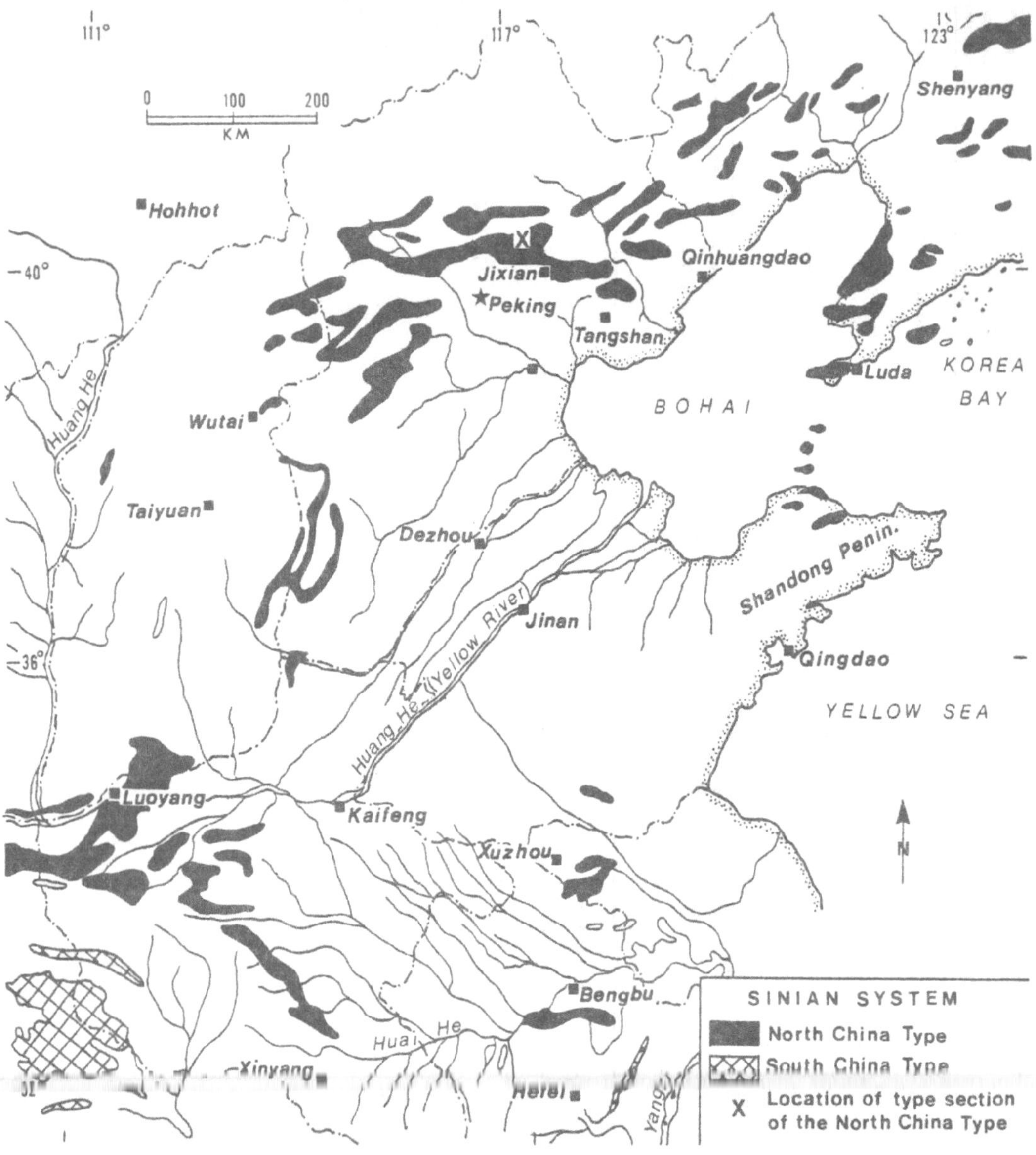

Fig. II–7. Outcrop distribution map of Sinian System in North China and northern South China (from Fang et al., 1979).

Table II-4. Stratigraphic correlation chart of the Sinian System in China

	SOUTHWEST CHINA			SOUTHEAST CHINA				NORTH CHINA		NORTHWEST CHINA		SOUTH	NORTH	WEST
	(1) WEST HUBEI	(2) EAST YUNNAN	(3) SOUTHWEST SICHUAN	(4) NORTH GUANGXI	(5) E. ANHUI-N. JIANGSU	(6) SOUTH ANHUI	(7) WEST ZHEJIBIANG	(8) LIAODONG	(9) HUAINAN	(10) QURUKTAGH	(11) NORTH QAIDAM	AUSTRALIA	EUROPE	AFRICA
CAMBRIAN	Cm_1 TIENZHLSHAN FM.	Cm_1 MEISHUCUN FM.	Cm_1 MAIDIPING FM.	Cm_1 QINGXI FM.	Cm_1 HUANGLISHU FM.	Cm_1 HUANGBO-LING FM.	Cm_4 HETANG FM.	Cm_1 JIANCHANG FM.	Cm_1 HOUJIASHAN FM.	Cm_1 XISHANBULAK FM.	Cm_1 XIAOGAOLU FM.			
600 UPPER SERIES	DENGYING FM. ***VENDOTAENIA*** ***CHARNIA*** ***DENGYINGENSIS*** 250–670 m	DENGYING FM. ***MICRHYSTRIDIUM*** ***LOPHOMINUSCULA*** 167–745 m	HONGCHUNPING FM. ***ACUS*** ***ACTINOPHYCUS*** ***BALIOS*** 940 m	LAOBAO FM. 131 m	DENGYING FM. 1,100 m	PIYUANCUN FM 80 m	XIFENGSI FM. 170 m	JINXIAN GROUP: XINMINCUN FM. ***BOXONIA*** 292 m; CUIJIATUN FM. 79 m; MAJIATUN FM. ***PRAESOLENO-PORA*** 199 m; SHISANLITAI FM. 155 m; YINGCHENGZI FM. 894 m	GOUHOU FM. 116 m; JINSHANZHAI FM. ***MULTISIPHONIA*** 23 m; XUHUAI GROUP: WANGSHAN FM. 377 m; SHIJIA FM. ***MULTISIPHONIA*** 902 m; WEIJI FM. 224 m; ZHANGQU FM. 132 m	HANGELHAOK FM. 465 m; SHUIQUAN FM. ***TAENIATUM*** 331 m; YUKENGOL FM. 583 m	ZHOUJIESHAN FM.; SABELLITIDIDS 22 m; HONGTIEGOU FM. 110 m; HEITUPO FM. 50 m; HONGZAOSHAN FM. ***MICRHYSTRIDIUM*** >100 m	MARINOAN: POUND QTZT.; WILPENA GROUP	BREIVIK FM.; STAPPOGIEDDE FM.	UPPER KUNDELUNGU GROUP
	DOUSHANTUO FM. ***MICRHYSTRIDIUM*** ***LOPHOSPHAERID-IUM*** 152–230 m	WANGJIAWAN FM. ***LOPHOSPHAERID-IUM*** ***MARGOMINUSCULA VERRUCOSA*** 180–360 m	GUANYINYA FM. ***PALAEOMICRO-CYSTIS*** 47 m	DOUSHANTUO FM. 49 m	DOUSHANTUO FM. 796 m	LANTIAN FM. 65 m	SANLITING FM. 68 m	WUHANGSHAN GROUP: GANJINGZI FM. 400 m; NANGUANLING FM. 400 m; CHANGLINGZI FM. SABELLITIDIDS ***MICRYSTRIDIUM*** 159 m	JIUDINGSHAN FM. 243 m; YIYUAN FM. 242 m; SIDINGSHAN FM. 144 m; JIULIQIZO FM 510 m	ZHAMOKTI FM. 793 m	SHIYINGLIANG 150 m			
680 LOWER SERIES	NANTUO FM. ***TRACHYSPHAERID-IUM*** ***LAMINARITES ANTIQUISSIMUS*** 90–150 m	NANTUO FM. 0–33 m	LIEGULIU FM. 0–204 m	NANTUO FM. 9€7 m	SUJIAWAN FM. 1,000 m	MAJIN FM. 88 m	LEIGONGWU FM. 127 m			TEREECKEN FM. 2,685 m	KUBOMU FM. 200 m	680 STURTIAN	VERRANGIAN	LOWER KUNDELUNGU GROUP: SMALL CGL. TILLITE
	LIANTUO FM. ***LAMINARITES*** CF. ***ANTIQUISSIMUS*** 50–280 m	CHENGJIANG FM. 300–1,200 m	KAIJIANQIAO FM. 680 m	FULU FM. 675 m; CHANGAN FM. 962 m	ZHOUGANG FM. 166 m	XIUNING FM. 484 m	ZHITANG FM. 966–4,200 m	QIAOTOU FM. 565 m		ZHAOBISHAN FM. 570 m; BAIYISI FM. 1,560 m	MAHUANGGOU FM. 400 m	TORRENSIAN	LISPAK	GREAT CGL. TILLITE
800	Pt_1 SANDOUPING GROUP	$Pt_{2,3}$ KUNYANG GROUP	Pt_3 ? SUXIONG FM.	Pt_3 GONGDONG FM.	Pt_3 ZHANGBALING GROUP	Pt_3 PULING FM.	Pt_3 SHANGSHU GROUP	Pt_3 XIHE GROUP	Pt_3 HUAIAN GROUP	Pt_3 PARGANGTAG GROUP	Pt_2 DAKDABAN GROUP	850 WILLOURAN		

From Wang and Qiao (1984)

(Figure II–6) is generally applicable to other parts of the Yangzi Platform. Thickness ranges from 1,000 to 3,000 m. Tillite and glaciofluvial beds are omnipresent. Lava flows appear in the basal part of the Lower series in western Sichuan Province, and somewhat higher (upper part of the Lower series) in southern Anhui Province. Bone coal, or sapromyxite, is common, and is mined for use as a fuel, especially for thermal power plants. Thin-bedded chert is common in the upper part of the section. The lower contact generally is unconformable; the upper contact is gradational or slightly disconformable with the Cambrian. Fossils are abundant.

The Sinian of western areas of China, well exposed along the northwestern and northeastern margins of the Tarim basin lies unconformably on all older units, and is disconformable with overlying phosphate-bearing Lower Cambrian (Table II–4). The Lower series here consists of about 2,000 m of grayish-green sandy shale, siltstone, sandstone, and conglomerate. The Upper series contains in its lower part a section, 300 to 900 m thick, of sandstone, siltstone, sandy shale, tuff, and terrestrial lavas (extruded either subaerially or on a shallow sea floor). The upper part of the Upper series consists of 160 to 350 m of massive dolomite, oncolitic dolomite, and minor siltstone. Microflora is abundant throughout.

In the Kuruktag (tag = mountains) of the easternmost Tarim basin north of Lop Nur, the Sinian, exceptionally thick, ranges from 3,000 to 8,000 m (Table II–4). Within the Lower series are three separate glacial (diamictite) units interbedded with thick terrigenous clastics and volcanic rocks (bottom two thirds of the Sinian). The upper third (= Upper series) comprises mainly a thick section (1,000 to 2,000 m or more) of carbonates (Huang Jiqing, 1978; Wang and Qiao, 1984; Zhang et al., 1984).

In eastern and southeastern Yunnan Province, the Sinian consists of molasse-type sandstone and glaciogenic red conglomerate (Lower series), 250 to 2,500 m thick; and of oolitic and cross-laminated dolomite, bituminous limestone, siliceous limestone, and birdseye dolomite, 600 to 900 m thick (Upper series). The Lower (Chengjiang) and Upper (mainly Dengying) series are separated by an unconformity (Table II–4; Li, 1984; Wang and Qiao, 1984). Beds of gypsum and halite, common on the Sichuan platform (Sichuan basin) just east of the Yunnan fold belt (Liu Hung-yun et al., 1973), form seals for Sinian gas reservoirs. Sources of the gas in the Sinian carbonates of the Dengying Formation are in the Sinian itself.

A generalized paleogeographic map of the Sinian is presented on Figure II–8.

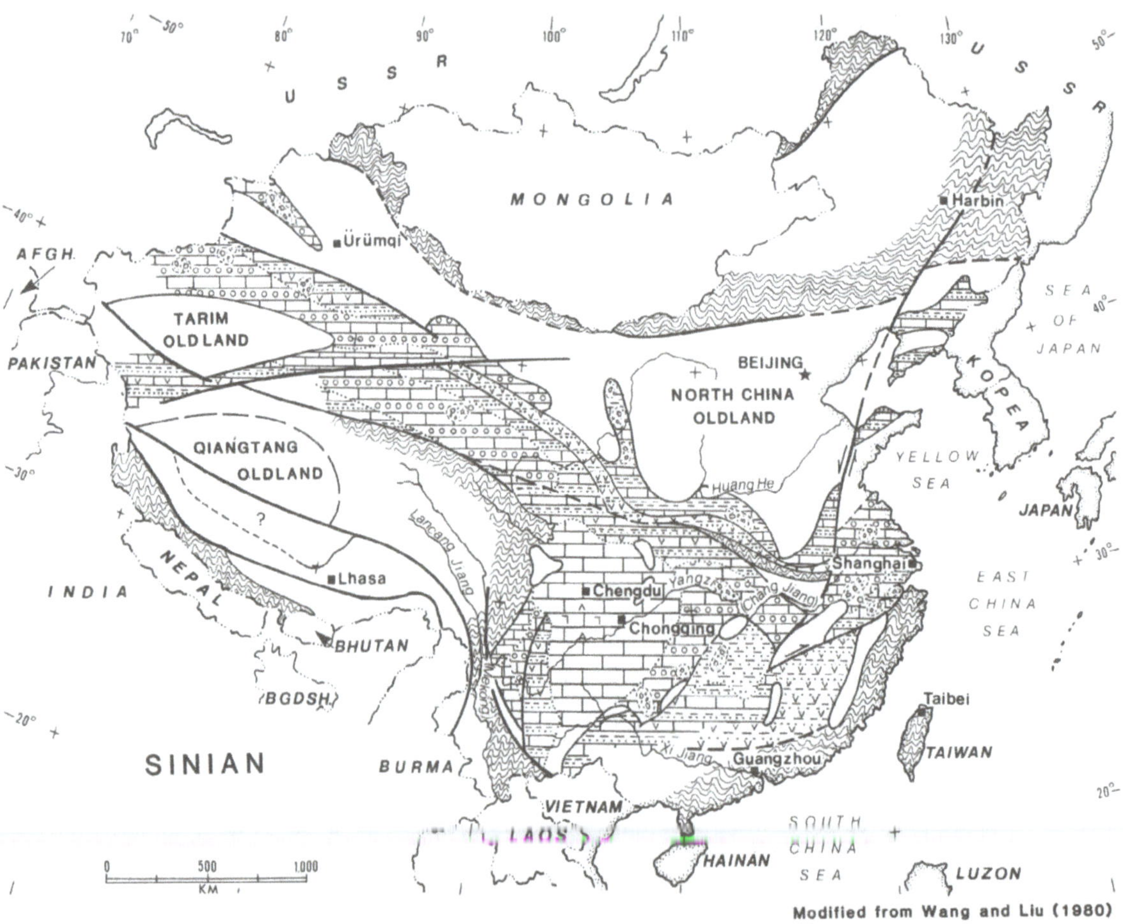

Fig. II–8. Paleogeographic and lithofacies map of Sinian, China.

Magmatic Activity

Changjiang Intrusive Episode

This episode, with a radiometric date of about 700 Ma, included moderate tectogenesis in Central-South China. Granite and other intrusive rocks are found at the disconformity between Zz_1 and Zz_2.

Post-Changjiang

Post-Changjiang deposits contain diamictites, and from these has arisen an inference of extensive glaciation in the late Sinian. On this inference, glaciation took place in Central-South China, in the southern part of North China, and in the Tian Shan of Northwest China. The radiometric date of this episode is 700 to 600 Ma. A smooth and flattened topography emerged. Transgression of marine waters followed, as evidenced by the rich biota of Zz_3 and Zz_4. Erosion took place in North China (Fang et al., 1979).

The end of the Sinian is marked by unconformities. In North China, unconformity is associated with moderate crustal movement. In Central-South China, unconformity is associated with a regression of marine waters.

PALEOZOIC

General

Paleozoic sequences are well developed and widely distributed in China. Richly fossiliferous marine deposits predominate in the lower Paleozoic. In contrast, continental non-marine sedimentary rocks are much more conspicuous in the upper Paleozoic.

The Compilation Group (1976) divided the Paleozoic of China into three tectonic belts and eight sedimentary regions (Figure II-9). The three tectonic belts are: (1) the Northern Tectonic belt (Junggar-Hinggan), and the north of the Tian Shan and the Yin Shan; (2) the Middle Tectonic belt (Tarim-North China), between the Tian Shan and the Yin Shan on the north and the Kunlun Shan and the Qin Ling on the south; and (3) the Southern Tectonic belt (Qinghai-Xizang-Central-South China), south of the Kunlun Shan and the Qin Ling. The eight sedimentary regions (Figure II-9) are: (1) Junggar-Hinggan; (2) Tarim; (3) North China; (4) Kunlun-Qilian-Qin Ling; (5) Xizang (Tibet)-West Yunnan; (6) Yangzi (Yangtze); (7) Southeast; and (8) Himalayas.

Fig. II-9. Paleozoic sedimentary regions of China (from Compilation Group, 1976).

Junggar-Hinggan

This region (Northern Tectonic belt) stretches across the whole of northern China along latitudes north of the Tian Shan, the Yin Shan, and the drainage basin of the Songhua Jiang. Among its geographic features are the Junggar depression of Xinjiang Uygur Autonomous Region in Northwest China and the mountainous Da Hinggan Ling (Great Khingan Range) of northeastern Nei Monggol Autonomous Region and western Heilongjiang, Jilin, and Liaoning Provinces in Northeast China. Marine waters invaded this region from the north, and their deposits were thick and complex. Scattered outcrops of lower Paleozoic strata consist mainly of slightly metamorphosed marine terrigenous-clastic rocks, with intercalated carbonates and terrigenous clastics. A facies change in the Upper Silurian brings purplish-red terrigenous-clastic strata into the lower Paleozoic column. The lower Paleozoic fauna consists chiefly of pelagic agnostid trilobites, archaeocyathids, graptolites, and brachiopods. Studies made in the 1970s suggest that the so-called 'Bei Shan complex' of western Gansu Province belongs mainly to the Ordovician and the Silurian.

The upper Paleozoic consists mainly of marine to paralic strata. Present is an abundant benthos of brachiopods and corals of the Boreal Pacific realm, as well as an Angaran (northern) flora. Other constituents of the column are terrigenous-clastic and pyroclastic rocks. The Upper Devonian is extensive, with continental to littoral deposits. In contrast, the Upper Permian consists mainly of continental strata with an Angaran flora, although with local interbeds of marine strata.

Tarim

In this region of the Middle Tectonic belt, Paleozoic rocks crop out around the margins of the Tarim basin and along the northern slopes of the western Kunlun Shan, both in southwestern Xinjiang Uygur Autonomous Region. In both areas the column is relatively complete. The Cambrian to Middle Ordovician consists mainly of carbonates and terrigenous-clastic strata intercalated with volcanic rocks. The Lower Cambrian consists of phosphatic carbonaceous shale, silicilith, and limestone, with a rich benthos of trilobites and brachiopods; cephalopods also are numerous. Parts of the Silurian and the Devonian contain neritic to littoral red sandstone and shale interbedded with limestone. The Carboniferous and Lower Permian consist of paralic carbonates, sandstone, and shale, locally with beds of coal. The upper Permian consists of continental clastics.

North China

This region covers a large area, bounded on the north by the Yin Shan, on the south by the Qin Ling and the Huang He (Yellow River), and on the west by the Helan Shan. Parts of the Silurian, Devonian, and Carboniferous are absent. The Cambrian and the Ordovician consist mainly of neritic limestone and dolomite, intercalated with sandstone and shale. In the western part of Shandong Province and in the Yin Shan of Nei Monggol Autonomous Region, relatively complete and richly fossiliferous columns contain type sections of the Cambrian and the Ordovician for China. Benthic trilobites of the Cambrian are so abundant that they yield a total of nineteen fossil zones. During the 1970s, beds with the trilobite *Paleolenus* were discovered in the middle lower Cambrian Tsanglangpu Formation across wide areas of North China, as well as in the southern part of Northeast China. Above the Cambrian, actinocerid cephalopods characterize the Ordovician. The Middle and Upper Carboniferous and the Permian of the Taiyuan and Ningwu areas of central and northern Shanxi Province, respectively consist mainly of continental terrigenous clastics interbedded with marine limestones and volcanics. In the Upper Permian, a rich Cathaysian flora is present.

Kunlun-Qilian-Qin Ling

This region also is part of the Middle Tectonic belt, and its Paleozoic deposits are thick and widespread. The lower Paleozoic column consists mainly of slightly metamorphosed marine carbonates and terrigenous-clastic strata interbedded with volcanic rocks. The upper Paleozoic sequence consists of paralic deposits. The Upper Devonian and the Upper Permian consist of continental deposits with plant remains. The Paleozoic of this region attains an aggregate thickness of 10,000 to 20,000 m.

Xizang (Tibet)-West Yunnan

This region west of the Jinsha Jiang (Jinsha River) lies within the Southern Tectonic belt. There the lower Paleozoic part of the column lies mainly in western Yunnan Province. The Cambrian consists of a thick sequence of flysch and flyschoid deposits. The Ordovician and the Silurian include both graptolitic shale and shelly limestone rich in cephalopods and corals. Graptolites also occur in the shelly facies. From western Yunnan Province, the fossiliferous Cambrian and Ordovician continue northwestward into eastern Xizang (Tibet) Autonomous Region. Devonian, Carboniferous, and Permian, widespread through the later region, contain a lithology mainly of marine sandstone, shale, and carbonates, interbedded locally with extrusive volcanics and beds of coal.

Yangzi (Yangtze)

This region also lies within the Southern Tectonic belt, and the development of its Paleozoic column is much more complete than in many parts of China. The Cambrian and the Ordovician consist mostly of neritic limestone, sandstone, and shale. The basal Cambrian faunas contain abundant trilobites, hyolithids, and archaeocyathids. The Ordovician contains cephalopods such as *Vaginoceras* and *Sinoceras*, together with numerous graptolites.

The Silurian consists mainly of graptolitic shale and shelly sandstone. Intercalations of redbeds in the upper part are rich in graptolites, brachiopods, and corals. The Devonian and the Carboniferous, well developed in the southwestern part of this region, consist primarily of marine limestone, sandstone, and shale, locally intercalated with siliciliths and with volcanic rocks. These strata, however, mostly are absent in the southeastern part of the Sichuan basin. The Maoshan Formation of the lower Yangzi, long placed in the Upper Silurian, more recently has been placed largely in the Lower and Middle Devonian, following a discovery of antiarchs (ostracoderms).

The Permian is widespread, with a lower part in neritic limestone and an upper part in paralic limestone and coal. The Permian also contains the widely distributed Emei Shan Basalt (at the boundary between Lower and Upper Permian) in the upper Yangzi, a unit associated with several major world-class mineral deposits (e.g., Panizhihua). In central Guangxi Zhuang Autonomous Region, intermediate to silicic volcanic rocks occur.

Many taxa contribute to the rich upper Paleozoic fauna in this region. The Devonian contains primarily brachiopods, corals, primitive ammonoids (including the wocklumerid 'goniatite' fauna), tentaculities, and fish. The marine Carboniferous and Permian contain abundant fusulinids, corals, and brachiopods, whereas the continental Permian contains the well-known Cathaysian flora with *Gigantopteris* as its representative.

Southeast

This region southeast of the Yangzi includes the islands of Hainan and Taiwan. Its lower Paleozoic consists of a thick series of slightly metamorphosed sedimentary rocks. Cambrian and Ordovician sequences of this series contain mostly flysch and flyschoid deposits, graptolitic beds, siliciliths, carbonaceous shale, and coal. Australian and Atlantic taxa are associated with endemic Chinese taxa. Devonian and Lower Carboniferous columns consist mainly of paralic deposits. In the Guangxi Zhuang Autonomous Region, *Monograptus uniformis* and *M. yukonesis* indicate the base of the Devonian. The upper Paleozoic is similar to that of the Yangzi region, both in lithofacies and biotas. Upper Carboniferous and Permian contain interbedded coal measures. On Taiwan, crystalline limestone of the Upper Carboniferous and the Permian contain fusulinids and corals in the lower part of the Nanao Group.

Himalayas

This is the last of the regions within the Southern Tectonic belt, and its Paleozoic is well developed. Relatively detailed studies on the northern slope of Qomolangma Feng (Mt. Everest) show a column mainly of neritic sandstone, shale, and limestone from Ordovician through Permian. Among rich faunal assemblages, that of the 'transitional beds' across the Silurian-Devonian boundary contains the monograptids *Neomonograptus himalayensis* and *Monograptus thomasi*. This same fauna occurs in western Sichuan Province, in the western part of the Junggar basin of the Xinjiang Uygur Autonomous Region, in Yunnan Province, and in the Guangxi Zhuang Autonomous Region. Evidently the fauna is widespread also in Northwest and Southwest China, especially in deposits of the Qin Ling, the Qilian Shan, and the Kunlun Shan.

Faunal Zones

Figures II–10 and II–11 show the major faunal zones of the Paleozoic of China and their correlatives in other parts of the world.

CAMBRIAN SYSTEM

Distribution and Facies

With the exception of Taiwan, northern Nei Monggol Autonomous Region (Inner Mongolia), and the Junggar basin of the Xinjiang Uygur (Sinkiang-Uighur) Autonomous Region, the Cambrian is ubiquitous in China (Figure II–12). Its strata are mainly marine, and they contain rich faunas, characteristics that led Lu (1962) to the recognition of five Cambrian 'regions' or subprovinces, each with a distinctive fauna and each with a distinctive lithology. A later study of faunas (Lu, 1981) suggested the existence of three Cambrian 'realms' of trilobites (Table II–5). Wang et al. (1981) presented a tabulation of Cambrian brachiopod assemblages (Table II–6). Huo Shicheng et al. (1989) reported on distribution of the Cambrian bradoriida (marine small bivalve), and on the Cambrian stratigraphic successions throughout China. The main conclusion of their study is that during Cambrian time, though the lithologic successions vary, all seas were connected with each other.

We discuss briefly the five 'regions' described by Lu (1962).

Northern Part of Northeast China and Southern Part of Nei Monggol

The Cambrian sequence, about 4,000 m thick in Da Hinggan Ling (Greater Khingan Range), consists of partly metamorphosed carbonates. The spores in these rocks suggest a close relationship with similar strata in Siberia (U.S.S.R.).

North China-Southern Part of Northeast China

Here the Cambrian consists of shale, carbonates (limestone and dolostone), argillaceous limestone, and sands-

System	Series	Formation	Fossil Zonation or Representative Fossils	Correlation with foreign countries
Silurian	Upper	S_3	*Pristiograptus tumescens* zone *Bohemigraptus bohemicus* *Pristiograptus nilssoni* zone	Pridolian Ludlovian
	Middle	S_2	*Monograptus flexilis* zone *Cyrtograptus rigidus* zone *Monograptus riccartonensis* zone *Cyrtograptus murchisoni* zone	Wenlockian
	Lower	S_1	*Oktavites spiralis*- *Glyptograptus* *Persculptus* zones	Llandoverian
Ordovician	Upper	O_6	*Dicellograptus szechuanensis* zone *Dicellograptus complanatus* zone	Ashgillian
		O_5	*Pleurograptus lui* zone *Sinoceras chinense* (Foord) zone	Caradocian
	Middle	O_4	*Nemagraptus gracilis* zone *Glyptograptus teretiusculus* zone	Llandeilian
		O_3	*Pterograptus elegans, Amplexograptus confertus, Glyptograptus austrodentatus* three zones	Llanvirnian
	Lower	O_2	*Didymograptus hirundo* zone *Didymograptus deflexus* zone *Dichograptus separatus* zone	Arenigian
		O_1	*Clonograptus tenellus* zone *Dictyonema flabelliforme* zone	Tremadocian
Cambrian	Upper	€$_9$	*Tellerina-Calvinella* zone *Quadraticephalus-Dictyella* zone *Ptychaspis-Tsinania* zone	Trempealeauan
		€$_8$	*Kaolishania* zone *Changshania* zone *Chuangia* zone	Franconian
		€$_7$	*Drepanura* zone *Blackwelderia* zone	Dresbachian
	Middle	€$_6$	*Damesella, Amphoton, Crepicephalina* *Liaoyangaspis* four zones	Albertian
		€$_5$	*Bailiella, Poriagraulos abrota, Sunaspis, Kochaspis hsuchuanensis* four zones	
		€$_4$	*Shantungaspis* zone	
	Lower	€$_3$	*Redlichia murakamii*- *Hoffetella* zone	Wacoubian
		€$_2$	*Megapalaeolenus, Palaeolenus, Drepanuroides, Malungia* four zones	
		€$_1$	*Eoredlichia, Hyolithus* two zones	

Fig. II–10. Fossil zonation of lower Paleozoic in China and correlation with other parts of the world.

System	Series	Formation	Fossil Zonation or Representative Fossils	Correlation with foreign country
Permian	Upper	P_4	*Palaeofusulina* zone	Tatarian
		P_3	*Codonofusiella* zone	Kazanian
	Lower	P_2	*Yabeina* zone *Neoschwagerina* zone	Kungurian
		P_1	*Parafusulina* zone: *Cancellina* subzone, *Misellina* subzone	Artinskian Sakmarian
Carboniferous	Upper	C_4	*Pseudoschwagerina* zone	Asselian Gzhelian Kasimovian
			Triticites simplex zone	
	Middle	C_3	*Fusulina*-*Fusulinella* zone	Moscovian Bashkirian
			Gastrioceras-*Eostaffella subsolana* zone	Namurian
	Lower	C_2	*Yuanophyllum* zone *Kueichouphyllum sinensis* zone	Viséan
		C_1	*Pseudouralinia* zone *Cystophrentis* zone	Tournaisian Etroeungtian
Devonian	Upper	D_8	*Yunnanella*-*Yunnanellina* zone	Fammenian
		D_7	*Cyrtospirifer sinensis* zone	Frasnian
	Middle	D_6	*Stringocephalus* zone *Parabornharatina* zone	Givetian
		D_5	*Acrospirifer houershanonensis*-*Utaratuia sinensis* zones	Eifelian
	Lower	D_4	*Euryspirifer paradoxus*-*Trepezophyllum cystosum* zone *Otospirifer daleensis* zone *Subcuspidalla trigonata* zone	Emsian
		D_3	*Euryspirifer tonkinensis*-*Xystriphyllum nobilis* zones	Siegenian
		D_2	*Orientospirifer nakaolingensis* zone	
		D_1	*Polybranchiaspis*-*Yunnanolepis* zones	Gedinian

Fig. II-11. Fossil zonation of upper Paleozoic in China and correlation with other parts of the world.

tone. These deposits accumulated in a wide stretch of shallow-marine waters under conditions of high energy. Lithofacies change vertically more than laterally. Fossils are abundant, and are related closely to Indo-Pacific types.

Along the southeastern margin of the Eerduos basin, Cambrian strata crop out from Lishi in Shanxi Province to Hancheng in Shaanxi Province. At Hejin, in southwestern Shanxi Province, Zhang Wei (1983) designated a type section with 56 genera of trilobites distributed among 23 families. In this type section the Cambrian-Ordovician boundary falls within an interval of dolomitic limestone, and between the Cambrian trilobite *Calvinella* and the Ordovician trilobite *Aotiaspis* cp. *karaipsis.* Lithology and fauna suggest to us a transitional boundary.

Northwest China

The Lower Cambrian consists of volcanic rocks that

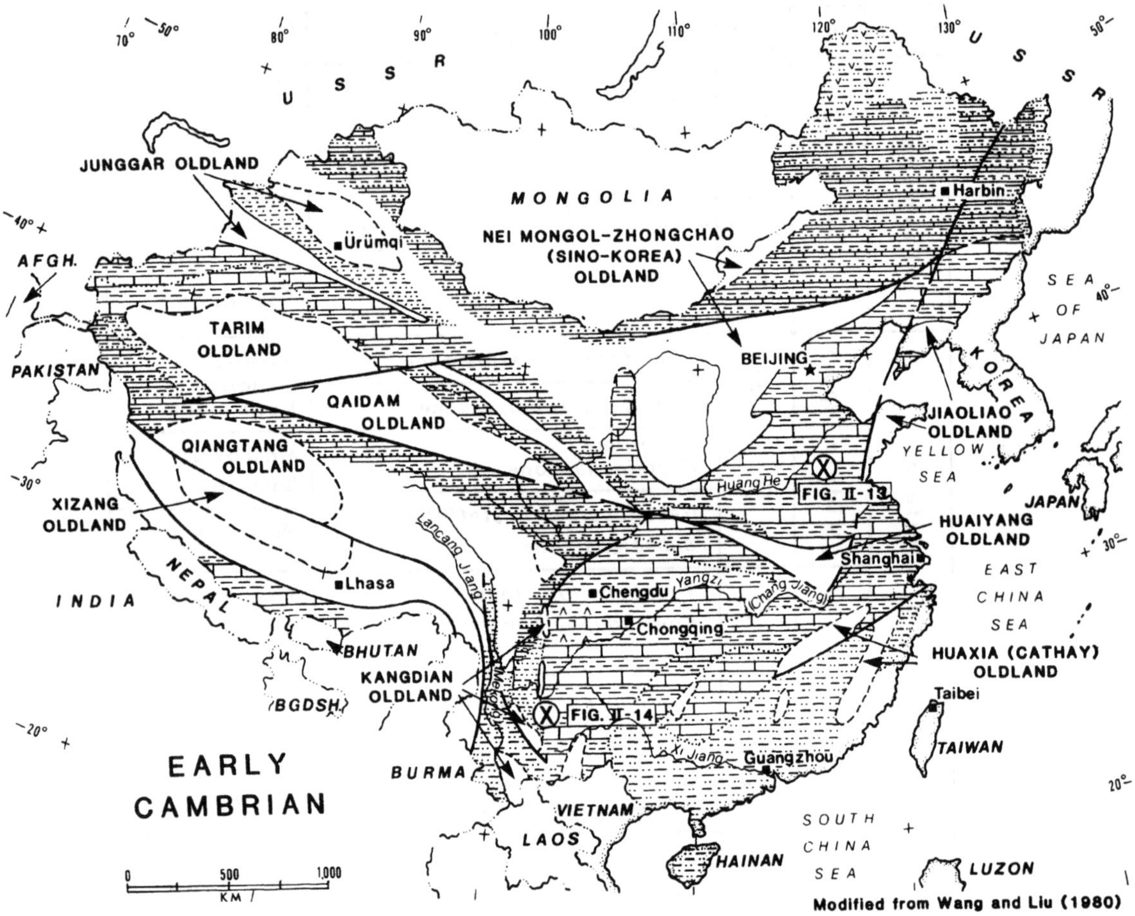

Fig. II-12. Paleographic and lithofacies map of Early Cambrian, China.

alternate with marine deposits. The Middle and Upper Cambrian deposits and faunas of the Tian Shan and of the northern part of the Qilian Shan are almost identical with those in Central-South and East China, whereas the faunas of the southern part of the Qilian Shan are similar to those of North China and Siberia (U.S.S.R.)

Central-South and East China

The Lower Cambrian consists of medium- to fine-grained terrigenous clastics with some siliceous and carbonate deposits. The Middle and Upper Cambrian

Table II-5. Cambrian realms of Trilobites in China (from Lu, 1981)

I. Oriental Realm:

- I. North China Province
  - A. North China–southern East China Subprovince
  - B. Yangzi–Qinghai–Sichuan–Xizang Subprovince
- II. Southeast China Province
  - A. Jiangnan–Northwest China Subprovince
  - B. Zhu–Jian Subprovince
- III. Transitional mixture of North China and Southeast China Provinces

II. Occidental Realm:

- I. North American Province (or "Pacific" Province)
- II. Atlantic Province
- III. Transitional mixture of North American Province and Atlantic Province

III. Transitional or mixed faunas

Intermediate areas between Oriental and Occidental faunas

Table II-6. Cambrian brachiopod assemblages in China (from Wang et al., 1981)

Stage	Assemblage
Upper Cambrian	
Fengshan Stage	*Mesonomia–Huenella* Assemblage
Changshan Stage	*Eoorthis–Palaeostrophia* Assemblage
Middle Cambrian	
Hsüchuang Stage	*Wimanella–Wynnia* Assemblage
Lower Cambrian	
Maochuang Stage	*Paterina–Cambrotrophia* Assemblage
Manto Stage	*Nisusia–Eoconcha* Assemblage
Tsanglangpu Stage	*Kutorgina–Israelaria* Assemblage
Chiongchussu Stages	*Diandongia* Assemblage
Meischucun Stage	*Ocruranus–Scambocris* Assemblage

Table II-7. Correlation of Lower Cambrian biostratigraphy in Central and Southwest China (from Yuan and Zhang, 1981)

Series	Stage	Substage	Zones or assemblages	Western Region: East Yunnan		Western Region: South Shaanxi (West)	Central Region: North Guizhou	Eastern Region: North Siehuan (Chengkou)	Eastern Region: West Hubei
Lower Cambrian	Lungwangmiao Stage		*Redlichia guizhouensis* Zone	Lungwangmiao Formation		Kongmingdong Formation	Qingxudong Formation	Shihlungtung Formation	
			Redlichia murakamii–Hoffetella Zone						
	Tsanglangpu Stage	Wulongjing Subst.	*Megapalaeolenus* Zone	Tsanglangpu Formation	Wulongjing Member	Yangwangbian Formation	Chingtingshan Formation	Tianheban Formation	
			Palaeolenus Zone					Yingzuiyan Formation	Shihpai Fm.
		Hongjingshao Substage	*Paokannia–Sichuanolenus* Zone		Hongjingshao Member				Shuijingtuo Formation
			Metaredlichioides–Chengkouia Zone						
			Drepanuroides Zone				Minghsingssu Formation	Liangshuijing Formation	
			Yunnanaspis–Yiliangella Zone			Xiannüdong Formation			
			Malungia Zone						
	Chiungchussu Stage		*Eoredlichia* Zone	Chiungchussu Formation		Kuojiaba Formation	Niutitang Formation		
			Parabadiella–Mianxiandiscus Zone						
	Meishucun Stage		*Siphogonuchites–Zhijintes–Sachites* assemblage	Meishucun Formation		Kuanchuanpu Formation		?	Huangshandong Member
			Anabarites–Circotheca–Protohertzina assemblage						
Late Precambrian				Töngying Formation					

Table II-8. Comparison and range of important Cambrian Archaeocyathids and Trilobites of Central China and Southwest China and of Siberia (from Yuan and Zhang, 1981)

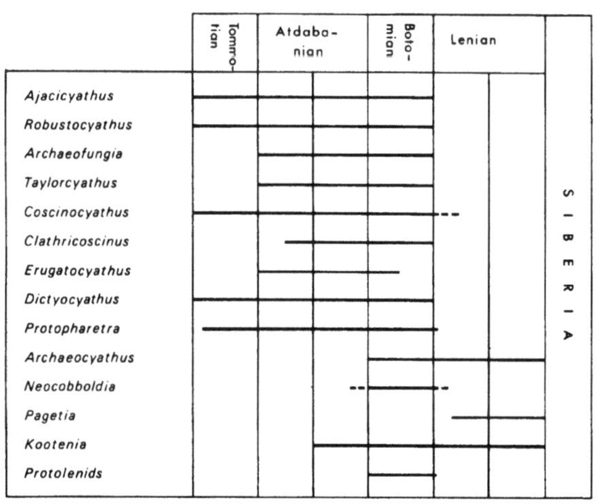

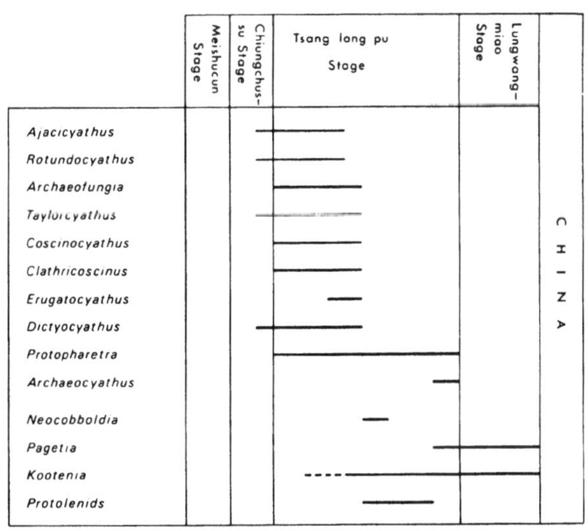

section is composed mostly of thin layers of black limestone. Faunas are more abundant above the Lower Cambrian. Most of the taxa are planktonic (e.g., the trilobite *Agnostus*, and abundant cephalopods; Chen and Qi, 1981). A few taxa are benthic. These are closely related to Atlantic faunas.

Central-Southwest China

This region contains an unusually well-developed Lower Cambrian sequence of shale, sandstone, and oolitic limestone. During the Middle and Late Cambrian a large marine basin formed, and this received thick deposits

Table II–9. Tentative correlation of Lower Cambrian of some areas of the World (from Yuan and Zhang, 1981)

	China		Siberia	South Australia	Morocco	North America
Lower Cambrian	Lungwangmiao Stage		Lenian	No. 7	Niveau d'Ouriken d'Ourmast	*Bonnia-Olenellus* Zone
					Asrir	
	Tsanglangpu Stage	Wulongjing Substage	Botomian	Clastics	Issafence	
				No. 4		
		Hongjingshao Substage	Atdabanian	Ajax Limestone	Amouslekien	*Nevadella* Zone
					Ouneinien	*Fallotaspis* Zone
	Chiungchaussu Stage		Tommotian	No. 2 No. 1 Parachilna Fm.	Adoudounien Supérieur	Pre-trilobite Zone
	Meishucun Stage			Uratanna Fm.		
Late Pre-cambrian	Töngying Formation		Judoma Suite	Pound Quartzite	Adoudounien Inférieur	

Note: Units not drawn to scale.

of dolomitic limestone, dolostone, and limestone. Benthic Indo-Pacific faunas, abundant on the margins of the basin, are less abundant in the central part. Yuan and Zhang (1981) presented a correlation of Lower Cambrian biostratigraphy (Table II-7), a fauna of Lower Cambrian archaeocyathids and trilobites (Table II-8), and a tentative Lower Cambrian global correlation (Table II-9).

Also, Zhang Senqui (1989) studied the archaeocyathids distribution of Lower Cambrian in China and he concluded that all the regions south of the Tian Shan-Da Hinggan belt belong to the Oriental Realm. The Tian Shan-Da Hinggan is considered part of the Intermediate Realm together with the Siberian Platform.

Blank Areas

Several blank areas shown on Figure II–12 in western China such as the Qinghai-Xizang (Tibet) Plateau, are covered by thick sections of younger rock. Cambrian shallow-water facies have been penetrated locally in the Tarim basin, but is absent in other wells. The distribution of Cambrian strata in these blank areas of western China therefore still remains to be determined.

Type Sections

North China

As noted above, Zhang Wei (1983) designated a type section of the Cambrian for North China at Hejin in southwestern Shanxi Province. Another type section (Fang et al., 1979) lies in Shandong Province (Figure II-13). An angular unconformity separates the Lower Cambrian from the underlying Archean, and the total thickness of the Cambrian section is 810 m. Obviously a long period of erosion took place after the Luliang movement (Table II-2). Only $€_3$ (part of the upper third) is present in the Lower Cambrian of this type section, with purplish-red shale (containing halite pseudomorphs) and interbedded argillaceous limestone. In the present Sichuan basin, evaporite beds are present (Wang and Liu, 1980). This section and fauna are nearly identical to equivalent sections on the Indian shield, in Kashmir, Iran, and Oman (Wolfart and Kursten, 1974; Wolfart, 1983).

The Middle Cambrian forms a transitional sequence from the predominantly purple shale below to limestone ($€_4$, $€_5$) above and to oolitic limestone ($€_6$) at the top. The latter is notably rich in trilobites. The Upper Cambrian consists of basal conglomerate and of intraformational edgewise limestone breccia interbedded with thin layers of limestone and purplish-red shale. Basal conglomerates indicate the presence of depositional hiati. Calcareous cement has indurated the conglomerates. A major discovery in recent years has been the finding of a rich cephalopod fauna in the Upper Cambrian of North China, the oldest such fauna known in the world (Chen Jun-yuan et al., 1979a, 1979b).

Southwest China

The type section of the Cambrian in Southwest China

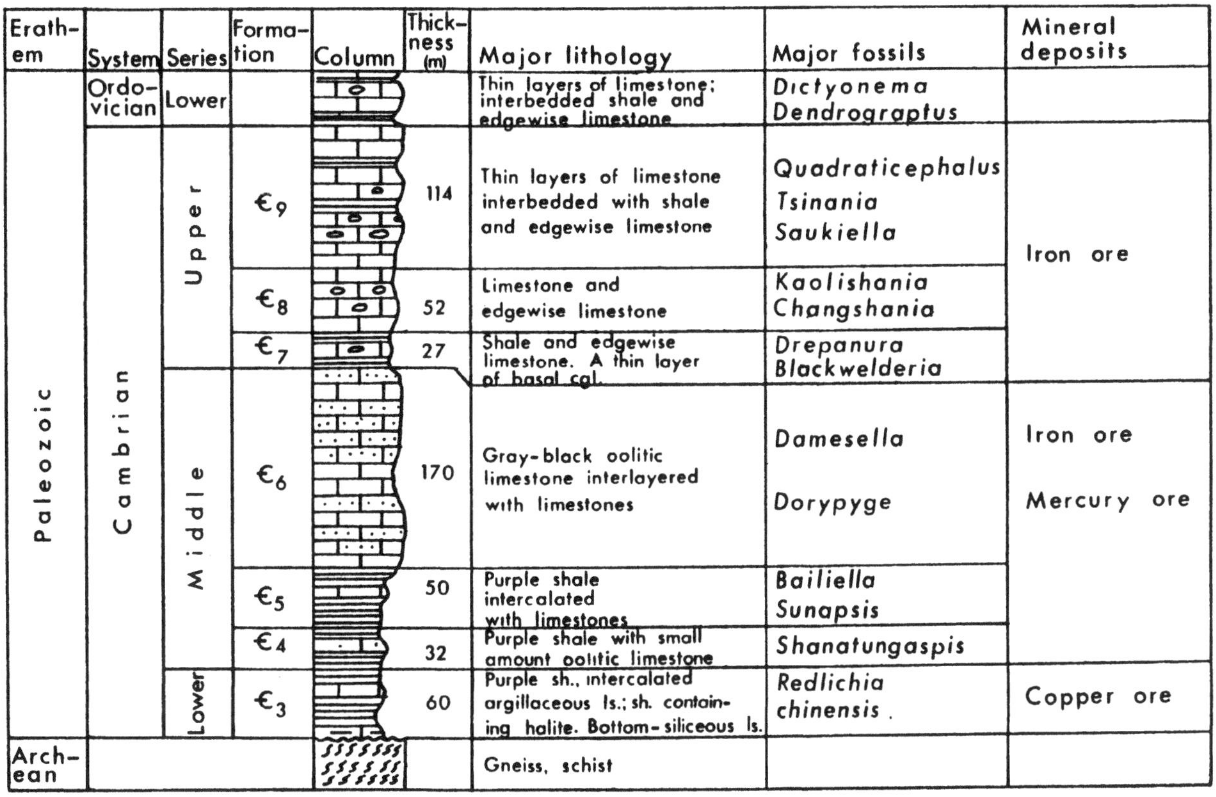

Fig. II-13. Type stratigraphic column of Cambrian System in Shangdong Province, North China. Location is shown on Figure II-13 (from Fang et al., 1979; mineral deposits after Lu, 1962).

lies in the eastern part of Yunnan Province where the Lower Cambrian is best developed (Figure II-14). Lu (1962) discussed the problem of the lower boundary of the Cambrian in this region. A thin layer below $€_1$ consists mainly of dark-gray, fine-grained, terrigenous-clastic strata and chert containing the enigmatic *Hyolithus* with a phosphatic bed at the base. Some geologists considered this phosphatic bed to be latest Sinian (latest Proterozoic); others considered it Early Cambrian; still others considered it transitional between the two. In some areas the layer in question appears to be either conformable or only slightly disconformable on Zz_4 (uppermost Sinian). (For an excellent discussion of the age, see Misch, 1942).

Chang (1980) also discussed the problem of the lower boundary of the Cambrian in eastern Yunnan Province, especially in the column exposed at Jinning (Junyang). Below the trilobite *Parabadiella* there occurs a shelly fauna with hyolithids, brachiopods, gastropods, monoplacophorids, tommotidiids, conodontophorids, rostroconchidiids, and merismoconchidiids. These and other taxa constitute the Meishucunian, an evident correlative of the Tommotian of Siberia (U.S.S.R.). Chang placed the lower boundary of the Cambrian at the base of the shelly Meishucunian in the column at Jinning (Kunyang; Jiang Zhiwen et al., 1989).

The $€_1$ and the $€_2$ divisions of the Cambrian are well developed in Southwest China (Fang et al., 1979), from where they wedge out gradually toward the northeast (Figure II-12). Even in Southwest China, however, the lower member of $€_1$ covers only a limited area. The $€_2$ division of South-Central and East China, moreover, and the $€_3$ division of North China, overstep the $€_1$ onto older Proterozoic, and, in many places, onto Archean. This relation indicates a gradual marine transgression from southwest to north during the Early Cambrian (Zhang Sengui, 1989).

Paleoecology and Paleogeography

Lu et al. (1965) published one of the most complete studies of the paleogeography and lithofacies of the Cambrian. His maps now are largely out of date, and regrettably they contain numerous errors. One point, however, is clear from his maps: the distribution of land and sea during Cambrian time changed greatly in response to differing types and differing amounts of crustal movement after the close of the Sinian (late Proterozoic). One of the principal phenomena illustrated by the maps of Lu et al. (1965) is a diverse topography in the region of what is now China.

Figure II-12, a lithofacies and paleogeographic map of the Early Cambrian, is modified from the work of Wang and Liu (1980). During $€_1$, earliest Early Cambrian, marine waters began to invade a flat-lying area

System	Series	Formation	Column Section	Thickness (m)	Major lithology	Major fossils	Mineral deposits
Silurian	Middle				Argillaceous ls. and sh. Basal conglomerate		
Cambrian	Middle	$€_5$		28–323	limestones, argillaceous ls., dolomitic ls., interbedded sandy shale	*Manchuriella* *Protahedinia yunnanensis*	Mercury ore
		$€_4$		14–67	sandy shale interbedded with dolomitic ls.	*Kutsingocephalus* *Douposiella*	
	Lower	$€_3$		48	gray argillaceous and dolomitic limestones	*Redlichia* sp *Hoffetella transversa*	
		$€_2$		155	Upper: gray-green sandy sh. interbedded with thin sandstone layers. Lower: gray thick quartz ss. interbedded with sandy shale	*Palaeolenus* *Redlichia mai*	
		$€_1$ Upper member		127	Dark gray-black shale and sandy sh. interbedded with thin fine sandstone	*Yunanocephalus* *Eoredlichia*	
		$€_1$ Lower member		60	Gray-black thin layer of fine sandstone and siltstone, phosphatic layer at the bottom	*Hyolithes*	Phosphate ore
Sinian	Z_z				Fault(?) Limestone		

Fig. II-14. Composite stratigraphic column of Cambrian System in eastern Yunnan Province, Southwest China. Location is shown on Figure II-12 (from Fang et al., 1979; mineral deposits after Lu, 1962).

in both Southwest and Central-South China during a transgression from the Indo-Pacific realm (Lu et al., 1965; Wang and Liu, 1980). The advance took place in two directions; east-northeastward; and northwestward. Chinese geologists refer to the transgressed area as the Shang Yangzi (Upper Yangtze) Sea. The east-northeastern branch of this sea was bounded by three connected ancient land areas, here called 'oldlands.' The Huaiyang oldland lay mainly on the east, Huaxia (Cathay) mainly on the southeast, and Nei Monggol-Zhongchao (Sino-Korea) mainly on the north (Figure II-12). The predominantly argillaceous deposits of $€_1$ time (Figures II-14, II-15, and II-16) are interbedded with sandstone in the northeast, and probably reflect the existence of shoal and coastal environments. The source area of the terrigenous clastics presumably was in the highland of the Huaxia oldland in the southeast (Figure II-12).

Land in the northwest stood noticeably low (Lu et al., 1965). The northwestern branch of the sea extended to the present Qilian Shan and Tian Shan, thereby connecting with Tethys. The resulting Tian Shan-Qilian Shan geosyncline was bounded by a postulated, but unproved, Junggar oldland on the north (Figure II-12) and by a second postulated, also unproved, oldland on the south (Xizang (Tibet)-Tarim oldland). The Tian Shan-Qilian Shan geosyncline was filled mainly with volcanic rocks interbedded with sandstones in the central part of the basin, and with carbonates in the western part. The carbonates indicate the existence of a stable and relatively shallow depositional environment in the west.

During $€_2$, the middle part of the Early Cambrian (Figures II-12 and II-15), the sea transgressed from the Shang Yangzi region toward the east-northeast, thereby separating the Huaxia-Huaiyang oldlands from the Nei Monggol-Zhongchao oldland (Lu et al., 1965). The widespread distribution of carbonate rocks and of the trilobite *Paleolenus* in these areas may indicate the former presence of a neritic environment. Reduced volcanic activity, and more carbonate deposition with the trilobite genus *Kingaspis*, characterized the geosyncline of Tian Shan and Qilian Shan. The presence of *Kingaspis* indicates an increasing water depth and suggests the

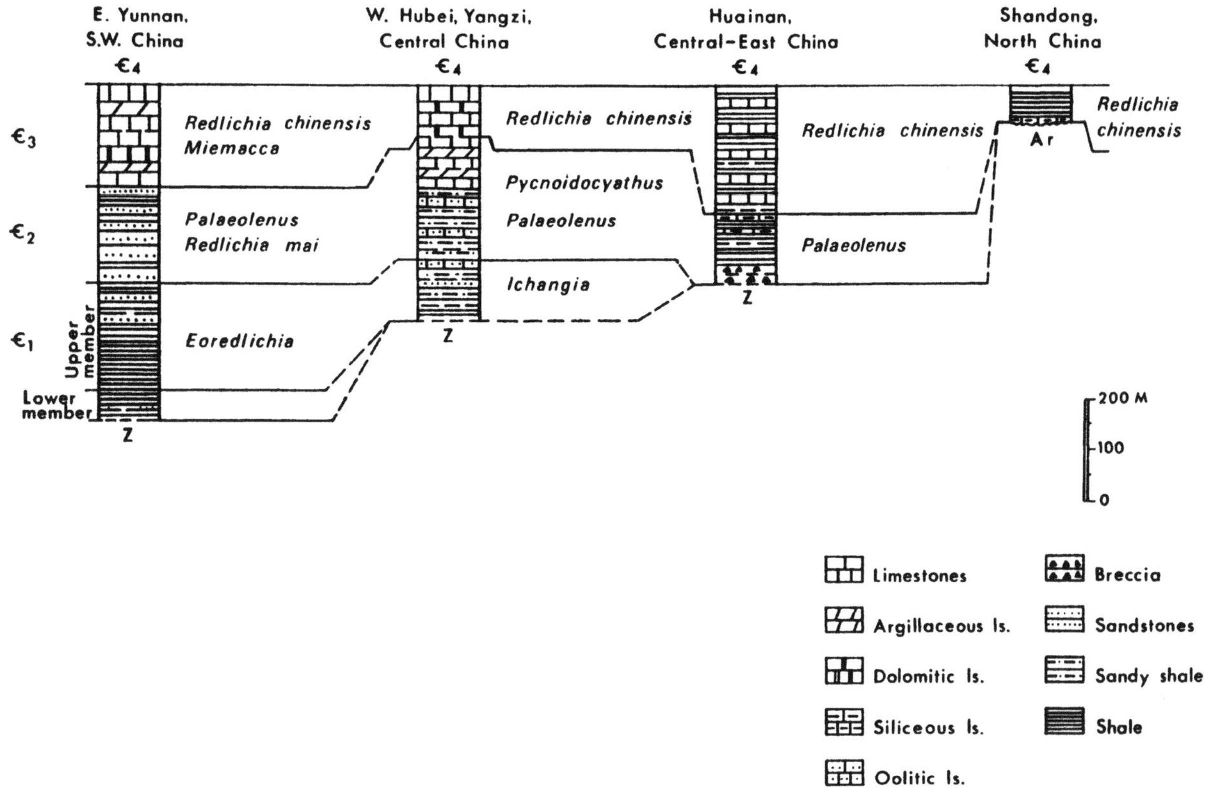

Fig. II-15. Stratigraphic correlation of Lower Cambrian from Southwest China to North China (after Geological College of Chengdu, 1974 *in* Fang et al., 1979).

possibility that these waters were connected with those of middle and western Siberia (U.S.S.R.). The sea also transgressed into the Tethys of the present Himalayas region. Abundant archaeocyathids, correlative with those in Australia, Siberia, and the Mediterranean, appeared in Central-South China (Table II-8).

During €$_3$, the latest part of the Early Cambrian, continued subsidence and marine ingression brought about a gradual diminution of the Nei Monggol-Zhongchao and Huaxia oldlands (Lu et al., 1965; Figure II-12). Most of northern China became part of a shallow continental shelf whose neritic sediments include purplish-red shales with a few sandstone and limestone beds. Halite pseudomorphs, present in some areas, indicate a high rate of evaporation typical of a hot arid climate. Carbonates predominate in the more subsident sectors. In Central-South and East China, a comparative stagnation of waters and an onset of reducing conditions took place (Figure II-12), with deposition of black carbonaceous pyritic shale. In Central-South and Southwest China, a stable basin with somewhat deeper marine water was characterized by a warm humid climate. Sediments deposited in this basin now form thick carbonates with some oolitic and edgewise intraformational

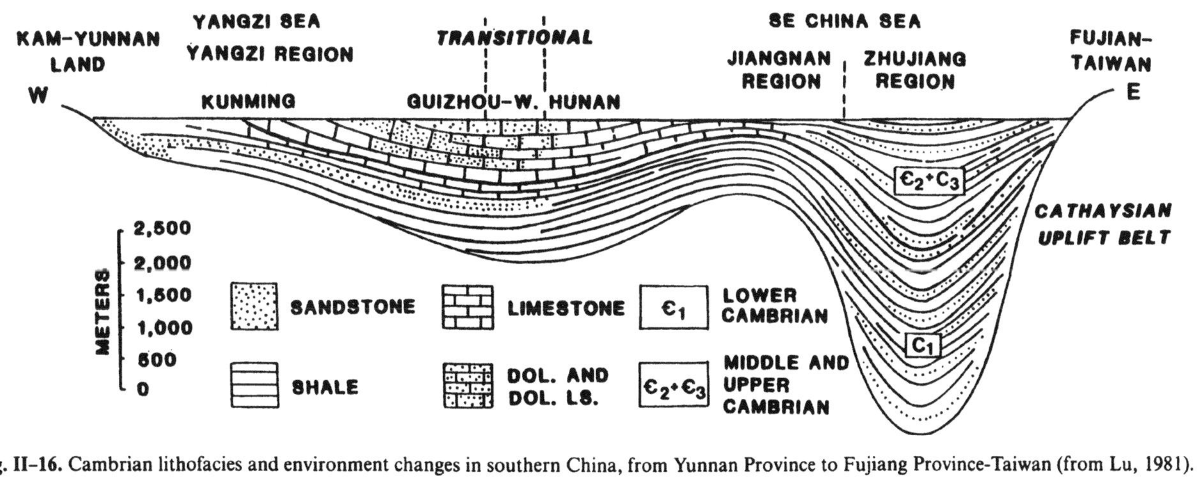

Fig. II-16. Cambrian lithofacies and environment changes in southern China, from Yunnan Province to Fujiang Province-Taiwan (from Lu, 1981).

limestone conglomerate. The €_3 faunas of North China, northern Central-South China, and Southwest China are similar. The fauna common to southern Central-South China and East China is quite distinctive, and different from faunas in the north.

During €_6, in the middle part of the Cambrian, a maximum marine transgression took place (Lu et al., 1965). Neritic carbonates formed widely, whereas terrigenous-clastic deposition was mainly in areas that fringed exposed oldlands and islands. In North China and in the southern part of Northeast China, a change from predominantly purple shale (€_4 and €_5) to thick carbonate units with oolitic limestone (such as €_6) characterizes the Middle Cambrian. An increase in the area and depth of marine water, as well as a more abundant trilobite fauna, accompanied marine transgression and a warmer climate. In Central-South and Southwest China, carbonates also formed in transgressing marine waters and under a warm climate. In southern Central-South China, coarser-than-normal terrigenous clastics accumulated near the source areas of the Huaxia oldland, while finer grained terrigenous clastics, carbonates, and deposits with carbonaceous constituents, accumulated farther from the oldland (Figure II–16). In Northwest China, volcanism continued.

At the same time (€_6), the diversity of marine taxa reached its peak in an environment whose favorable elements were adequate water depth, good circulation, and a warm climate. This is particularly evident in North, Northeast, Southwest, and northern Central-South China. The character of the trilobite fauna in these regions is close to that of the typical Indo-Pacific fauna. In southern Central-South and East China, on the other hand, a reducing environment predominated and was not favorable to the growth of benthos (Figure II–17). In these regions a pelagic fauna with the trilobite *Agnostus* is closely related to Atlantic fauna. In Northwest China, faunas are mixed, but the Atlantic type predominates.

Some general diastrophism took place between €_6 and €_7 (Middle and Late Cambrian), but in North China diastrophism took the form of local uplift (Lu et al., 1965), whereas in the southern part of Northeast China, gradual subsidence occurred.

During €_7, in North China, edgewise limestone conglomerate accumulated as a result of wave erosion along beaches. Reddish rims around the components of these conglomerates may indicate oxidation under shoal conditions of high energy, changing water depth, and a warm climate. The purplish-red color of accompanying shale also may indicate a warm climate. In Central-South China and Southwest China, the depositional environment was similar to environments of North China, and to environments of the southern part of Northeast China. Volcanism continued in the orogenic belts of Northwest China. Terrigenous clastics with flyschoid depositional patterns accumulated near the Huaxia oldland (Figure II–12) whereas fossiliferous carbonates dominated more distal sectors.

The depositional environment of €_8 was closely similar to that of €_7 (Lu et al., 1965). Marine transgression, however, was slightly stronger, and it increased still more in €_9, with a corresponding increase in depositions of carbonate. Dominance of the Indo-Pacific and Atlantic faunas persisted throughout the Late Cambrian.

Figure II–18 shows, in generalized form, the paleozoogeographic provinces of China during the Cambrian, as interpreted by Lu (1981).

ORDOVICIAN SYSTEM

Chang (1962) asserted that the best and most complete sequences of the Ordovician in the entire world occur in China. These sequences consist of richly fossiliferous carbonates. Marine transgression in China reached a maximum during the Ordovician, and the population of marine invertebrates reached a peak.

Distribution and Facies

Chang (1962) recognized eight distinct regions of Ordovician rocks, based on their fossil assemblages and their sedimentary tectonics. These regions are shown on Figure II–19, a paleogeographic map of China for the

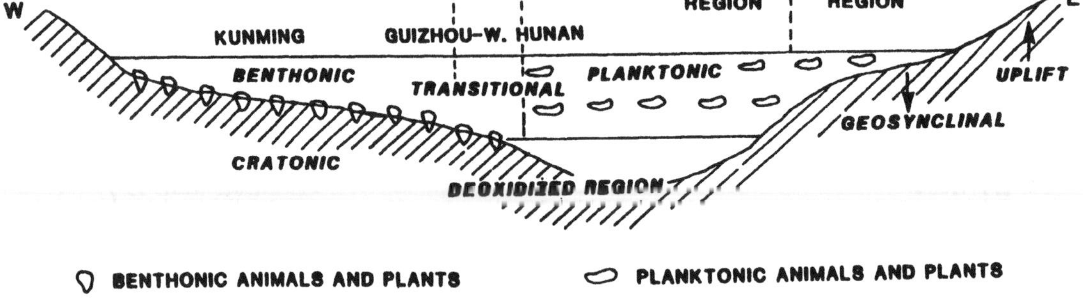

Fig. II–17. Ecology of Cambrian Seas in southern China (from Lu, 1981).

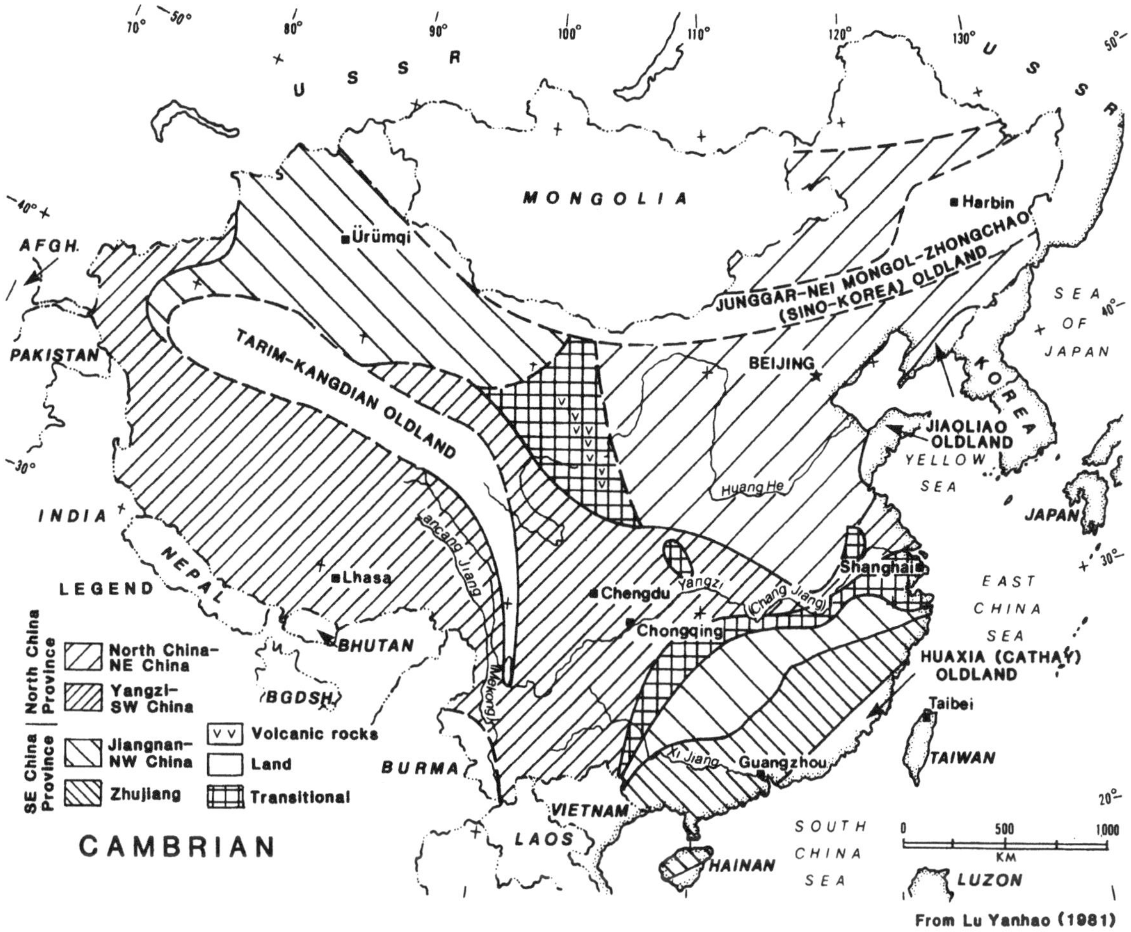

Fig. II–18. Cambrian paleozoogeography map of China.

Early Ordovician (Wang and Liu, 1980). These regions are: (I) Da Hinggan Ling; (II) North China (southern part of Northeast China); (III) Northwest China; (IV) Qin Ling; (V) Central-Southwest China; (VI) southern Central-South and East China; (VII) western Yunnan Province; and (VIII) Xizang (Tibet) Autonomous Region. Most regions, especially (I), (IV), (V), (VII), and (VIII) were geosynclinal depositional regimes with fossil assemblages of southern affinities similar to those found in Europe. The depositional environment of (II) was that of a platform, with more northern fossil assemblages related somewhat closely to Pacific faunas. Within (VI), a parageosyncline is present with a southern type of fauna. In (III), the best-known regime is geosynclinal, with a southern type of fauna, although fossils of both the northern type and the platform occur. Places that appear blank on Lu's paleogeographic map, such as the Tarim platform, yield sections of Ordovician platforms facies in a few exploratory wells, while in other exploratory wells the Ordovician section is missing. The extent of Ordovician strata in these blank areas thus still is unknown.

Type Sections

Central-South China

A complete Ordovician sequence without faunal breaks of any consequence crops out at Yichang in the Yangzi Gorges of Central-South China (Figure II–20; also see Figure II–10). Ordovician boundaries with the underlying Cambrian and the overlying Silurian in this region are mainly transitional. The Lower Ordovician contains principally carbonates interbedded with a few thin layers of shale rich in trilobites, brachiopods, cephalopods, and graptolites. The lower and middle parts of the Middle Ordovician consist predominantly of nodular limestone with polygonal patterns, rich in orthoceratid cephalopods. The upper part of the Middle Ordovician consists of black shale with abundant graptolites.

The Upper Ordovician sequence closely resembles that of the upper part of the Middle Ordovician. Sheng (1974) stated that a layer rich in the trilobite genus *Dalmanitina* (widely distributed in northern Central-South China and Southwest China) represents the uppermost part of the Upper Ordovician. The sequences on the west consist mainly of terrigenous-clastic strata, whereas the sequen-

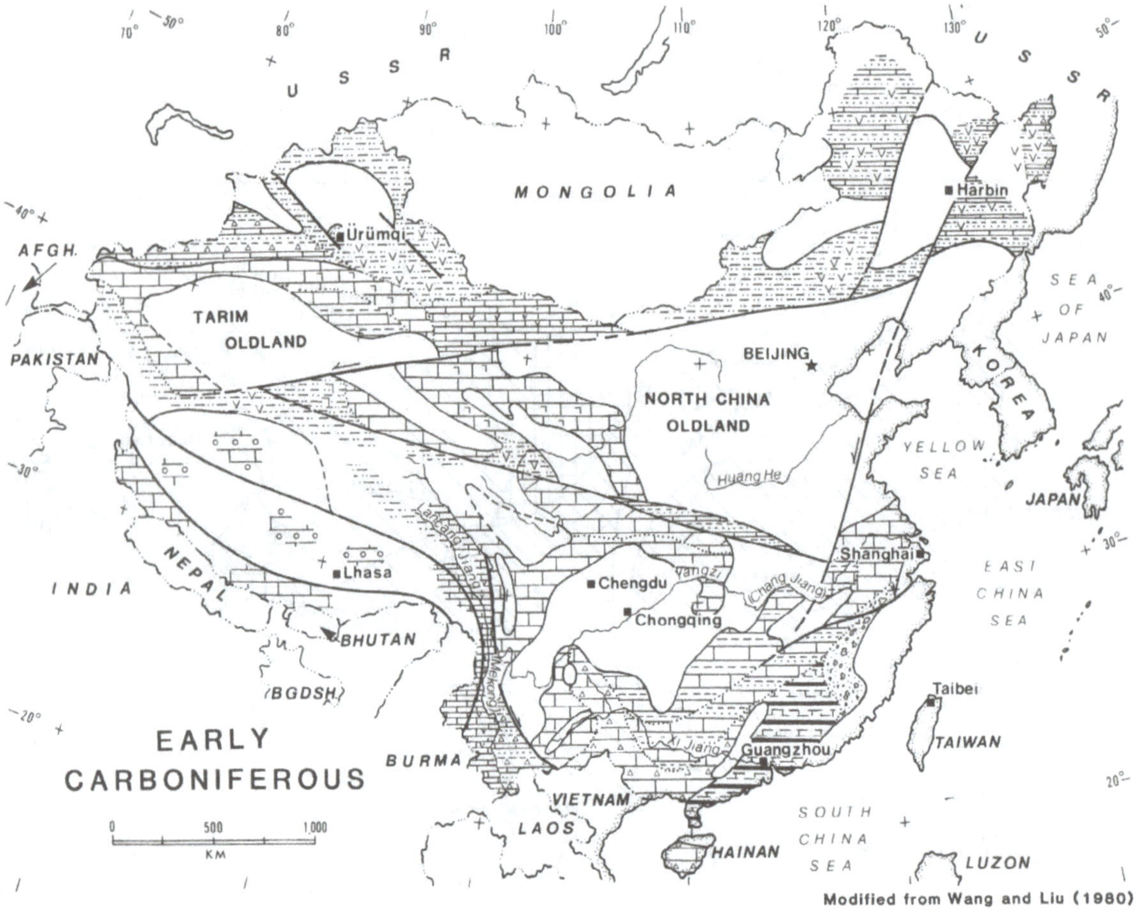

Fig. II-19. Paleogeographic and lithofacies map of Early Ordovician, China.

ces on the east consist entirely of carbonate rocks. The terrigenous clastics indicate provenance from an ancient highland on the west, whereas the carbonates indicate a comparative deepening of marine waters on the east.

North China

Most of the Ordovician sequence in North China consists of limestone (Figure II–21) deposited principally in a transgressive sea. Much of this has metasomatized to dolomite today. The entire area sank slowly during the Ordovician, and in some areas this submergence had continued unbroken from the latest Cambrian. As noted in a preceding section, the section at Hejin in southwestern Shanxi Province is unbroken by hiati. In dolomitic limestone embracing the Cambrian-Ordovician boundary, the trilobite *Calvinella* indicates latest Cambrian; the trilobite *Aotiaspis* cf. *karaipsis* indicates earliest Ordovician. In other sectors a faunal change takes place, even though no lithologic change is apparent. Evidence for discontinuity at the Cambrian-Ordovician boundary in some areas is provided by a basal conglomerate. Whereas *Saukia* sp. and other trilobites are abundant in Є$_9$ of the Cambrian, graptolites and brachiopods become dominant in O_1 of the Ordovician. Interlayered limestone, edgewise limestone conglomerate, and shale that are present in the section appear to have formed while slow transgression and regression took place in the epicontinental sea of the latest Cambrian and the earliest Ordovician.

In O_2 (Figure II–21), the latest Early Ordovician, a lower richly fossiliferous limestone grades upward into a dolostone with sparse fossils. The Middle Ordovician consists of thick limestone beds with gastropods and cephalopods, interbedded with dolostone, dolomitic limestone, and/or cherty limestone. A disconformity separates the Lower Ordovician from the Middle Ordovician. At the end of the Ordovician, uplift affected North China and persisted until the Middle Carboniferous. Despite the consequent great stratigraphic hiatus, only a disconformity separates Ordovician from Middle Carboniferous across the Sino-Korean platform, an area of tens of thousands of square kilometers. On this platform, the youngest Ordovician usually is Middle Ordovician, but locally is Caradocian (early Late Ordovician, Mu En-zhi et al., 1986). No agreement exists concerning reasons for the absence of pre-Carboniferous, post-Caradocian strata on this platform.

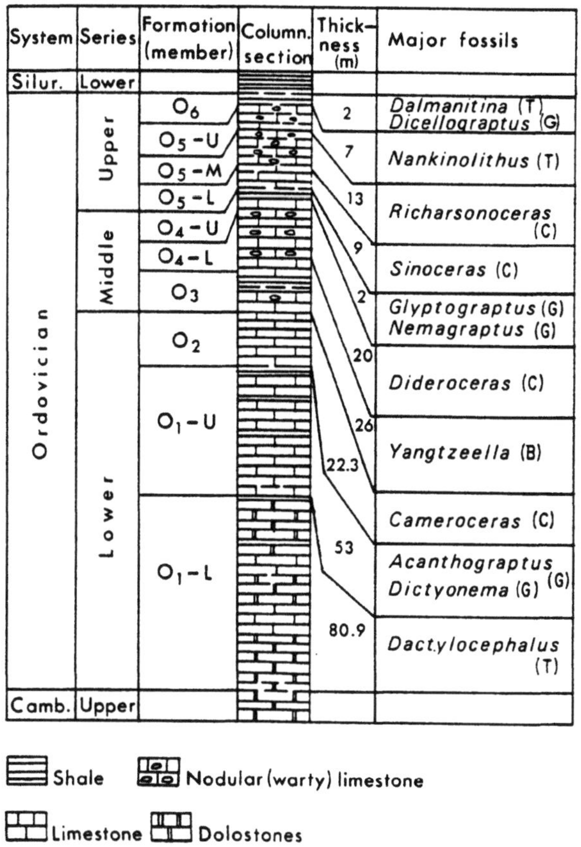

Fig. II-20. Stratigraphic type section of Ordovician System in Yichang (Yangzi Gorges), Hubei Province, Central-South China. Location is shown on Figure II-19 (from Fang et al., 1979).

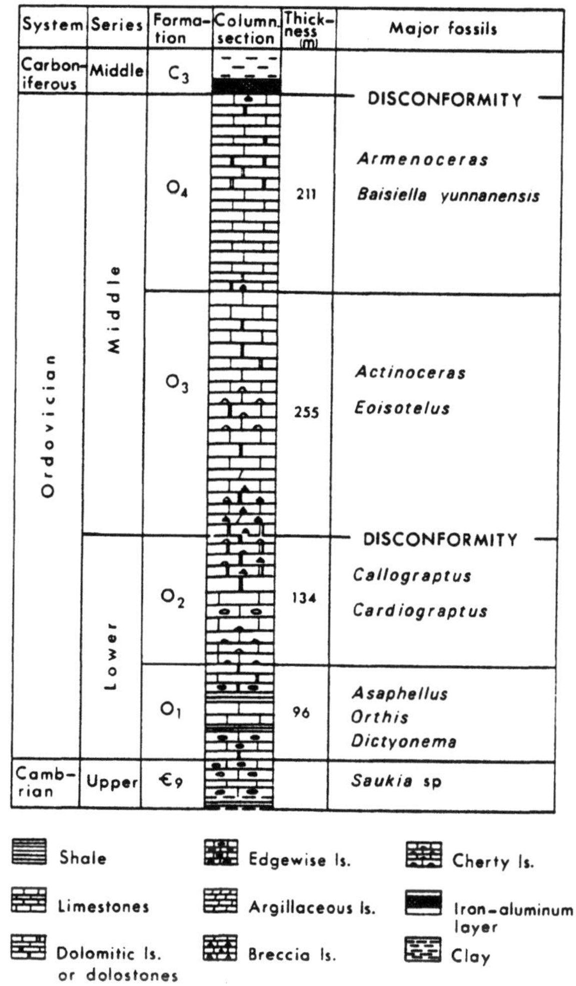

Fig. II-21. Stratigraphic type section of Ordovician System in Tangshan, Hebei Province, North China. Location is shown on Figure II-19 (from Fang et al., 1979).

Faunal Boundaries

The base of the Ordovician corresponds with the lowest appearance of the graptolites *Rhabdinopora flabelliforme* and *Staurograptus*, the trilobites *Leiostegium* and *Yosimuraspis*, the conodont *Cordylodus intermedius* (Chen Jun-yuan et al., 1988).

An (1981) proposed a Cambrian-Ordovician boundary based upon the transition from paraconodont types to euconodont types (Table II-10). The two lower zones, with *Proconodontus* and *Cordylodus proavus*, contain paraconodont types believed to represent the Cambrian. The two upper zones, with *Drepanodus simplex* and '*Acodus*' *oneotensis-Acantodus costatus*, contain euconodont types believed to be of Ordovician age.

Sheng (1980) discussed the boundaries between subdivisions of the Ordovician in China in terms of graptolite zones. The zone of *Didymograptus abnormis* or of *Azygograptus suecius* and the zone of *Oncograptus* presumably represent respectively the top of the Lower Ordovician and the base of the Middle Ordovician. Graptolites are the index fossils for the boundary between the Middle Ordovician and the Upper Ordovician. In faunal terms, the boundary lies at the base of the *Climacograptus wilsoni* zone or at the base of the *C. spiniferus* zone. He (1978) recognized the coral assemblages of *Borelasma* and *Sinkiangolasma* as the uppermost faunal zone of the Ordovician in the Yangzi region. For the Ordovician-Silurian boundary in southern South-Central China, the faunal indicator is the highest occurrence of the trilobite *Dalmanitina* (Sheng, 1980).

Paleoecology and Paleogeography

Because no major crustal movement took place between the Cambrian and the Ordovician in China, both the distribution and the geological characteristics of marine basins during the Ordovician were very similar to those during the Late Cambrian (Lu, 1975). After continuous transgression during the early Ordovician, marine waters reached their maximum extent during the Middle Ordovician. At the end of the Middle Ordovician, a crustal uplift reduced the areas of marine basins and brought on a concomitant increase of land by the Late Ordovician.

Table II-10. Approximate correlation of Cambrian- Ordovician Conodont zones in China and other parts of the world (from An, 1981)

China			Iran (Müller, 1973)		Queensland, Australia (Druce and Jones, 1971)			Utah, U.S.A. (Miller, 1976)		
Ordovician	Yehli Formation		Ordovician	7	Ordovician	Warentian	*Chosonodina herfurthi–Acodus* Zone	Ordovician	*Cordylodus proavus* Zone	
		"Acodus" oneotensis–Chosonodina herfurthi Zone (Zone 4)					*Cordylodus rotundatus–C. angulatus* Zone			*Clavohamulus hintzei* Subzone
							Cordylodus prion–Scolopodus Zone			
				6		Datsonian	*Cordylodus oklahomensis–C. lindstroemi* Zone			
		Drepanodus simplex Zone (Zone 3)		5			*Drepanodus simplex–Oneotodus bicuspatus* Zone			*Hirsutodontus simplex*
										Clavohamulus primitus Subzone
							Cordylodus proavus Zone			*Fryxellodontus inornatus* Subzone
Cambrian	Fengshan Formation	*Cordylodus proavus* Zone (Zone 2)	Cambrian	4	Cambrian	Payntonian				*Hirsutodontus hirsutus* Subzone
								Cambrian	*Proconodontus* Zone	*Oistodus minutus* Subzone
		Proconodontus Zone (Zone 1)		3						*Proconodontus notchpeakensis* Subzone
										Proconodontus muelleri Subzone

Early Ordovician (Tremadocian and Arenigian)

In the basin of northern Central-South China and in Southwest China (region V of Figure II–19), a littoral zone was present along the western margin near the Kangdian oldland. The depth of marine waters increased gradually toward the east and southeast. Accordingly, Lower Ordovician lithofacies change from reddish to purplish sandstone on the west, where a notably dry coastal climate is presumed to have prevailed, to alternate shale and limestone deposited in shallow to neritic waters at the middle of the basin (largely a carbonate shelf), and, farther east, to limestone deposited in neritic to deep water (Figure II–22). In present-day southeastern Shanxi, southern Hebei, northern Henan, and western Shandong Provinces, beds of anhydrite and halite formed in large lagoons on this carbonate shelf (Wang and Liu, 1980). The marine shelfal carbonate basins of North China (Region II) and Northeast China (Region I) had become largely isolated from southern and southwestern China, but they connected with the ancestral Boreal Pacific Ocean on the east.

Data are limited for the rest of China, but a carbonate-shelf regime existed in the Tarim region of Northwest China, on the Gangdise block east and west of Lhasa in Xizang (Tibet) Autonomous Region, and in Qomolangma Feng (Mt. Everest) of the Himalayas block (Mu Anzi et al., 1973; Yin Jixiang et al., 1983). Here, nearly 900 to 1,070 m of carbonates was deposited on the shelf of an epeiric sea from the Arenigian to the middle Caradocian. These figures are approximately the same as those of central China, of the Yangzi platform, and of Nei Monggol Autonomous Region, from which this area was never far removed. Nei Monggol Autonomous Region, northern Xinjiang Uygur Autonomous Region, northern Gansu Province, and northernmost Heilongjiang Province were the sites of terrigenous-clastic deposition, possibly in an environment of shelf margin or of slope. Mafic volcanic breccias are present in southern Gansu Province and in northernmost Heilongjiang Province (Lai, 1984).

Middle Ordovician (Llanvirnian and Llandeilian)

The maximum transgression of the Ordovician sea took place during the middle of the Middle Ordovician. The Kangdian oldland on the west (Figure II–19) diminished considerably. In northern Central-South China and in Southwest China, terrigenous-clastic deposits accumulated in the west (Figure II–23), finer grained than those of the Early Ordovician. An alternation of shale and limestone accumulated in the middle of the area, and limestone formed on the east (Figure II–23). Deposition of carbonates may have taken place in a distinctly quiet

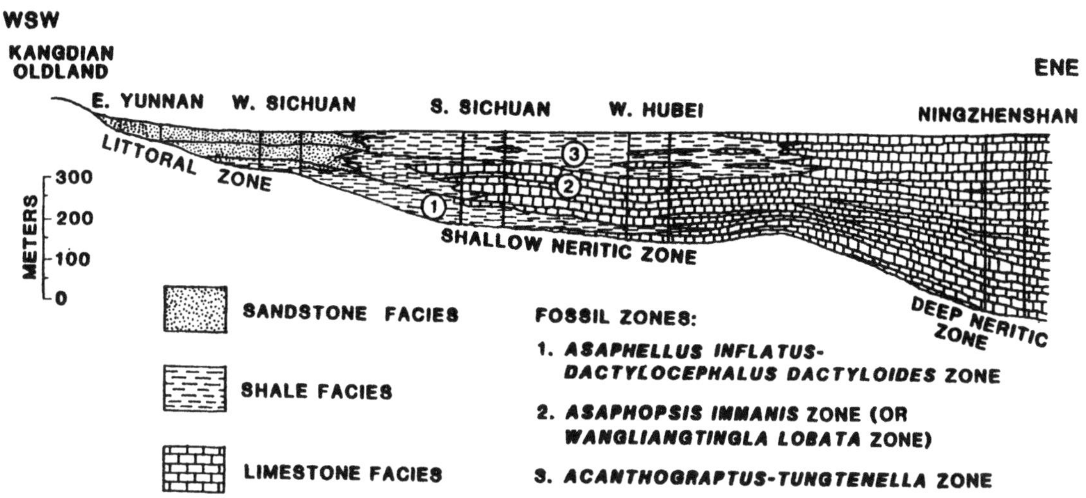

Fig. II-22. Profile of Lower Ordovician facies and environmental changes in Central-South China. Location is shown on Figure II-19 (from Lu, 1975).

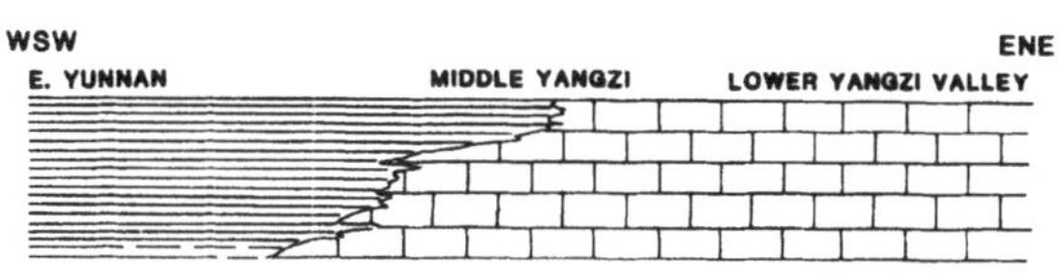

Fig. II-23. Profile of Middle Ordovician facies changes in Central-South China. Location is shown on Figure II-19 (from Lu, 1975).

neritic environment. The basins of North China, Northwest China, and Northeast China also were inundated by marine waters, and thereby they probably acquired marine connections not only with each other but also with basins in northern Central-South China, Southwest China, and East China. Marine waters in the geosynclines of the Tian Shan and the Qilian Shan of Northwest China may have covered marginal areas of the Xizang-Tarim oldland south of these geosynclines and the Junggar oldland north of them (Figure II-19). The Middle Ordovician sequence of North China consists of a few layers of dolostone and dolomitic limestone, indicative of a hypersaline environment that evidently was hostile to most flora and fauna.

At the end of the Middle Ordovician much of China rose epeirogenically, and thereby isolated many basins. In basins where stagnant conditions prevailed, black shale with abundant graptolites formed locally. All of North China stood above sea level.

Late Ordovician (Caradocian and Ashgillian)

Marine waters continued to regress during the Late Ordovician, and three depositional environments mark three stages in this regression. At the beginning of the Late Ordovician, carbonate deposition gradually decreased and gave way to terrigenous muds (now yellow to green shale) with abundant shelly benthic fauna. In the middle of the Late Ordovician, sandy muds with shelly benthic fauna predominated, but some graptolites also are present (plankton or nekton). At the end of the late Ordovician, the sea floor deepened, and a reducing environment in the deepest parts of the basin diminished the benthos, but graptolites from surface waters accumulated in black muds at the bottom of deep basins.

Paleobiogeography

Lu et al. (1976) recognized four Ordovician biogeographic provinces in China, based upon the ecological habitats of trilobites, graptolites, and nautiloid cephalopods. Lai Caigen (1989) also divided China into four Ordovician biogeographic provinces only based on Cephalopods, thus his province boundaries somewhat differ from Lu et al.'s. Here we discuss only Lu et al.'s Ordovician biogeographic provinces.

Faunal Province I

This province, in North China, is notable for its predominant cephalopod zones in carbonates. Trilobites, graptolites, and ostracods also occur. The important cephalopod zones are based upon (1) Tremadocian *Protocameroceratidae*; (2) Arenigian *Ordosoceras*, *Polydesma*, *Manchuroceras*, and *Coreanoceras*; (3) Llanvirnian *Tofongoceras*, *Discoactinoceras*, and *Steroplasmoceras*; and (4) Caradocian *Gonioceras*. Much of the fauna consisted of benthons or of demersal nektons in a relatively clean shallow-marine environment, and most forms in this fauna are similar to North American forms.

Faunal Province II

This province in the Yangzi River valley is faunally notable for its graptolite zones, mostly in black shale, and mostly correlative with forms in Europe, the Middle East, and North America. In carbonates, trilobites and a few brachiopods and cephalopods also are present. The cephalopods *Richardonoceras* and *Sinoceras* characterize two zones in the Caradocian. The faunas contain both benthons and planktons. In the latest Ordovician, a semi-isolated basin with relatively quiet water occupied the Yangzi region. Its predecessors in the Middle Ordovician and in the Early Ordovician may have connected with other basins.

Faunal Province III

This province covers two regions: (1) South-Central China and East China and (2) Northwest China. Environments similar to those in other regions provide similar fossil zones. In general, ceratopygid and olenid trilobites predominate. Graptolites usually are not abundant, except in the Llanvirnian, in which stage they are related mostly to forms in Australia. As carbonates decrease, strata of terrigenous clastics increase and enter the column as layers of black shale and chert. In South-Central China and East China, Jiangnan and Zhujiang form two subprovinces, based upon sedimentological characteristics. The Zhujiang (Pearl River) Subprovince contains a thick lithofacies of flysch with important zones of graptolites, whereas the Jiangnan (south of the Yangzi), devoid of flysch, contains important trilobites. Spinose trilobites, preserved with graptolites in black shale, may represent the plankton or the nekton of muddy waters.

Faunal Province IV

This province constitutes a transitional or intermediate center where a mixture of faunas of Provinces I, II, and/or III occurs in varying degree. The two regions of this faunal province are Southwest China and the northern part of South-Central China.

Lu (1981) considered that Occidental trilobites originated in the Americas, especially in North America, and that Oriental trilobites originated in south China, Southeast Asia, and Oceania. Occidental trilobites could have migrated from west to east, whereas Oriental trilobites could have migrated from east to west. In the places of their meeting and their juncture, an intermediate fauna could have formed. In Asia, Oriental trilobites seem to have dispersed most conspicuously in the Cambrian and in the Ordovician.

In the above terms, migration of Occidental fauna would have been a common occurrence. For example, the trilobite *Dalmanitina*, found in the Middle Ordovician of the Occidental realm flourished in China during the Late Ordovician (Lu and Wu, 1979). After the Ordovician, when marine interconnections persisted for long periods, endemism in trilobites in their respective Oriental and Occidental realms disappeared progressively.

SILURIAN SYSTEM

Distribution

Silurian rocks are widely distributed in China (Figure II–24), the exceptions being North China, many parts of Central-South China and East China, and Taiwan (Mu Enzhi, 1962). The absence of Silurian in North China resulted from major uplift and erosion between the Middle Ordovician and the Middle Carboniferous. In East China and Central-South China, as well as on Taiwan, Silurian has been discovered in few places, and its development generally is a result of local factors, especially topography, orogeny, and epeirogeny. Blank areas on Figure II–24, particularly in Southwest China and Northwest China, may be filled in part as additional field mapping and subsurface drilling are conducted.

An important summary of the Silurian of China was published by Mu Enzhi (1962). Since the time of that publication, additional studies have appeared, including the excellent summaries by Wang and Liu (1980), Wang and Ho (1981), and Wang Yu et al. (1984). A monographic work on the Silurian of all China appeared in 1986 (Mu En-zhi et al., 1986).

Facies

In general, three Silurian platform facies exist: (1) a facies of graptolitic shale through entire sequences, as in parts of Hunan and Guandong Provinces and Guangxi Zhuang Autonomous Region; (2) a facies of shelly limestone that is highly fossiliferous (e.g. Xinjiang, Xizang, etc.); and (3) a mixture of shelly and graptolitic facies, where the lower part contains mostly graptolitic shale and the upper parts contain shelly limestone (all other regions, outside of the geosynclinal complexes). The striking contrast between the shelly and the graptolitic facies in the lower Paleozoic of China is wholly reminiscent of a similar contrast in northwestern Europe. There, the contrast has attracted the attention of geologists and paleontologists for nearly a century.

Between the platforms lie thick geosynclinal associations of flysch with sparsely fossiliferous flysch and paraflysch, associated with siliciliths, mafic to intermediate (locally silicic) volcanic rocks, and a few deep water carbonates. At the margins of subsiding troughs, deposition of shale and sandstone along shelving shorelines gave way to deposition onshore and to the accumulation of continental deposits. The Guangxi movement (late Caledonian tectogenesis) brought the Silurian to a close, and affected almost all of the area of modern China. Tectogenesis was especially severe in the geosynclinal troughs. Granitic intrusions accompanied and followed folding. The Guangxi movement, which corresponds to the last Caledonian phase in Europe, took place mainly in Central-South China (Hunan-Guangdong-Guangxi) and Northwest China (Tian Shan, Qilian Shan, Qin Ling, Kunlun Shan, etc.), and it affected all of China's major

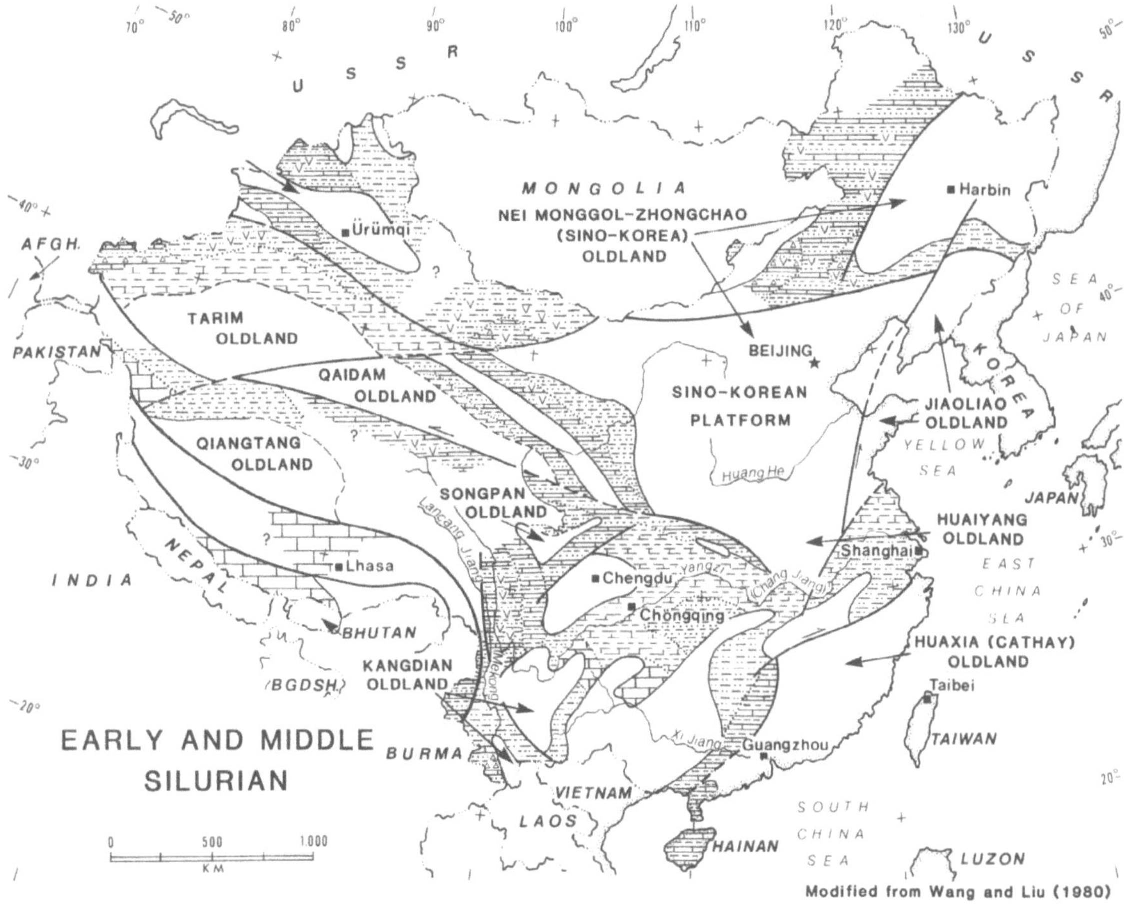

Fig. II–24. Paleogeographic and lithofacies map of Early-Middle Silurian, China.

Paleozoic geosynclines (Huang Jiqing, 1978, 1980, 1984a, 1984b).

Type Section

The type section for the Silurian in China is at Yichang, in the region of the Yangzi Gorges of Hubei Province (Figure II–25; also see Figure II–10). In that locality the Ordovician-Silurian boundary is transitional. The three Silurian divisions present a complete depositional cycle of marine transgression and regression. The lithologic changes from dominant graptolitic shale in the lower division to predominant shelly and coralline limestone in the middle division reflect changes of environment, from conditions of quiet and stagnant-water in a basin to strongly oxidizing conditions on a broad neritic shelf. In the upper division, the deposition of shale, mudstone, siltstone, and sandstone indicates the former existence of a coastal environment associated with marine regression. A disconformity caused by the Guangxi movement separates the Silurian from the overlying Devonian.

Distribution, Paleoecology and Paleogeography

Figure II–24 is the Wang and Liu (1980) paleogeographic and lithofacies map, also used by Wang and Ho (1981). Wang and Ho divided China into six sedimentational-biogeographic domains, some of which were divided into subdomains or 'regions.' These are shown on Figure II-26. Following Wang and Ho (1981), and Mu Enzhi et al. (1986), we describe briefly the main domains.

Northern Mobile Domain

Wang and Ho (p. 60) also called this the 'northern geosynclinal domain.' It includes the Altay-Argun geosynclinal complex and the Da Hinggan Ling mobile zone. The Da Hinggan region of Northeast China is characterized by associations that grade from poorly fossiliferous graptolitic shales to unfossiliferous graywacke rich in volcanic materials, mainly silicic to intermediate in composition, with some mafic volcanics. Terrigenous-clastic flysch, argillaceous flysch, paraflysch, and other flysch-like sediments are abundant, with locally abundant beds of siliciliths (Zhang and Tang, 1989). The fauna belongs to the Boreal realm.

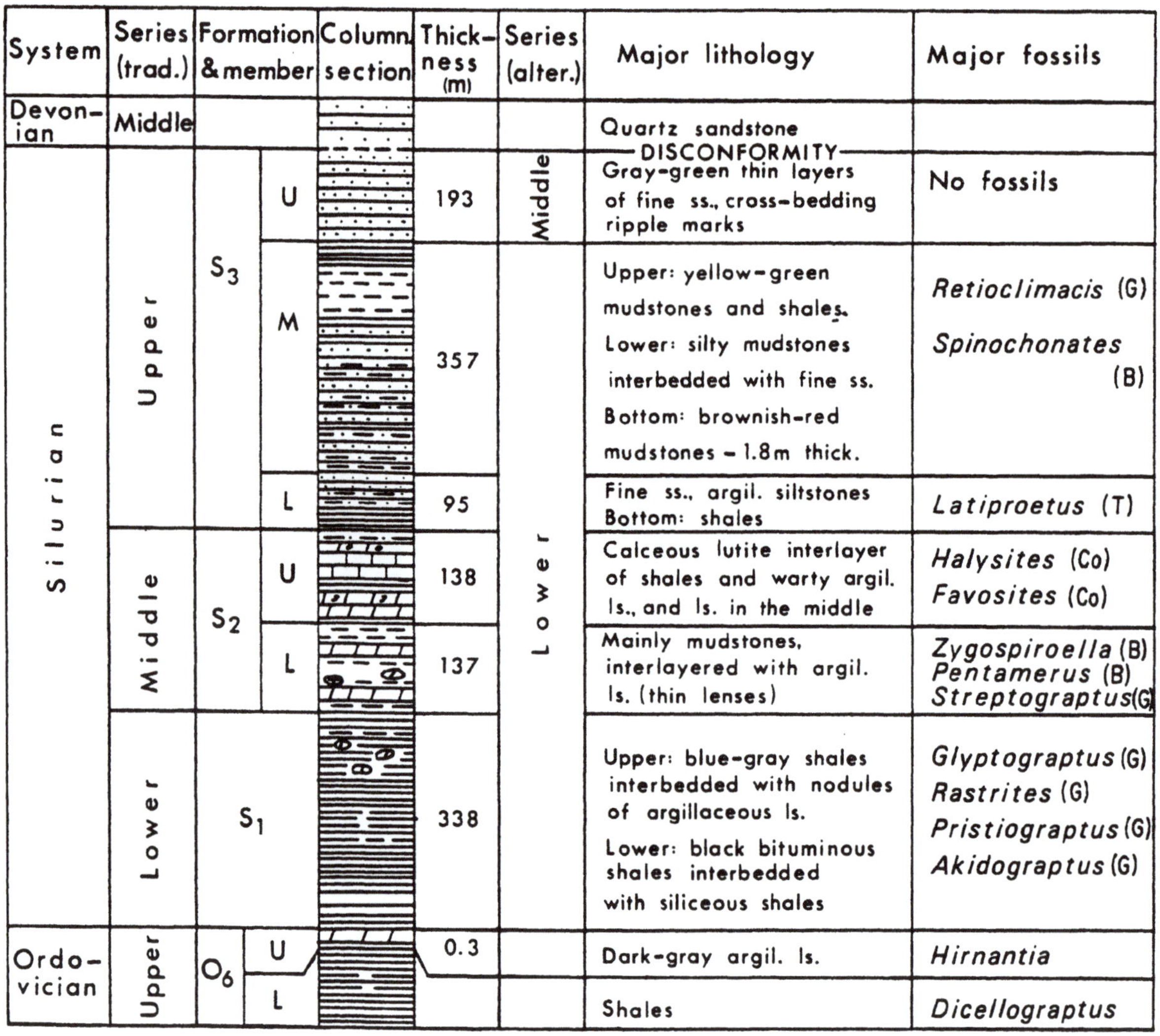

Fig. II–25. Stratigraphic column of Silurian in Yichang (Yangzi Gorges), Hubei Province, Central-South China (from Fang et al., 1979).

In eastern Jilin Province, in an area isolated from the Da Hinggan Ling section on the west, and where the Upper Silurian Erdaoguo Formation is present, it consists of shallow-water shelf limestones. This section overlies a graptolitic facies of Lower and Middle Silurian shale.

In the Altay and Bei Shan regions of western Nei Monggol Autonomous Region, western Gansu Province, and northern Xinjiang Uygur Autonomous Region, the section is dominated by unfossiliferous argillaceous flysch and paraflysch, with graptolitic shale and small areas of clastic flysch and silicilith. Intermediate to mafic volcanigenic rocks are abundant west of the Junggar basin and in western Nei Monggol. On the southern flank of the Altay Shan, the lower part of the section is graptolitic and shaley, whereas the upper part is shelly and carbonate-rich (Mu En-zhi et al., 1986). The fauna is Boreal.

Northern Stable Domain

The Sino-Korean platform (Nei Monggol-Zhongchao oldland) was exposed during the Silurian, as was a part of the Tarim platform (proved by well records). Around the margins of the Sino-Korean platform light-gray-green sandstone, siltstone, and phyllitic to schistose shale were deposited. Fossils are scarce. The Middle Silurian consists of carbonates with corals. The Silurian does not lap onto the platform.

In the area of the Junggar basin, terrigenous-clastic, shallow-water, greenish-gray sandstone, siltstone, and shale contain brachiopods and corals.

The Tian Shan is underlain by thick sections of mildly metamorphosed deep-water argillaceous flysch, paraflysch, thin-bedded carbonates, silicilith, andesite porphyry, and andesite tuff. Corals are present. The Lower Silurian is absent, except locally (Mu En-zhi et al., 1986). The reminder of the Silurian consists of a lower graptolitic facies and an upper carbonate facies (Mu En-zhi et al., 1986).

Along the northern margin of the Tarim basin, up to 6,000 m of andesitic volcanic rocks, argillaceous flysch, and paraflysch grade southward into shaly sandstone, siltstone, and silty-to-sandy shale. The Lower

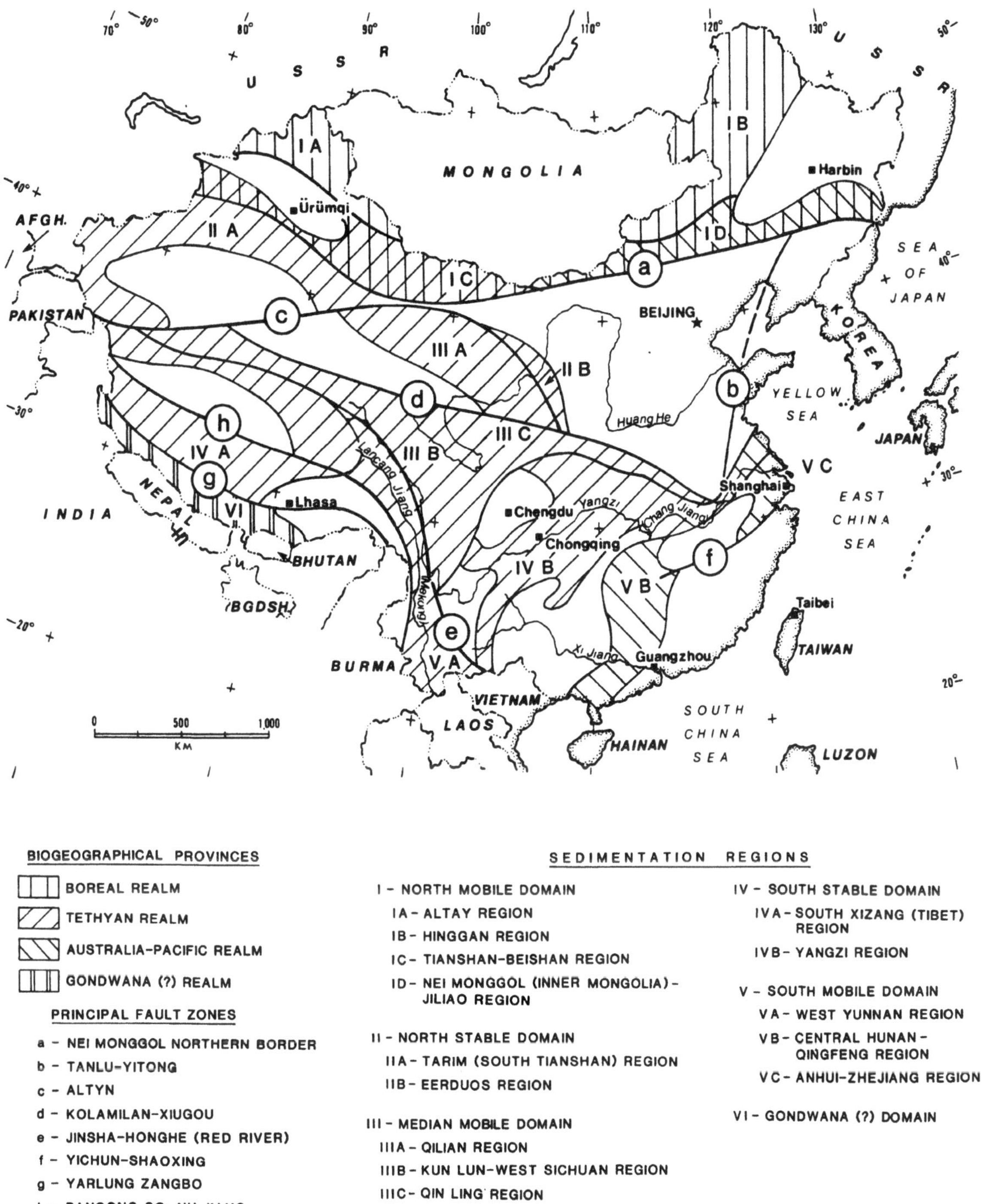

Fig. II-26. Silurian sedimentation and biogeographical domains of China (from Wang and Ho, 1981).

Silurian consists largely of graptolitic shale and slate; the Middle Silurian comprises mainly shelly limestone and limestone; the Upper Silurian is in shelly facies interbedded with sandstone and shale, together with scarce volcanic rocks and redbeds. The Lower Silurian west of the Tarim basin, includes, in addition to gray graptolitic shale, thick units of unfossiliferous mudstone (Mu En-zhi et al., 1986).

Median Mobile Domain

This domain of northwestern and north-central China consists of the Kunlun, Qilian, and Qin Ling geosynclinal complex. Unusually thick sections of terrigenous-clastic and argillaceous flysch are interbedded with mafic and intermediate volcanigenic rocks. In the Qilian Shan, the Lower Silurian consists of graptolitic shale, which characterizes both the Lower and Middle Silurian of the

Qin Ling (Chen Xu, 1984). The Middle Silurian of the Qilian Shan consists of coralline limestone with interbedded shale, and the Upper Silurian includes shale, pink siltstone, fine-grained sandstone, and flysch. Some corals and trilobites are present in shallow-water facies. The Qin Ling section is partly metamorphosed. Volcanic rocks occur mainly in the west, from the Qilian Shan to the Kunlun Shan.

Southern Stable Domain

The best studied sections of this domain occur on the Yangzi platform. Fossiliferous strata younger than early Wenlockian (early Middle Silurian) are absent, except in northeastern Yunnan Province, where Upper Silurian predominates. Lower Silurian consists largely of a graptolitic facies, with shale dominant in the west, and shale with siltstone in the east, succeeded by a Lower to Middle Silurian shelly or 'mixed' facies and a graptolite-shale facies. A shelly unit above the 'mixed' unit includes sandstone, mudstone, shale, and minor limestone. In general, the carbonate content decreases from west to east. Above the fossiliferous Middle Silurian, scattered outcrops of unfossiliferous (late Wenlockian) redbeds are preserved. The succeeding Devonian is everywhere unconformable (Mu En-zhi et al., 1986). Rough topography on the platform, especially in the southern part, caused (1) many local hiati to develop, and (2) the development of erratic, abrupt, facies changes.

In Xizang (Tibet) Autonomous Region, a broad stable platform existed (Wang and Ho, 1981; Wang Hongzhen, 1983). North of the Indus-Yarlung fault zone, the Lower Silurian strata contain limestone with a shelly fauna of corals and cephalopods. The Middle and Upper Silurian strata consist of siliceous limestone and sandstone, mainly with a fauna of brachiopods and other shelly elements.

In western Yunnan Province near the Burma border, paraflysch is present in thick masses. Lateral facies changes are numerous and abrupt.

East and south of the shelf section of the Yangzi platform (Central-South China and East China), occurrences of the Silurian are erratic. Where found, the Silurian includes gray shale, some black shale, thin-bedded siltstone and sandstone, and carbonates. The shale is overwhelmingly in graptolitic facies (Mu En-zhi et al., 1986). The Upper Silurian consists mainly of sandstone with only scattered fossils. The graptolitic shale-siltstone facies characterizes southern Hunan Province, Guangxi Zhuang Autonomous Region, and Guangdong Province. Early Silurian is absent in southern Guizhou Province and eastern Yunnan Province, but the Middle Silurian and younger beds are of shallow-water origin. In Jiangxi, Anhui, and Zhejiang Provinces, the Silurian consists mostly of sandstone and shale. Graptolites are present in the lower part of the section; brachiopods, trilobites, and pelecypods occur in the middle and upper part. No geosynclinal facies are present (Mu En-zhi et al., 1986).

Gondwana Domain

This domain occupied the present site of the Himalayas. The section is dominated by shallow-water shelf carbonates, but shelf sandstone, siltstone, and shale also are present. The Silurian section on the northern flank of Qomolongma Feng (Mt. Everest), for example, although incomplete, includes a lower section of 136 m of graptolitic black shale, shallow-water marine sandstone and gray shale, overlain by Middle and Upper Silurian that is predominantly shallow-water limestone, capped by sandy and clean platform-type limestones (Mu En-zhi (Anzi et al., 1972). The fauna is similar to that in equivalent section on the Yangzi platform toward the east.

Detailed studies of all Cambrian through pre-tillite Permian sections in the Qinghai-Xizang Plateau and in the Himalayas have led to the nearly unanimous conclusion by Chinese geologists that this area was always a part of Asia (see excellent summary by Mu En-zhi, Boucot, et al., 1986). The close affinities, in places near-identities, of the shallow- and warm-water faunas and lithofacies with those of the Yangzi platform, Nei Monggol, Siberia, and Western Europe leave little room for doubt. The Indian Block, at least that part of it where the Himalayas is now located, was always a part of Asia (Saxena, 1978; Crawford, 1979; Gansser, 1981; Chang and Pan, 1981; 1984; Waterhouse, 1983; Bhat, 1984; Stocklin, 1984; Mu En-zhi et al., 1986).

Biogeographic Realms

Wang and Ho (1981) recognized three Silurian biogeographic realms, based mainly on rugose corals and partly on brachiopods. These realms are: (1) Boreal; (2) Tethyan; and (3) Australo-Pacific (Figure II–26). Associated with them are two belts with mixed faunas: one with Boreal and Tethyan elements; and another with Boreal and Australo-Pacific elements. Table II–11 gives details of assemblages in China that contain rugose corals and brachiopods.

Boreal Realm

Regionals 1_A and 1_B belong to the Boreal realm (Figure II–26). Their representative faunas include the coral *Tungussophyllum* and the brachiopods *Tuvaella* and *Tunguspirifer*. The brachiopod genera of this realm are widely distributed from the Hinggan and Mongolia to the Altay, the Sayan (U.S.S.R.), and Siberia (U.S.S.R.).

Tethyan Realm

This is the largest of all three realms, with regions II_A, II_B, III_A, III_B, III_C, IV_A, IV_B, V_A, and VI (Figure II–26). Early and Middle Silurian assemblages of rugose corals occur in the Yangzi region (Table II–11), where they indicate the presence of epicontinental marine

Table II-11. Silurian assemblages of brachiopods, cephalopods, and rugose corals in China. The Llandoverian and Wenlockian corals occur in the Yangzi region, the Ludlovian and Pridolian corals in the Nei Monggol region.

GEOLOGICAL TIME	BRACHIOPOD ASSEMBLAGES IN SOUTH CHINA (Wang Yu et al., 1981)			RUGOSA CORAL ASSEMBLAGES (Wang and Ho, 1981)	CEPHALOPOD ASSEMBLAGES IN SOUTHWEST CHINA (Chen et al., 1981)
PRIDOLIAN	UNNAMED ASSEMBLAGE			*MUCOPHYLLUM-PSEUDOMICROPLASMA* ASSEMBLAGE	*YUNNANOCERAS* ASSEMBLAGE
	PROTATHYRISINA PLICATA-SCHIZOPHORIA HESTA ASSEMBLAGE				?
LUDLOVIAN	*PROTATHYRISINA UNIPLICATA-ATRYNIDEA QUJINCANSIS* ASSEMBLAGE			*WEISSERMELIA-ALTAJA* ASSEMBLAGE	*EUTHYOCYCLOCERAS* ASSEMBLAGE
WENLOCKIAN				*MICULA-KETA KETOPHYLLUM* ASSEMBLAGE	*HEYUNCUNOCERAS* ASSEMBLAGE *SICHUANOCERAS* ASSEMBLAGE
	SALOPINA-XINANOSPIRIFER ASSEMBLAGE			*KYPHOPHYLLUM-IDIOPHYLLUM* ASSEMBLAGE	
LLANDOVERIAN	*HALIVKINIA-NUCLEOSPIRA* ASSEMBLAGE				*YICHANGOCERAS* ASSEMBLAGE
	PARACONCHIDIUM-VIRGIANELLA ASSEMBLAGE	*PENTAMERUS-EOSPIRIFER* ASSEMBLAGE	*STRICKLANDIA-MERCIELLA* ASSEMBLAGE	*KODONOPHYLLUM-MAIKOTTIA* ASSEMBLAGE	*SONGHANOCERAS* ASSEMBLAGE
	BOREALIS-KRITORHYNCHIA ASSEMBLADGE			*DINOPHYLLUM-RHABDOCYCLUS* ASSEMBLAGE	

waters connected with the troughs of Qin Ling and Sanjiang. Faunas of this whole geogeographic province show a pronounced endemism.

Australo-Pacific Realm

Regions V_B and V_C of southern South-Central China and East China belong to this realm (Figure II-26). A few of its species of rugose corals, of Late Silurian age, relate to Australian forms.

Mixed Faunas

As noted above, Wang and Ho (1981) designated two belts with mixed faunas:

Mixed Tethyan and Boreal – Silurian rugose corals of Tian Shan-Bei Shan (I_C) correlate with both Russian, Uralian, and western European forms.

Mixed Boreal and Australo-Pacific – Silurian corals of Nei Monggol-Jiliao (I_D) correlate with both northern Asian forms and Australian forms.

DEVONIAN SYSTEM

As a result of uplift in a 'Caledonian' tectogenesis (Guangxi movement) at the end of the Silurian, marine waters receded from much of the Chinese mainland, and apparently remained absent from the Sino-Korean platform throughout the Devonian. In Northwest China (the Xinjiang Uygur Autonomous Region) and in Southwest China (Xizang Autonomous Region and Yunnan Province), the sea persisted from the Silurian into the Devonian (Wang and Yu, 1962). The greatly enlarged landmasses of Early Devonian time in China, however, suggest that a complex topography was present, and that it consisted of numerous dissected uplifts, plateaux, platforms, and intermontane basins. Marine waters soon returned, and as deposition resumed with them a new marine fauna proliferated.

Distribution and Facies

Wang and Yu (1962) recognized five major geographic divisions of the Devonian (Figure II-27): (1) northern part of Northeast China and Nei Monggol Autonomous Region; (2) Northwest China; (3) Central-South China; (4) Southwest China; and (5) East China.

Cai and Li (1980) recognized four major geographic divisions of the Devonian: (1) Central-South China; (2) western Yunnan-Xizang (Tibet); (3) Qilian Shan, and (4) Junggar-Hinggan.

Yang et al., (1981) recognized eight major geographic divisions of the Devonian: (1) Junggar-Hinggan; (2) southern Tian Shan; (3) Qilian Shan; (4) Longmen Shan; (5) Central-South China; (6) East China; (7) western Sichuan-northern Xizang (Tibet); and (8) Himalaya-western Yunnan.

The description that follows uses the five divisions of Wang and Yu (1962), but also includes five tabulations of corresponding Devonian sequences from Yang et al. (1981).

Northern Part of Northeast China and Nei Monggol

This was a region of geosynclinal deposition (Altay-Argun geosynclinal complex), thus one of thick Devonian sequences, mainly hard sandstone and sandy shale interbedded with agglomerate, andesite, and rhyolite. Complete sections generally are absent because of complex structure produced by folding and faulting. A close relation to faunas of the Soviet Union (U.S.S.R.) and of eastern North America indicates a transgression of marine waters from the north.

Cai and Li (1980) noted the presence of alternate marine and non-marine deposits in the Middle and Upper Devonian together with a richness in plants of Angaran affinities. The Middle Devonian contains the plant *Protolepidodendron scharyanum*.

For the sector of Da Hinggan Ling ('Great Khinghan Range'), Yang et al. (1981) described a column that appears in Table II-12.

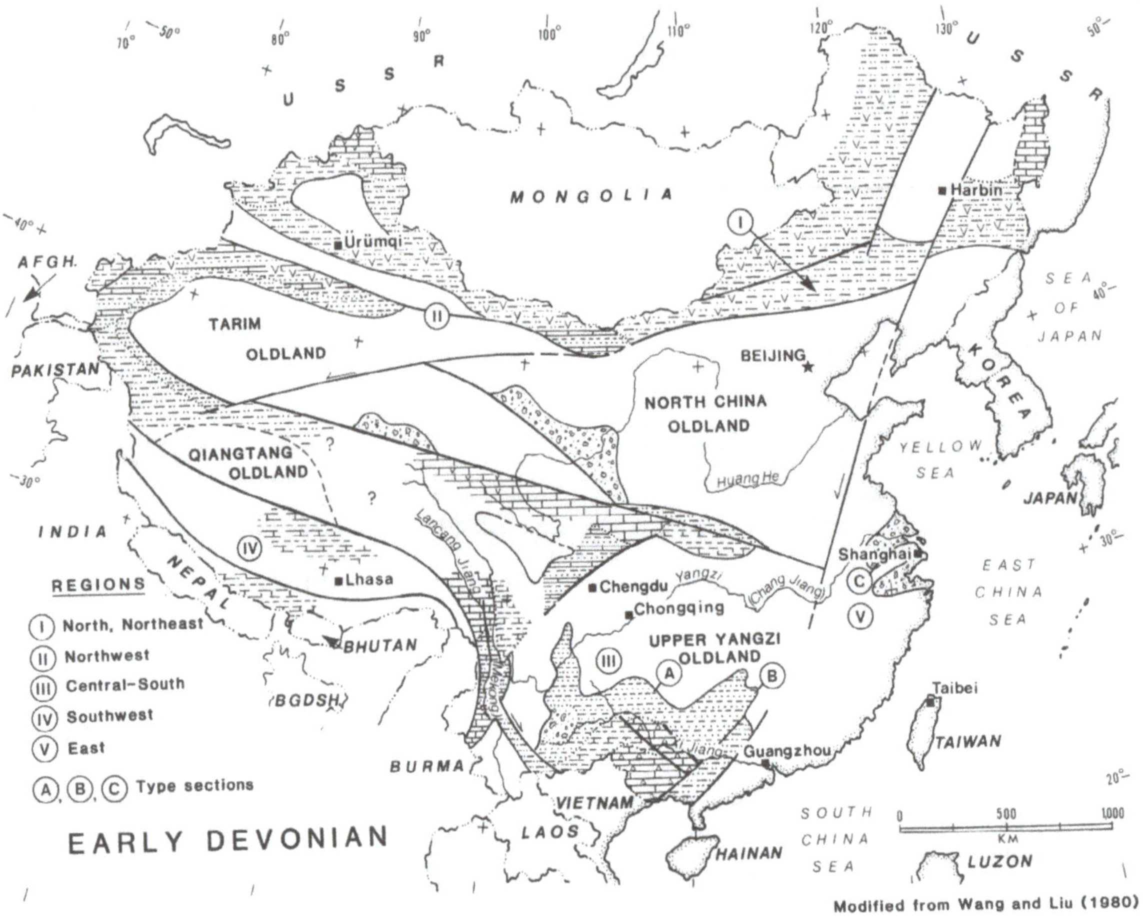

Fig. II-27. Paleogeography and lithofacies map of Early Devonian, China.

Table II-12. Devonian sequence - Da Hinggan Ling region - Northeast China

DEVONIAN	Upper	Shangdamingshan Formation	Andesitic porphyry, upper part; sandstone, lower part; zones of *Platyclymenia wolcotti* and *Cheiloceras subpartitum* (ammonoids)	60 m
		Ziadamingshan Formation	Tuffaceous sandstone and intermediate to silicic lava; *Temnophyllum tomense*, *Macgeea solitaria*, (corals) *Peneckiella* (brachiopod)	183 m
	Middle	Hebaoshan Formation	Limestone, conglomerate; *Endophyllum abditum* and *Temnophyllum ornatum* (corals)	137 m
		Beikuang Formation	Calcareous siltstone, intercalated with slate; sandstone; *Breviphyllum lenense (corals)*, *Cayugaea subcylindrica*	140 m
	Lower	Unuoer Formation	Bioclastic limestone, siltstone; *Lyrielasma*, *Leptoinophyllum* sp., *Tryplasma hercynica*, *Amplexiphyllum hamiltonae* (corals); *Wilsonella grandis*, *Howellella amurensis* (brachiopods)	N.A.
		Luotuoshan Formation	Siltstone, sandstone intercalated with limestone lenses; *Cymostrophia alpha*, *Protathyris praecursor*, *Howellalla* sp., and *Ancolitoechia* sp. (brachiopods)	90 m

Northwest China

As in Northeast China and Nei Monggol, this was a region of geosynclinal deposition (Altay-Argun geosynclinal complex), thus a region characterized by thick sequences. These, however, are commonly metamorphosed. In addition, lithofacies differs greatly from area to area. In the Altay, columns consist of phyllite, sandstone, quartzite, silicic volcanics, and a few interbeds of lenticular fossiliferous limestone. In the Tian Shan, schist, shale, and argillaceous limestone are the main rock types. Southward in the Tian Shan, only red fine-grained sandstone and conglomerate are present in shoreline facies. Continental terrigenous-clastic beds are intercalated. In the Qilian Shan, the lithology consists of quartzite and argillaceous siltstone. In the Kunlun Shan,

Table II-13. Devonian Sequence in Kiziltag, southern Tian Shan - Northwest China

DEVONIAN	Upper	Harzirbulak Formation	Intermediate to silicic volcanic rocks intercalated with limestone in upper part, terrigenous clastics in lower; *Yunnanella* sp. and *Yunnanellina* sp. (brachiopods); *Tabulophyllum* sp. (coral)	2,000 to 2,500 m
	Middle	Saermin Formation	Terrigenous clastics intercalated with andesitic porphyry; limestone; *Stringocephalus obesus* (brachiopod); *Cyathophyllum* sp. (coral)	250 to 1,200 m
		Alatag Formation	Schist, metamorphosed limestone, volcanic rocks; *Favosites* sp., *Heliolites* sp. (corals) *Atrypa* sp. (brachiopod)	500 to 1,000 m
	Lower	Alpishemeibulak Formation	Lightly metamorphosed schist, marble, lenticular limestone; *Tryplasma* sp., *Favosites* sp., *Squameofavosites* sp. (corals)	2,600 to 6,500 m

argillaceous limestone and some shale layers constitute most of the section.

Yang et al. (1981) provided detailed descriptions of three Devonian sequences in the southern segment of the Tian Shan. Of these, the sequence at Kiziltag appears in Table II-13.

The Early Devonian fauna relates closely to that of the adjacent Soviet Union (U.S.S.R.), whereas the Middle Devonian and the Late Devonian faunas are almost identical with those of Central-South China. These relations indicate that the Early Devonian sea connected directly with a seaway in the Soviet Union, whereas the Middle Devonian and the Late Devonian seas spread uninterruptedly into Central-South China.

Central-South China

The environment of deposition in this region during the Devonian was mainly that of a platform. Exceptions are the parageosynclines of the present Qin Ling and northwestern Sichuan Province, and parts of Hunan and Guangdong Provinces and the Guangxi Zhuang Autonomous Region. Differing rates of subsidence, and consequent differences between geosynclines and platforms, resulted in marked changes of the section from area to area.

Rocks near the base of the section consist mainly of reddish-purple conglomerate and sandstone that grade upward into reddish and purplish shale and mudstone. The latter grade upward, in turn, into shale and mudstone interbedded with a few layers of limestone. Carbonates dominate the upper part of the section. All divisions of the section undergo lateral changes of facies. Some of the Devonian faunas of this region contain cosmopolitan genera such as the brachiopod *Stringocephalus*, the coral *Calceola*, and the ammonoid *Manticoceras*, all strongly represented in northwestern Europe. Endemic genera also occur, as for example the brachiopod *Yunnanella*. Rich plant remains in continental facies of the Devonian sequence constitute the 'South China type' of flora (Cai and Li, 1980).

Yang et al., (1981) provided detailed descriptions of five Devonian sequences in South China. The most complete of these, at Nandan in Guangxi Zhuang Autonomous Region, appears in Table II-14.

Table II-14. Devonian sequence of Nandan, Guangxi Zhuang Autonomous Region, Central-South China

DEVONIAN	Upper	Daihua Formation	Banded and nodular limestone; Zone of *Worklumeria spheroides* (ammonoid); Zone of *Clymenia laevigata* (ammonoid); and Zone of *Dzieduszyckia baschkirica* (brachiopod)	94 m
		Xiangshuidong Formation	*Upper member*: mudstone; *Uniconus* sp. (tentaculite) and *Richterina latior;*	98 m
			Lower member: thin-bedded siliceous strata; *Beloceras* cf. *acutum* (ammonoid)	80 m
	Middle	Luofo Formation	Black mudstone intercalated with argillaceous limestone; Zone of *Nowakia otomari* (tentaculite)	584 m
		Napiao Formation	*Upper member*: Dolomitic mudstone; Zone of *Nowakia sulcata* (tentaculite), and *Pinacites jugleri* (ammonoid);	120 m
			Lower member: black mudstone; Zone of *Anarcestes noeggerati* (ammonoid); Zone of *Nowakia cancellata* (tentaculite) and *Convoluticeras discordans* (ammonoid)	235 m
		Tangdian Formation	Purplish gray mudstone; Zone of *Nowakia barrandei* (tentaculite) and *Erbenoceras ellipticum* (ammonoid); Zone of *Nowakia praecursor.* Zone of *Nowakia zlichovensis* (tentaculites)	220 m
	Lower	Yilan Formation	Mudstone; *Dicoelostrophia annamitica* (brachiopod), *Gravicallymene maloungkaensis* (trilobite)	86 m
		Danlin Group	Gray-white sandstone.	700 m

Southwest China

This was a region of thick platformal deposition in which the principal rocks consist of siliceous limestone and dolostone, with local conglomerate and sandstone, in places interbedded with volcanic rocks in the lower part of the column. The fauna in this region relates closely to that of Central-South China, the Western Cordillera of North America, and central Europe. In Early Devonian strata, Mu and Ni (1975) discussed the occurrence of the graptolite *Monograptus* in black shale, the first such occurrence reported from China (Mu Anzi et al., 1973) and Mu and Ni (1975) discussed occurrence of the cosmopolitan graptolites *M. thomasi* and *M. yukonensis*. A rich warm-water shelf fauna characterizes most of the Devonian sequence on Qomolangma Feng (Mt. Everest), where it is 350 m thick (Yin Ji-xaing et al., 1983). Cai and Li (1980) noted the presence of paralic deposits that contain plant remains related to the flora of South China.

Yang et al. (1981) provided detailed descriptions of eight Devonian sequences in West China. A complete sequence from Southwest China appears in Table II-15.

East China

Yang et al. (1981) reported that some investigators classified basal red sandstones in this region as Late Silurian, whereas most workers assigned the vertebrates of these basal strata in continental facies to the Early Devonian. In the latter case Lower Devonian presumably lies conformably or paraconformably on Silurian. Until the early 1980s, workers had not found the Middle Devonian in East China (see 'Hiatus' on Table II-16). The Upper Devonian consists of light-colored quartzose sandstone with remains of ostracoderm fish and plants.

Yang et al. (1981) provided detailed descriptions of two Devonian sequences in East China. Table II-16 contains the more northerly of these two sequences.

Type Sections

Fang et al. (1979) discussed three types of depositional environments within the Devonian: (1) marine; (2) continental; and (3) marine-continental. All three type sections of the Devonian in Central-South China contain these three depositional environments. For illustration we show only one sequence (Figure II-28), and this is comparable to the sequence shown in Table II-14. We call our illustrated sequence in Central-South China, type section A (Figure II-27), a marine and estuarine column deposited in Guilin, Guangxi Zhuang Autonomous Region. D_1, the lowest unit of the Lower Devonian, consists of estuarine and deltaic deposits, and is unconformable on rocks affected by a 'Caledonian' tectogenesis. Carbonates with corals and brachiopods increase upward in the Lower Devonian from D_2 to D_4, an indication that a neritic environment developed during a marine transgressive phase. At the end of the Early Devonian, a moderate uplift in Central-South China caused marine waters to recede.

Middle Devonian D_5, D_6 witnessed a second transgression, and a corresponding expansion of neritic deposition. During the Late Devonian, two depositional

Table II-15. Devonian sequence - Lijiang, Yunnan Province, Southwest China

	Upper		Platy shale, siliceous limestone, and sandstone;	800 m
	Middle		Siliceous rock intercalated with platy shale; *Nowakia cancellata*	21 m
		Banmandaodi Formation	Blackish-gray siliceous rocks, limestone slate; *Nowakia barrandei*, and *Nowakia zlichovensis* (tentaculites); *Erbenoceras ellipticum* (ammonoid)	140 m
DEVONIAN	Lower	Alengzhoa Formation	Limestone intercalated with shale, conglomerate; *Nowakia acuaria* (tentaculite); *Monograptus yukonensis fangensis* (graptolite); *Lyrielasma chapmani* and *Tryplasma hercynica* (corals)	520 m
		Minying Formation	Limestone intercalated with shale; *Paranowakia bohemica* (tentaculite)	350 m

Table II-16. Devonian sequence, southern Jiangsu, western Zhejiang, and southern Anhui Provinces, East China

	Upper	Wutong	Yellow quartzite and gray shale; *Sinolepis macrocephala, S. wutungensis, Astrolepis sinensis, Holoptychius nankinensis*, and *Parholoptychius lungtanensis* (placoderm fish); *Leptophloeum rhombicum, Sublepidodendron mirabile*, and *Lepidodendropsis hirmeri* (plants); *Cymbosporitis scabellus* (spore); *Sphenophyllum lungtanensis* (coral)	50 to 800 m
DEVONIAN	Middle	Hiatus		
	Lower	Maoshan Group	Red, purple, and green sandstone, shale, and quartzite; *Galeaspidae, Polybranchiaspidae* (ostracoderm fish)	0 to 500 m

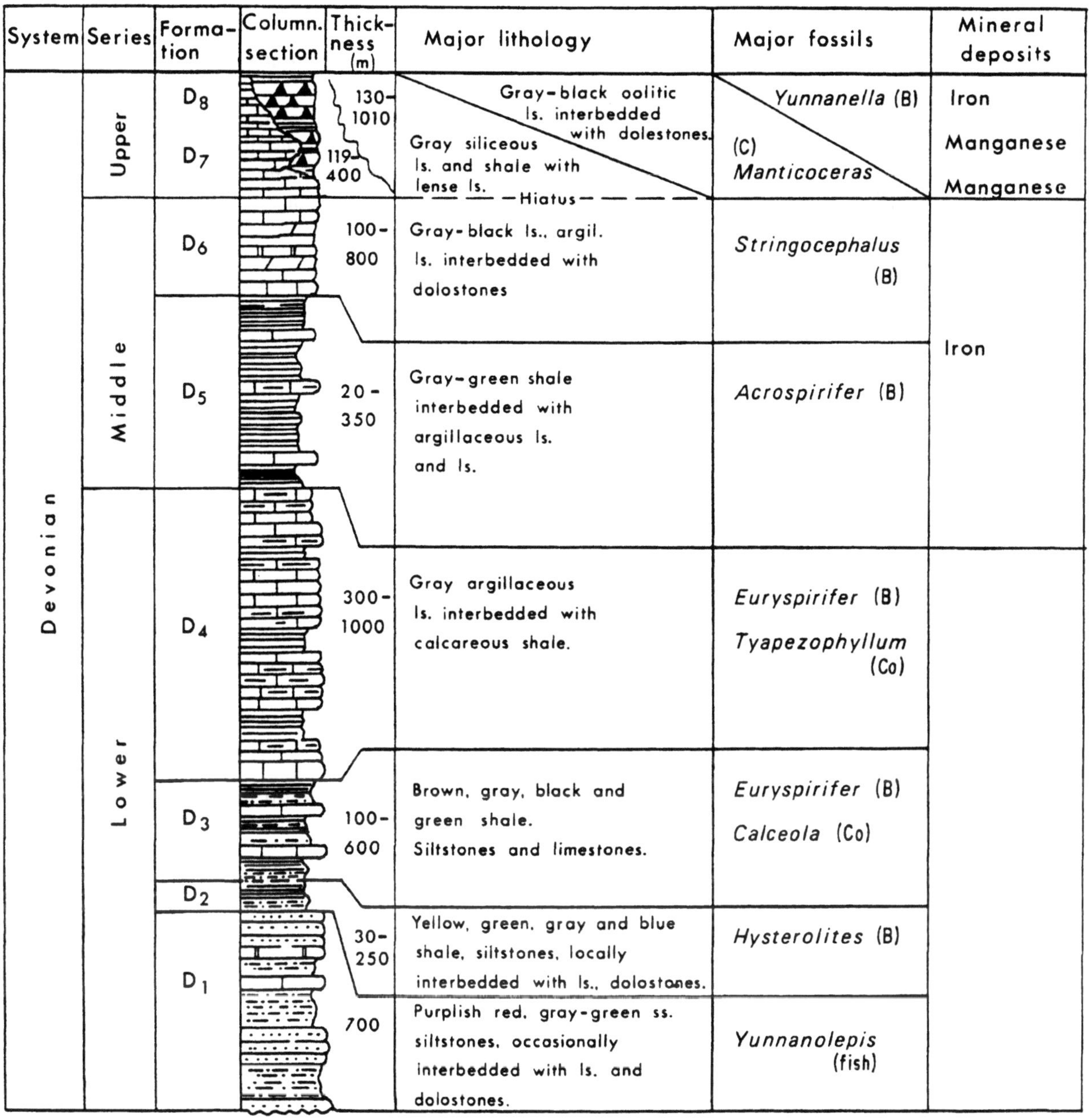

Fig. II-28. Stratigraphic type section of Devonian in Guilin area, Guangxi Zhuang Autonomous Region, China (from Fang et al., 1979).

environments developed. D_7 represents a quiet neritic environment with deposition of limestone bearing a fauna of nektonic cephalopods. D_8 represents an environment of active currents, and a regime of high temperature. Oolitic carbonates form part of the sequence. By the latest Devonian a marine regression resulted in the formation of a depositional hiatus. A disconformity is present at the base of overlying Carboniferous deposits.

Type section B (Figure II-27), in southern Jiangxi Province, consists of alternate marine and non-marine strata. Divisions D_1 to D_5 of the Lower and Middle Devonian are missing. D_6 of the upper Middle Devonian consists of sandstone and conglomerate interbedded with a silty mudstone whose flora includes *Lepidodendropsis* and *Protolepidodendron*. D_6 is strongly unconformable on folded and metamorphosed lower Paleozoic rocks. D_7 of the lower Upper Devonian consists mainly of non-marine clastics with the plant *Leptophloeum rhombicum* and the placoderm fish *Bothriolepis*, the latter probably an anadromous genus inasmuch as it is found in both non-marine and marine environments. Interbedded marine intervals of D_7 contain the brachiopods *Tenticospirifer* and *Lingula*. D_8 represents the uppermost Devonian, and it consists mainly of fine- to medium-grained terrigenous clastics with the plants *L. rhombicum* and *Sublepidodendron*. Marine intervals consist of limestone with the brachiopods *Yunnanella*, *Crytospirifer*, and *Tenticospirifer*.

Type section C (Figure II-27), a non-marine sequence,

lies in the lower valley of the Yangzi River. The Lower Devonian consists of purplish-red sandstone and sandy shale with ostracoderm fish. The Middle Devonian is missing. The Upper Devonian consists mainly of quartz sandstone interbedded with black shale and gray claystone whose flora contains the plants *Leptophloeum rhombicum* and *Sublepidodendron*, and whose fauna contains placoderm fish (see Table II–16). The environment of deposition appears to have been lacustrine and intermontane.

Paleoecology and Paleogeography

We summarize here a discussion of Devonian geography by Wang and Yu (1962), to which we have added on the basis of more recently collected data. Wang and Yu pointed out that the geography of the earliest Devonian was produced as a result of 'Caledonian' movements at the end of the Silurian, and that land areas then became important. Mountains and plateaux formed in many parts of China, although marine waters covered a basin on the site of the Tian Shan geosyncline. Folded rocks underlay most of China in the earliest Devonian, and initial Devonian deposits consisted characteristically of purplish-red terrigenous clastics with sparse fossils. The climate may have been hot and dry.

During the late Early Devonian, marine waters ingressed from three directions: (1) from the Indo-Pacific through the present Indochina Peninsula to Central-South China; (2) from the present Sea of Okhotsk into North China; and (3) from Tethys on the southwest into the area of the Himalayas and the Qinghai-Xizang (Tibet) Plateau. The faunas of North China relate closely to those of the Boreal Pacific realm; those of Central-South China are similar to those of Western Europe; and those of Xizang (Tibet) Autonomous Region are related to those of Western Europe, the Middle East, and the southwestern Pacific. Small-scale movements at the end of the Early Devonian caused local regressions and subsequent erosion. Small-scale movement between the Middle Devonian and the Late Devonian, in their turn, caused a local hiatus in the southern part of Central-South China. Transgression then reoccurred in the Late Devonian, with an increase in depth of waters, but with an expansion of area that reached barely into East China, owing to the relatively higher elevation in that area. The faunas, partly endemic, still show clear relations with Western European, Central Asian, Mongolian, and North American biotas. Faunas collected in the area of Qomolangma Feng (Mt. Everest) and in northern India also show direct relationship with Asia and Europe, but with no affinities to the Malvinokaffric realm of the southern continents (Gupta and Janvier, 1970; Wang Yu et al., 1984).

CARBONIFEROUS SYSTEM

Distribution and Facies

The Carboniferous System, extremely well developed in China, shows a wide distribution, excellent outcrops, and an abundance of fauna and flora, both endemic and cosmopolitan (Figure II–29). During the late 1970s and the first half of the 1980s, there has been some advocacy for a two-fold division of the Carboniferous to replace the three-fold division of the late 1950s (i.e. Lower Carboniferous, Middle Carboniferous, and Upper Carboniferous). The two-fold division would retain the Lower Carboniferous as the Fengeng Series, but would unite the Middle Carboniferous and the Upper Carboniferous in the Hutian Series. The latter constitutes a viable unit in terms of lithology and of marine faunas, but Li Xingxue and Zhang Linxin (1983) pointed out that a considerable difference exists between Middle and Upper Carboniferous floras in North China. In this study we retain the three-fold division of the Carboniferous in China.

Yang et al. (1962) observed a notable difference between the Carboniferous of North China and that of Central-South China. In the south, apart from the continental deposits of East China, the Lower Carboniferous occurs mainly in marine facies. In the north, the Middle and Upper Carboniferous exhibit alternate sequences of marine and non-marine strata. Patterns of distribution also changed through Carboniferous time. The pattern of distribution for the Lower Carboniferous remains very similar to that of the Devonian. The patterns for the Middle and Upper Carboniferous change gradually to that of the succeeding Permian. For the Carboniferous as a whole, the nature of deposition and the characters of fauna and flora have suggested a geographic division into (Figure II–29): (1) Central-South China; (2) North China; (3) Northwest China; (4) Xizang (Tibet)-western Yunnan; and (5) Tian Shan-Hinggan (Boreal Depositional Area). Using Martinez-Diaz (1983, gen. ed.), we note briefly the characteristics of these five regions.

Central-South China

The main lithofacies consist of marine carbonates, richly fossiliferous, with terrestrial intercalations through the Middle and Upper Carboniferous. Coal measures are associated with the terrestrial intertongues. On Taiwan, the Carboniferous (carbonate lenses in flysch-like metamorphics) may be present in the Central Range (Chang and Liu, 1984), but the claim of Chang and Liu remains to be proved.

North China

This region shows a notable absence of Lower Carboniferous. Lithofacies of the Middle and Upper Carboniferous is mainly clastic, with intercalated limestone,

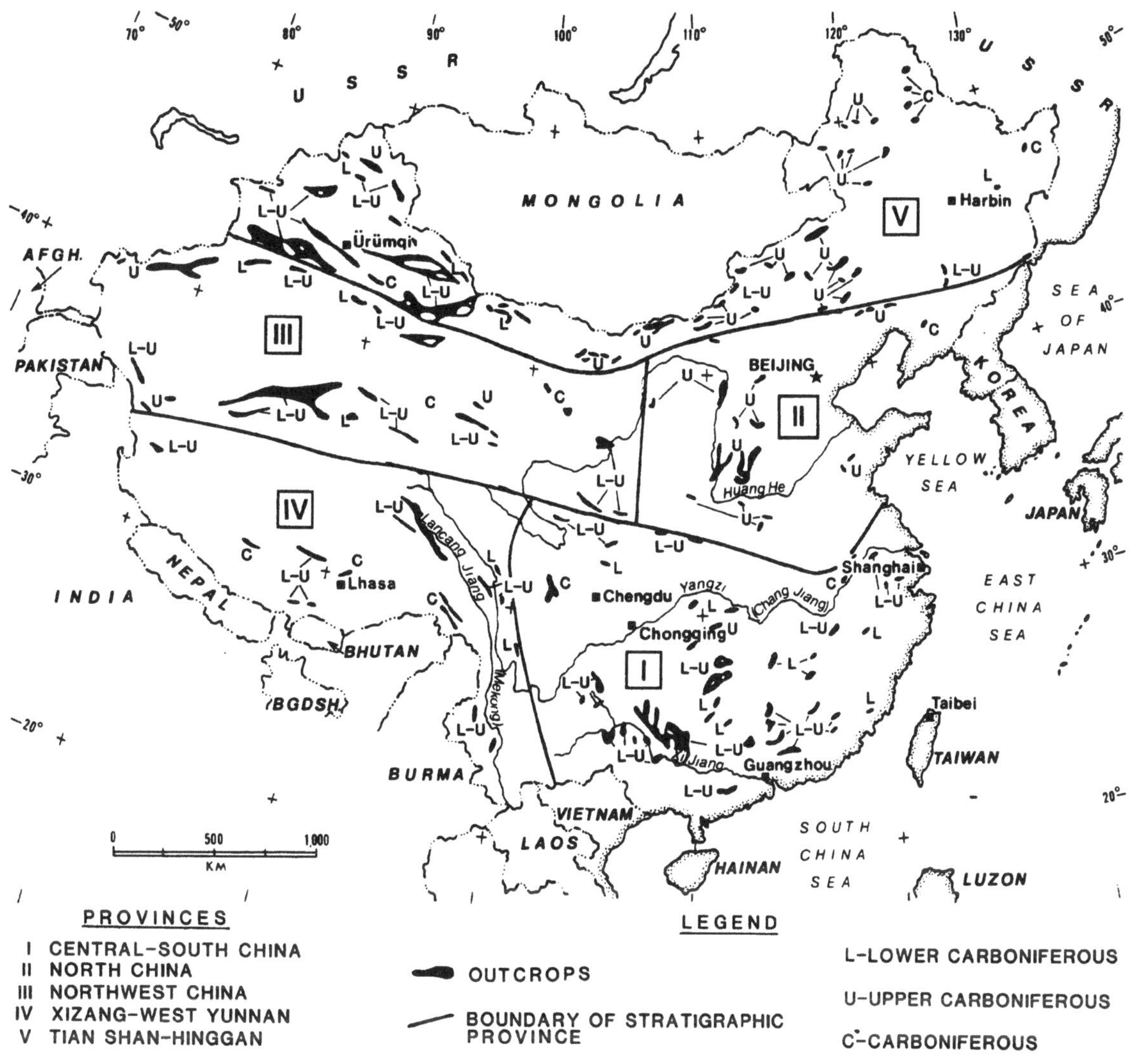

Fig. II-29. Outcrop distribution and stratigraphic provinces of Carboniferous in China (from Yang Shihpu et al., 1983).

shale, and coal. A plentiful marine and non-marine fauna, together with plants, is present.

Northwest China

Extensive carbonates and terrigenous clastic rocks occur here, together with fauna and flora that resemble those of Central-South China.

Kunlun Shan-Xizang (Tibet)-Western Yunnan

Sandstone, shale, sandy shale, siltstone and carbonate rocks of platform facies predominate south of the Kunlun Shan geosynclinal zone. Basaltic volcanics are common on the Gangdise block (western Xizang), and again in western Yunnan Province. They are interbedded with neritic platform facies. North of the Gangdise block, up to 7,000 m of Carboniferous and Lower Permian is known. In the Kunlun Shan geosyncline, the rocks consist predominantly of flysch with eugeosynclinal mafic submarine volcanics (Yin Ji-xiang et al., 1983). A pronounced division of floras and faunas appears, with Cathaysian floras extending onto the Gangdise block from the north and Gondwana floras across the Himalaya block from the south. *Glossopteris* was discovered north of Qomolangma Feng (Mt. Everest) for the first time by Hsu Jen (1973, 1976). A narrow trough of paraflysch with deep-water facies occupied an area near Rutog extending on both the Gangdise and Qiangtang block. This trough cuts across the major east-west fault zones of today (Hu Changming, 1984).

Tian Shan-Da Hinggan Ling (Boreal Depositional Area)

Along this latitudinal segment of northernmost China, the deposits are thick and geosynclinal, intercalated with volcanics; Lower Carboniferous is absent in places.

Lower Carboniferous: Distribution and Facies

The distribution of lithofacies in the Lower Carboniferous of China depends upon a framework of two

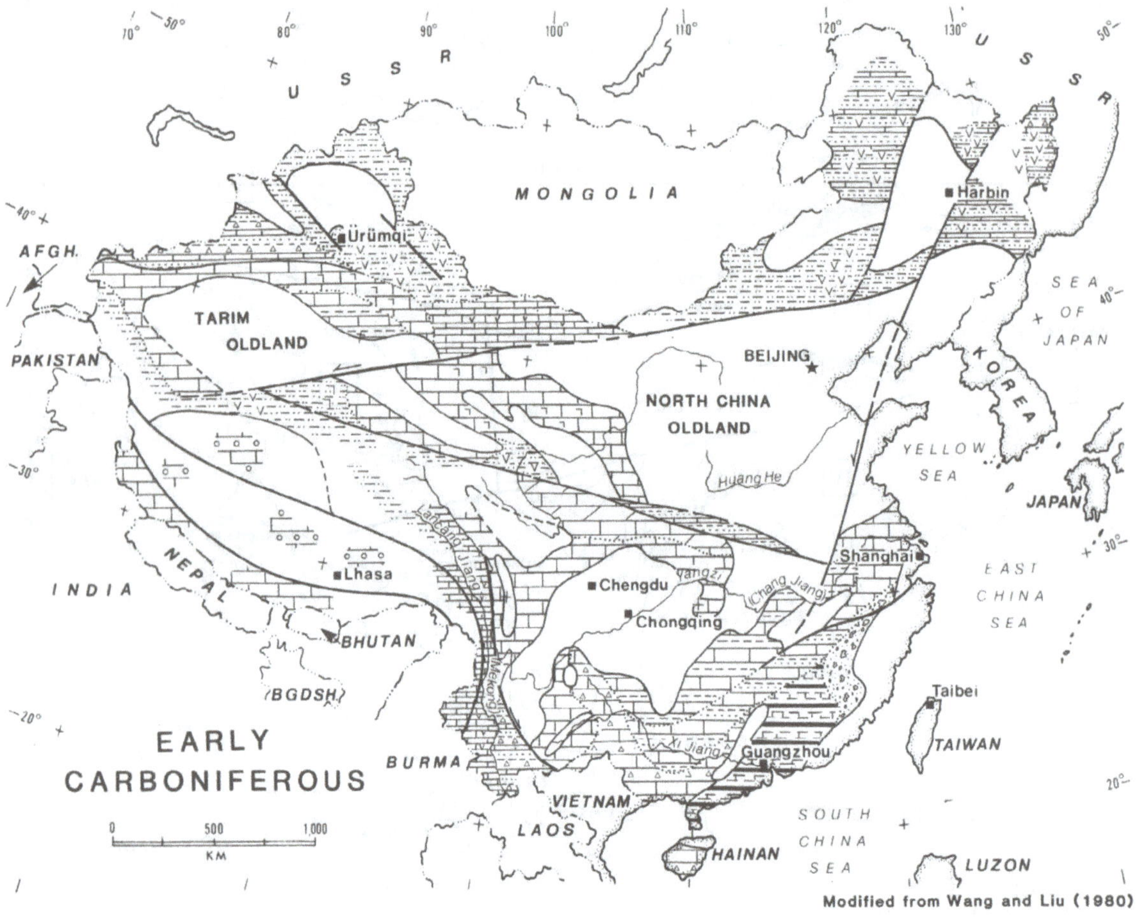

Fig. II-30. Paleogeography and lithofacies map of Early Carboniferous, China.

latitudinal tectonic zones (Figure II-30). The Northern line of the framework is formed by the Tian Shan and the Yin Shan, and the southern line, by the Kunlun Shan and the Qin Ling. Together, these two lines give rise to four latitudinal areas of deposition (Yang Shihpu et al., 1983).

Tian Shan-Da Hinggan Ling

Although the Lower Carboniferous is absent in some places along these latitudes on the northern slopes of the Tian Shan and the Yin Shan, generally an extremely thick succession of shallow-water terrigenous-clastic deposits, carbonates, and volcanic rocks is present in an extensive geosyncline. The coralline fauna is closely related to that of Central-South China, whereas the assemblage of brachiopods shows similarity to that of the U.S.S.R. on the north. North of the Tian Shan in the Junggar basin, the Lower Carboniferous, possibly marine, is present in the subsurface (Lin Longdong, 1984). It is, however, mildly metamorphosed and apparently in a geosynclinal (flyschoid) facies.

Western and Central China

Between the two latitudinal controls named above, a large area possibly stayed emergent. Deposition took place in a few areas, however, including the northern margins of the Tarim and Qaidam basins, northern Kunlun Shan, southern Dabie Shan, southern Tian Shan, and along the Qin Ling (Figure II-30). Basal beds may be absent or only poorly developed. Apart from these, the Lower Carboniferous sequence consists mainly of coal-bearing terrigenous clastics and carbonates. Gypsum and salt are present in lagoonal strata that crop out along the Qilian Shan and along the southern slope of the Tian Shan. Faunas show affinities with those of Europe.

Southern China

Uplifts - Cathaysian, Jiangnan, and central Yunnan - exist within this large area. Away from these uplifts, however, the Lower Carboniferous is widespread in the lowlands of the Yangzi River, as well as in those of the Pearl River (Zhujiang). The sequence, characterized by littoral and shallow-water deposits, is complete and thick. Limestone is the main element, and this alternates with sandstone and shale. Corals and brachiopods dominate an abundant fauna. The carbonates are oil-bearing in the Beibuwan (Gulf of Tonkin). Important coal measures developed in southeastern China, where some gas deposits have been traced to them (Han and Yang, 1979, 1980; Lee, 1984).

Xizang (Tibet)-Western Yunnan

The Lower Carboniferous of this large area remains poorly known compared with other areas of China. Some sequences contain great thicknesses of terrigenous-clastic strata, with minor carbonates. Others contain thick limestones and interbedded basalts. Coal measures appear to be absent. Gondwanan elements appear in extreme southern Xizang (Tibet) Autonomous Region, but most of the fauna is northern, and includes some North American taxa. On Qomolangma Feng (Mt. Everest), 126 m of gray marine shale with some marl is present, bearing a mainly northern fauna (Mu Anzi et al., 1973). On the opposite side of the present Qinghai-Xizang Plateau, several thousand meters of flysch and mafic volcanicgenic rocks accumulated in the Kunlun Shan geosynclinal basins (Yin Ji-xiang et al., 1983).

Southern Qinghai-Xizang Plateau

On the Qinghai-Xizang Plateau, the Lower Carboniferous is composed of shale, limestone, and siliciclastics in the Himalaya, whereas in northern Xizang (Tibet) Autonomous Region and southern Qinghai Province, it consists mainly of shallow-water carbonate.

In this region, pre-Carboniferous faunas possess a somewhat cosmopolitan character, although the assemblages contain species endemic to southeastern Asia, so that the region has a distinctly south or southeast-Asia flavor. The Cambrian *Redlichia* (trilobite) fauna, for example, is known from Iran, the Salt Range of Pakistan, the Tarim basin, Burma, Southwest China, and several other large areas. Ordovician faunas include numerous genera and some species from Europe, North America, and northern China. Several Silurian genera have a worldwide distribution. Numerous Devonian forms are found in various parts of China, the U.S.S.R., Europe, Australia, and North America.

Beginning in the Carboniferous, the present area of the southern Qinghai-Xizang Plateau was divided abruptly into three latitudinally oriented geogeographic zones (Figure II–31). The northern and central zones appear to be separated by the Bangong-Nujiang fault system; the central zone is separated from the southern by the Yarlung Zangbo (Indus) fault system.

Wen (1981) described the Lower Carboniferous fauna north of the Bangong-Nujiang fault system. Abundant fossils occur, and these include fusulinids, corals, brachiopods, and ammonoids. Present are genera such as *Eostaffella*, *Cystophrentis*, *Aulina*, *Kueichouphyllum*, *Yuanophyllum*, *Gigantoproductus*, *Schuchertella*, and *Praewaagenoceras*. These are closely similar to the faunas of South-Central China. South of the fault system there occurs a mixed Tethyan-Gondwanan fauna dominated by the brachiopods *Fusella*, *Ovatia*, *Marginirugus*, *Syringothyris*, and *Balakhonia*, clearly allied to the faunas of Australia and the Himalaya, yet different from wholly Gondwanan forms farther south.

All Wen (1981), Chang and Pan (1981) and Huang and Chen (1987) interpreted the faunal distribution of the Carboniferous, and the faunal and floral distributions of the Permian, as evidence that India-Himalaya-Northern Xizang (the area south of the Bangong-Nujiang fault system), separated and drifted southward from Asia in Early Carboniferous time. During the Mesozoic, the direction of drift is supposed to have reversed, with collision during the Late Jurassic, the Cretaceous, and the Paleocene-middle Eocene.

Middle Carboniferous: Distribution and Facies

Gao Lianda et al. (1983) defined the Middle Carboniferous column of China as a sedimentary succession whose lower boundary is the lowest appearance of the fusulinid *Pseudostaffella antiqua*, the brachiopod *Choristites*, and the ammonoids *Reticuloceras* and *Retites*; and whose upper boundary is the highest appearance of the fusulinids *Fusulina cylindrica* and *F. quasicylindrica*. Except for the uplifts of Yin Shan (Nei Monggol Autonomous Region), Huaiyang (Henan Province), Taiwan, and - possibly - central Xizang (Tibet) Autonomous Region, the distribution of the Middle Carboniferous across China is extensive. Sedimentation was controlled by the latitudinal structural zones of Tian Shan-Yin Shan and Kunlun Shan-Qin Ling, as well as in the southeast-trending structural zones of Hengduan Shan (Xizang-Yunnan) and A'nyemaqen Shan (Qinghai Province). Stratigraphy and lithofacies change across these zones, and five distinctive areas of deposition are established.

Central-South China

Marine carbonates are typical in many sequences of the Middle Carboniferous in this area. Columns commonly are complete, with varied and abundant fossils.

North China

The Middle Carboniferous sequence consists of alternate marine and terrestrial strata, the latter with coal measures. Fossil plants occur in the coal measures. Brachiopods and fusulinids characterize the marine strata. China's thickest Carboniferous (Middle and Upper) coal measures occur in North China and Northeast China (Figures II–32 and II–33; Li, 1962; Han and Yang, 1979, 1980; Gao Lianda et al., 1983).

Northwest China

Differences of tectonic development and of local paleogeography affect the Middle Carboniferous sequence

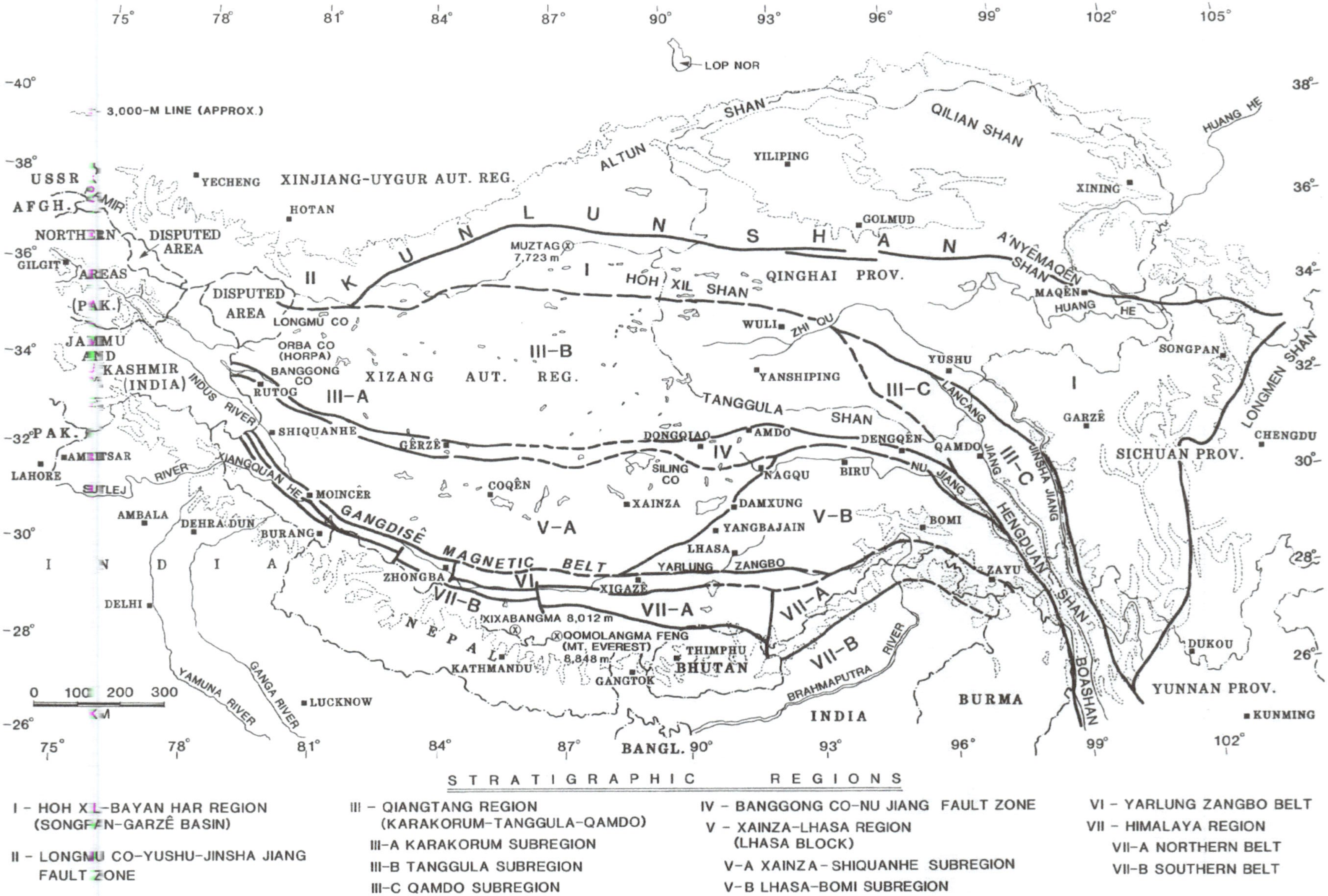

Fig. II-31. Stratigraphic regions of the Qinghai-Xizang (Tibet) Plateau, China.

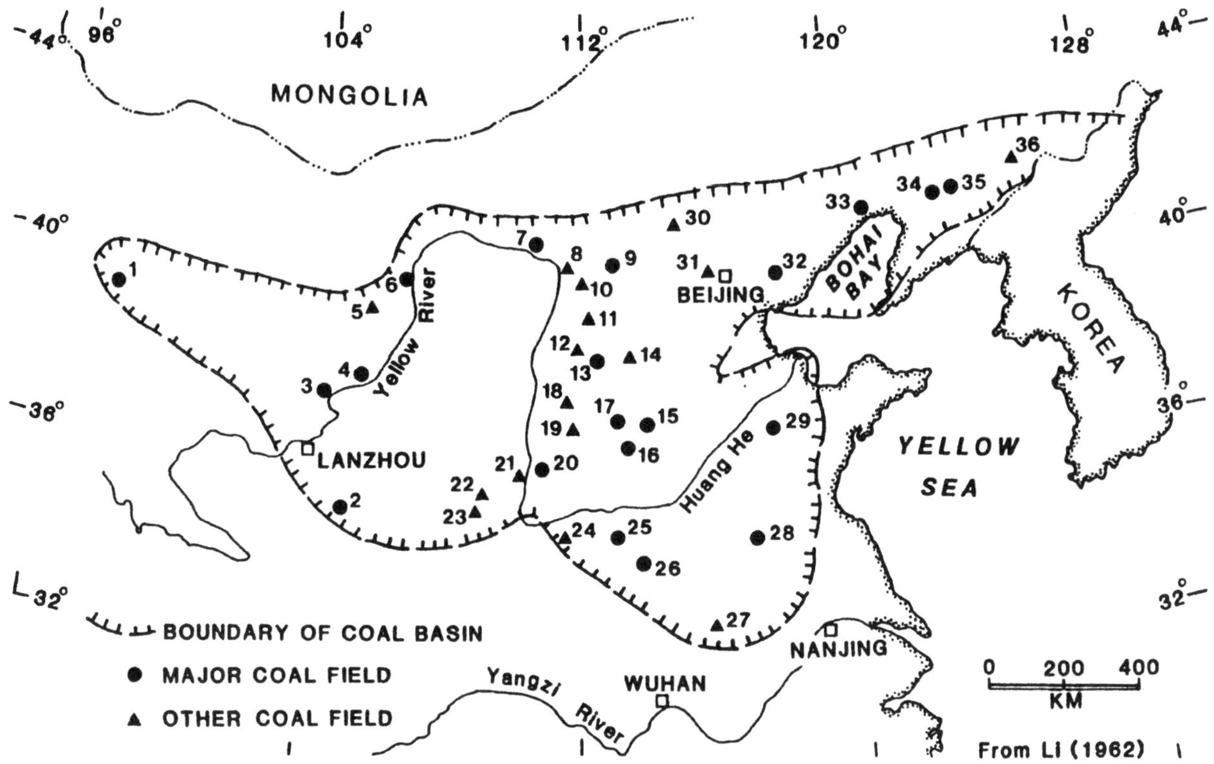

COAL FIELDS

1. NANSHAN	7. DAQINGSHAN	13. TAIYUAN	19. HUOXIAN	25. YUXIAN	31. XISHAN
2. TONGWEI	8. GINGSHUIHE	14. PINGDING	20. XIANGNING	26. PINGDINGSHAN	32. JIANPING
3. JINGTAI	9. DATONG	15. FENGFENG	21. HANCHENG	27. GUSHI	33. NANPIAO
4. ZHONGWEI	10. PINGLU	16. LUCHENG	22. TONGCHUAN	28. JAWANG	34. YANTAI
5. HELANSHAN	11. NINGWU	17. XIANGYUAN	23. YAOXIAN	29. ZIBO	35. BENXI
6. SHIZUISHAN	12. JINGLE	18. XIAOYI	24. YINGYANG	30. ZANGJIAKOU	36. TONGHUA

Fig. II-32. Distribution of major Carboniferous and Permian coal fields in North China.

of this area. In the Tarim basin, the sequence contains abundant fusulinids and brachiopods. Other areas exhibit marine deposits with ammonoids and alternate coal measures with plants. In the Junggar basin, Middle Carboniferous non-marine strata overlie the Lower Carboniferous (Lin Longdong, 1984).

Xizang (Tibet)-Western Yunnan

The Middle Carboniferous sequence of this region remains unclear, partly because of poor exposures and partly because of lack of detailed investigation. In western Yunnan Province, volcanic rocks are present in the sequence. In the Shenzha area of Xizang (Tibet) Autonomous Region, detrital deposits interpreted as glaciomarine are present. Thick eugeosynclinal sections are present in the Kunlun Shan with flysch and mafic volcanogenic rocks. On Qomolangma Feng (Mt. Everest), Mu Anzi et al. (1973) described 1,888 m of gray-black shale, with conglomerate, sandstone, and marl. The depositional environment is platformal, open-marine. The fauna is closely related to that of Western Europe.

Several recent publications on Xizang (Tibet) Autonomous Region have grouped the Middle and Late Carboniferous into one unit; hence we discuss the two together in the section on Late Carboniferous.

Tian Shan-Da Hinggan Ling

This area in general consisted of a mobile trough that received great thicknesses of detrital deposits with intercalated volcanic rocks and volcaniclastics. More stable sectors show a local development of neritic carbonates and of terrigenous clastics. Brachiopods and ammonoids dominate the fauna. Where flora is present some of its elements belong to the early Angaran realm.

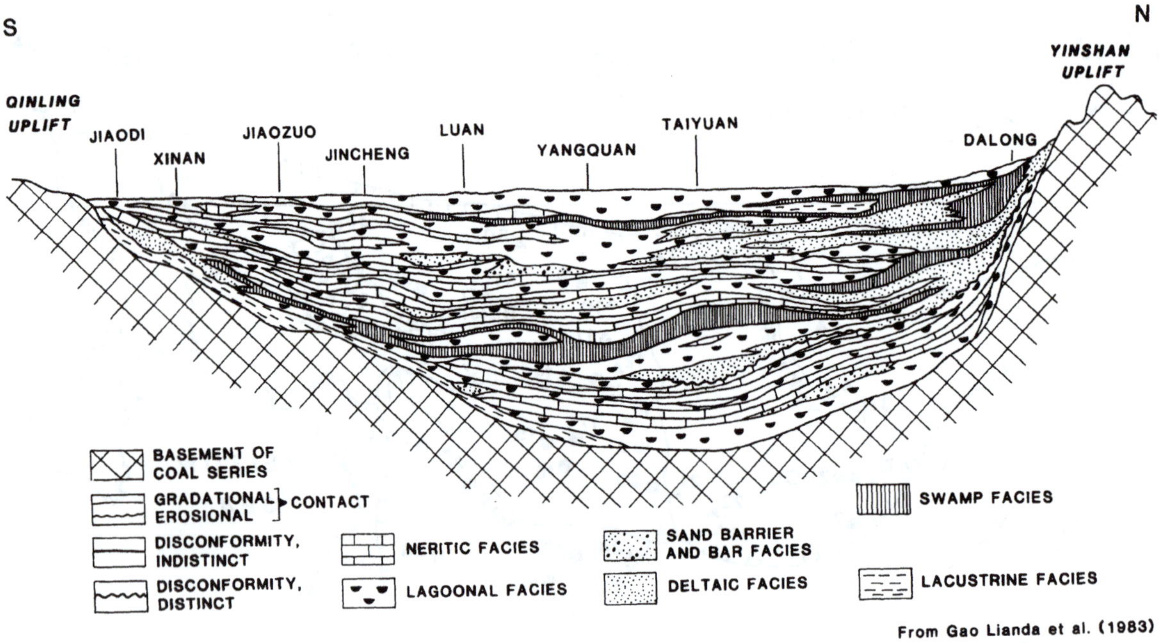

Fig. II-33. Generalized S - N lithofacies section of Carboniferous in North China.

Upper Carboniferous: Distribution and Facies

Li and Zhang (1983) named the main uplifts of the Upper Carboniferous as Yin Shan (Nei Monggol Autonomous Region), southern Kam-Yunnan (Xizang and Yunnan), and 'Paleoyangzi' (mainly Sichuan Province). The Kam-Yunnan uplift is the combined Xizang (Tibet)-Kangdian uplifts of Mu (1962). Marine deposits are extensive, but on the Sino-Korean platform they do not extend as far north as their predecessors of the Middle Carboniferous. Their northern boundary, instead, is a line linking the Datong coalfield (Shanxi Province), Beijing, and the Benxi coalfield (Liaoning Province). The five areas of the Middle Carboniferous persisted into the Late Carboniferous, although specific boundaries differ.

Central-South China

Limestone is a dominant element in the Upper Carboniferous of Central-South China, but shale, sandstone, and dolomite also are present. Fusulinids and corals are important contributors to the fauna.

North China

In this area the Upper Carboniferous lithofacies consist of mixed representatives of the platform. Siliciclastics are the major contributors, with intercalations of limestone and coal. Some of China's most important Paleozoic coal measures formed in the Middle Carboniferous, the Late Carboniferous, and the Early Permian (Figure II-32 and II-33; Li, 1962; Han and Yang, 1979, 1980). Marine and freshwater faunas are both widespread. The flora includes abundant leaves, trunks, and other large remains. Also present in the flora are sporo-pollen assemblages typical of the Middle and Upper Carboniferous of the Euramerican floral realm. On the whole, however, a definitely Cathaysian character was developing during the Late Carboniferous in North China.

Northwest China

Upper Carboniferous deposits show several lithofacies. In addition to marine carbonates, there are alternate non-marine and marine deposits, partly interbedded with thick volcanic layers. From the Qilian Shan eastward the sequences become increasingly like the sections of North China. To the northwest in the Junggar basin, the Upper Carboniferous consists of interbedded marine and non-marine dark-gray to gray-black dolomitic mudstone and agrillaceous tuff, or tuffaceous mudstone. These dark organic mudstones, formed in protected lagoons, are among the oil-source beds in the northwestern Junggar basin (Lin Longdong, 1984).

Xizang (Tibet)-Western Yunnan

Carbonate and terrigenous clastic rocks both occur in this area, and in western Yunnan Province and central Xizang (Tibet) Autonomous Region (between the Bangong-Nu Jiang and Yarlung Zangbo fault zones (Figure II-31), they are accompanied by andesitic and basaltic volcanic rocks, including flows and tuffs. Quartz sandstone, siltstone, silty shale, tuffaceous sandstone, and minor limestone predominate in the north; quartzite, sandstone, sandy shale, silty shale, sandy siltstone, and marl predominate in the south. Deposition took place on an open-marine shelf with small local depressions. In the Rutog area (Figure II-31), a deep-water Late Carboniferous to Early Permian trough with paraflysch formed, along a trend across the major east-west fault zones. A *Eurydesma* fauna flourished in this trough (Hu Changming, 1984). Diamictite is present with sedimentary breccias in the lower part. These diamictites have been interpreted to be glacial and glaciomarine deposits, but this is not proved. Recently, some Chinese geologists have expressed the opinion that the diamictites are nonglacial in origin (Yin Jixiang et al., 1983). Whatever their origin, they are found in the Himalaya, south of the Indus-Yarlung Zangbo fault zone, in the Gangdise block, between that fault zone and the Bangong-Nu Jiang fault zone, and in the region north of the Bangong-Nu Jiang fault zone. In all of these areas, Gondwana faunal and floral elements are present. The Gondwana elements comprise but a small part of the total flora and fauna (Waterhouse, 1983), and they disappear in a gradational manner northward (He and Weng, 1982, 1983; Liang Dingyi et al., 1983; Liu and Cui, 1983; Wang Hongzhen, 1983; Wang and Mu, 1983; Zhao Junpu, 1984). In the Gangdise block and in the northern Himalaya, beds with Tethyan faunas alternate with thin layers containing Gondwana elements, and thick zones of mixed Tethyan and Gondwana elements (Liang Dingyi et al., 1983; Wang and Mu, 1983). Abrupt lithologic and faunal changes are not associated with these major fault zones. At least 730 m is present (Mu et al., 1973).

In the Tingri County area north of Qomolangma Feng (Mt. Everest), the Late Carboniferous above a basal diamictite (Chataje Diamictite) contains the Gondwana fauna of the '*Stepanoviella*' (brachiopod) beds. The brachiopods include *Stepanoviella gracilis*, *Lissochonetes* cf. *qeinitzianus*, *Trigonotreta* cf. *narsahensis*, *Attenuatella convexa*, and '*Punctospirifer*' *jilongica*. The fauna is closely related to the Gondwana-Tethyan (mixed) fauna of Umaria on the Indian shield, in Madhya Pradesh, 650 km southeast of New Delhi (Wang and Mu, 1983).

North of the Indus-Yarlung Zangbo fault zone near Lhasa, the approximately equivalent section lies near the base of the Pondo Group, and was collected from Urulung, just north of Lhasa (Wang and Mu, 1983). This locality is on the Gangdise block between the Indus-Yarlung Zangbo fault zone on the south and the Bangong-Nu Jiang fault zone on the north (Figure II-31). Its fauna includes the brachiopods *Bandoproductus* (misidentified by some as *Stepanoviella*) *hemiglobicus*, *Chaoiella latisinuata*, *Leiorhynchoidea* sp., *Praeckmannella* sp. and *Syringothyris* cf. *nagmagensis*. According to Wang and Mu (1983), this fauna lived in warmer waters than the Tingri fauna south of the Indus-Yarlung Zangbo fault zone. That fault zone, however, is unrelated to the interpreted difference in water temperature, a statement proved by the Rutog section at the western end of the Chinese part of the Gangdise block. At Rutog, and north of it in the Gaggar Co-Duoma district, the basal unit of the Upper Carboniferous Horpa Co (Huoerpacuo) Group is the Cameng Formation, a marine diamictite unit. It is overlain by the Zhanjin Formation (Liang Dingyi et al., 1983; Liu and Ciu, 1983). The lower Zhanjin is a paraflysch, associated with slaty shale; the upper part also is paraflysch, but includes conglomerate, sandstone, and conglomeratic sandstone. In the lower part are the gastropods *Eurydesma perversum* (the first Chinese occurrence of *Eurydesma*), *Nuculopsis* sp., and *Mourlonia* cf. *freneyensis*. Above are *Eurydesma playfordi*, *E. mytiloides*, *E.* cf. *globosum*, *E. sumoratum*, *Schizodus* cf. *occidentalis*, *S.* cf. *meekanus*, and other forms. The Zhanjin also contains the brachiopod *Ambikella fusiformis* and the corals *Amplexocarinia* and *Cyathaxonia*.

The northern margin of the Gangdise block is the Bangong-Nu Jiang fault zone, north of which lies the Qiangtang block. On the Qiangtang block (Figure II-31), north of Rutog at Tulonggongba (Toinlunggomba), the basal (Late Carboniferous part) Longmu Co Group contains the warm-water fusulinids *Triticites altus Pseudofusulina ovata*, *Boultinia willsi*, and *Schwagerina quembeli pseudoregularis* (Wang and Mu, 1983).

Tian Shan-Da Hinggan Ling

Along this latitudinal trend in northernmost China, the Upper Carboniferous sequence shares the character of the whole Carboniferous sequence of this trend in being thick and geosynlinal, the latter feature involving the presence of many interbedded volcanics. The Xinjiang Uygur Autonomous Region (central) and Gansu Province exhibit a biofacies of ammonoids and brachiopods. The Bogda Shan of northern Xinjiang Uygur Autonomous Region exposes numerous outcrops of coarse-grained silicilastics with a marine ammonoid facies in some intervals, and with a continental plant-bearing facies in others. Farther east, in the Nei Monggol Autonomous Region, the Upper Carboniferous sequence contains limestone and siliciclastics, both of which contain a marine biofacies of fusulinids and brachiopods. Corals are also present.

Type Sections

Lower Carboniferous

Fang et al. (1979) illustrated and tabulated the lithology of a type section of Lower Carboniferous, and indicated its major fossils, all representative of southern Guizhou Province (Figure II-34).

The strata in this sequence are marine. A hiatus is present locally between the underlying Devonian and the base of the Carboniferous, but in many places the contact between the two systems is conformable, a fact which suggests no major regression at the time boundary. Above the base, Fang et al., used C_1 for a lower division, and C_2 for the upper division, up to the top of the Lower Carboniferous sequence.

As Figure II-34 shows, limestones form an important

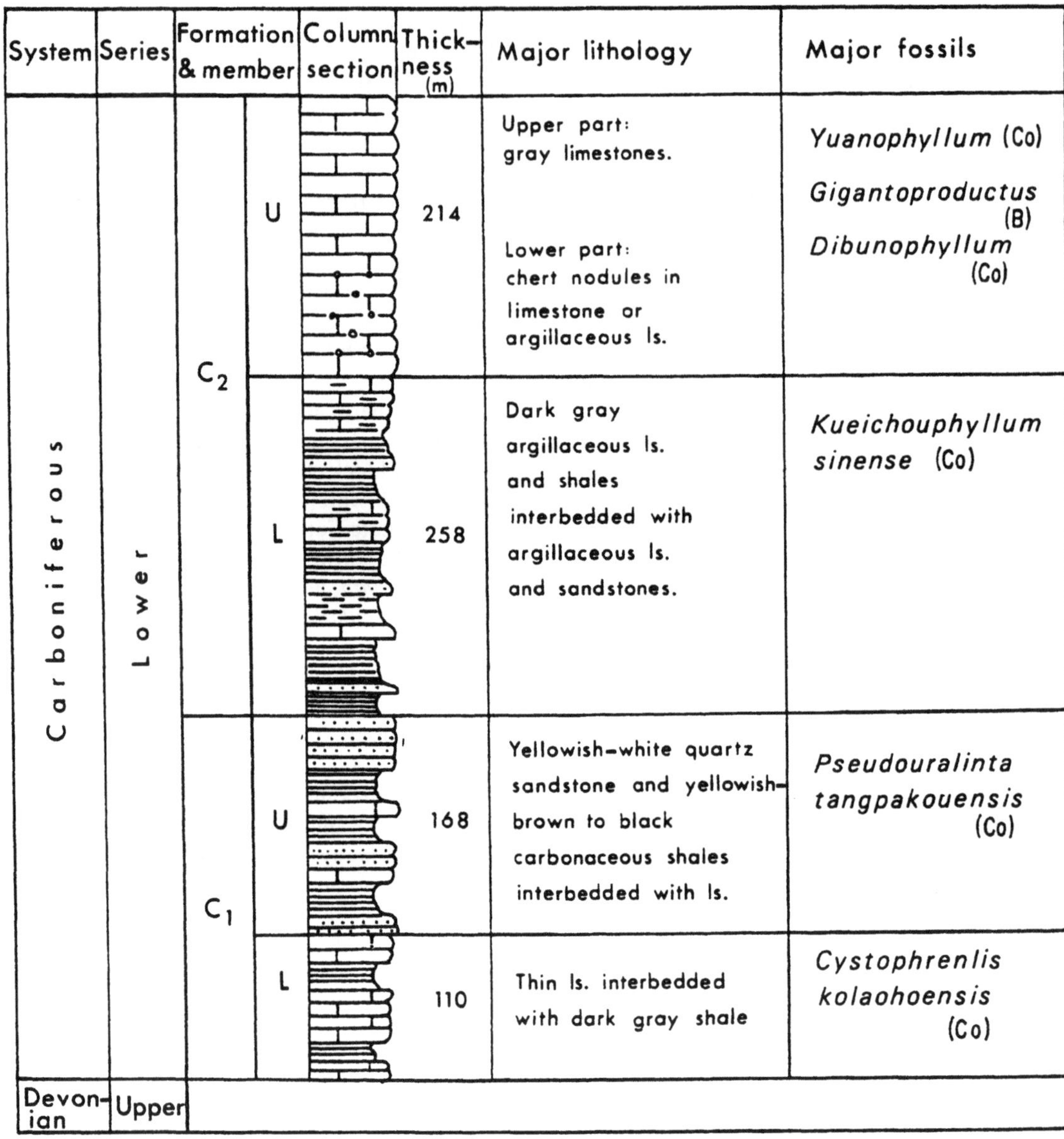

Fig. II–34. Lower Carboniferous stratigraphy in southern Guizhou Province, China (from Fang et al., 1979).

constituent of C_1, and they form also the best indicators of neritic facies in a regime of moderate transgression. Corals and brachiopods typify the fauna of C_1. Among the corals, characteristic forms are the solitary tetracorals with septa, tabulae, and dissepiments. The assemblage of fossils suggests a correlation with most of the Tournaisian stage of Europe (Figure II–11).

In C_2, clastics become relatively important, with a facies that suggests coastal environments and an episode of moderate regression. Limestones, the highest strata of C_2, signal the return of transgression and presumably a neritic environment. Corals and brachiopods typify the fauna. Among the corals, characteristic forms are colonial and rugose, the latter triple-zoned. All corals in C_2 exhibit septa, tabulae, dissepiments, and columella. The assemblage of fossils suggests a correlation with most of the Visean stage of Europe (Figure II–11).

A section from East China consists mainly of continental strata with a few marine layers. This conjunction of facies indicates the presence of an oldland, possibly a highland, invaded from the west by marine waters during transgressional maxima in Central-South China. Grayish-white and purplish-red terrigenous clastics are present, with a flora of the plants *Sublepidodendron*, *Rhodea*, *Asterocalamites*, *Mesocalamites*, *Archaeopteris*, and *Neuropteris*. A sparse marine fauna includes the brachiopods *Eochoristites*, *Echinoconchus*, and *Chonetes*.

The later investigations of Yang Shihpu et al. (1983) have yielded lithological and paleontological details of more than twenty sequences of Lower Carboniferous across their four latitudinal zones. For comparison with Figure II–34, and for information on chronostratigraphic series and on their relations with European zones, we present our tabulation of a sequence in Hunan Province, Central-South China (Table II–17).

Table II-17. Lower Carboniferous sequence - central Hunan Province - Central-South China

Lower Carboniferous	lower Namurian		Zimenqiao Formation	Limestone and argillaceous ls.; *Yuanophyllum kansuense, Neclisiophyllum yangtzeense* (corals); *Latiproductus edelburgensis* (brachiopod); *Eostaffella* (fusulinid); *Homoceras* cf. *subglobosum* (ammonoid)	15 to 160 m
			Zeishui Formation	Quartzose sandstone and shale with intercalated limestone and coal; *Archaeocalamites prolixus, Cardiopteridium spitsbergense, Triphyllopteris collombiana* (plants); *Arachnolasma sinense* (coral)	80 to 160 m
	Viséan	Datangian	Shidengzi Formation	Massive bedded limestone; *Thysanophyllum asiaticum, T. shaoyangense, Kueichouphyllum sinense* (corals); *Gigantoproductus giganteus* (brachiopod)	150 m
			Liujiadang Formation	Gray limestone with intercalated sandstone and shale; *Zaphrentites paralellus, Pseudouralinia tangpakouensis* (corals); *Hunanoproductus hunanensis, Fusella metatrigonalis, Maartiniella chinglungensis* (brachiopods)	140 to 300 m
	Tournaisian	Aikuanian	Menggongao Formation	Argillaceous limestone, massive bedded; *Cystophrentis kolaohoensis, Caninia cornucopiae* (corals); *Poulonia menggungaoensis; Cleiothyridina serra* (brachiopods)	60 to 90 m
			Shaodung Formation	Marl and limestone, thin-bedded; *Caninia dorlodoti, Ceriphyllum elegantum* (corals); *Sphensospira* sp. *Mesoplica* sp. (brachiopods)	70m

Yang Shihpu et al. (1983) noted three types of contact between the Lower Carboniferous and the Devonian or the pre-Devonian: (a) angular unconformity; (b) disconformity; and (c) conformity. Angular unconformity in Northwest China results from folding in Tianshan tectogenesis before the beginning of Carboniferous deposition. Lower Carboniferous deposition itself may be incomplete at the base. Disconformity occurs where overlap of the Lower Carboniferous took place on the eastern and western margins of the present South China Sea, also around Nanjing along the lower part of the present Yangzi River. To the west of these areas, in eastern Yunnan Province, rocks below the disconformity may belong either to Upper Devonian sequences or to Cambrian sequences. Conformable contacts occur where marine deposition took place continuously across the Devonian-Carboniferous boundary, and where Early Carboniferous taxa exhibit Devonian heritage. Three fossil facies provide part of the evidence for conformity. Evidence from the ammonoid facies occurs in Huishui County, southern Guizhou Province. Evidence from the coral-brachiopod facies occurs in central Hunan Province and in northern Heilongjiang Province. Evidence from the stromatoporoid-tabulate coral facies exists in Guizhou Province and in Sichuan Province.

Middle Carboniferous

Fang et al. (1979) illustrated the lithology and the paleontology of a type section of Middle Carboniferous and Upper Carboniferous at Taiyuan, Shanxi Province (Figure II-35). At this locality the systematic stage is C_3. The Benxi Formation at the base consists of two layers of mineral deposits, iron ore below and aluminum ore above. The iron ore is a massive weathered and unstratified residue of reddish-brown hematite and limonite disseminated in clay. This residue constitutes the Shanxi type of iron ore, to be discussed elsewhere. In the overlying clays, the aluminum ore is present, and is better stratified. Brachiopods and fragments of plants point to deposition in a subsiding coastal environment during the lower part of C_3. The middle part consists of fusulinid limestone of inshore marine facies. The upper part consists of sandstone and coal seams of supratidal and paludal facies.

Gao Lianda et al. (1983) presented lithological and paleontological details of eight sequences of the Middle Carboniferous across China. From these we have selected a sequence in Guizhou Province for tabulation (Table II-18), also as a supplement to the sequence in Figure II-34. The sequence in Guizhou Province is wholly carbonate and wholly marine. Its fauna of fusulinids and ammonoids yields data for correlation with equivalents in Europe, U.S.S.R., and North America.

Upper Carboniferous

The C_4 stage of the Upper Carboniferous shown in Figure II-35 is disconformable on the C_3 stage of the Middle Carboniferous. This relationship is an expression of a brief period of erosion before the beginning of deposition of C_4. Within the C_4 succession itself, sandstone deposits are the basal strata in each of three members. An accompaniment of coal and of limestone gives a partly cyclothemic character to each member, and the middle limestone represents a peak of marine incursion during the deposition of C_4. The water remained shallow, and the fusulinid fauna suggests a depth no greater than that of the photic zone (marine 'meadows'). Among the

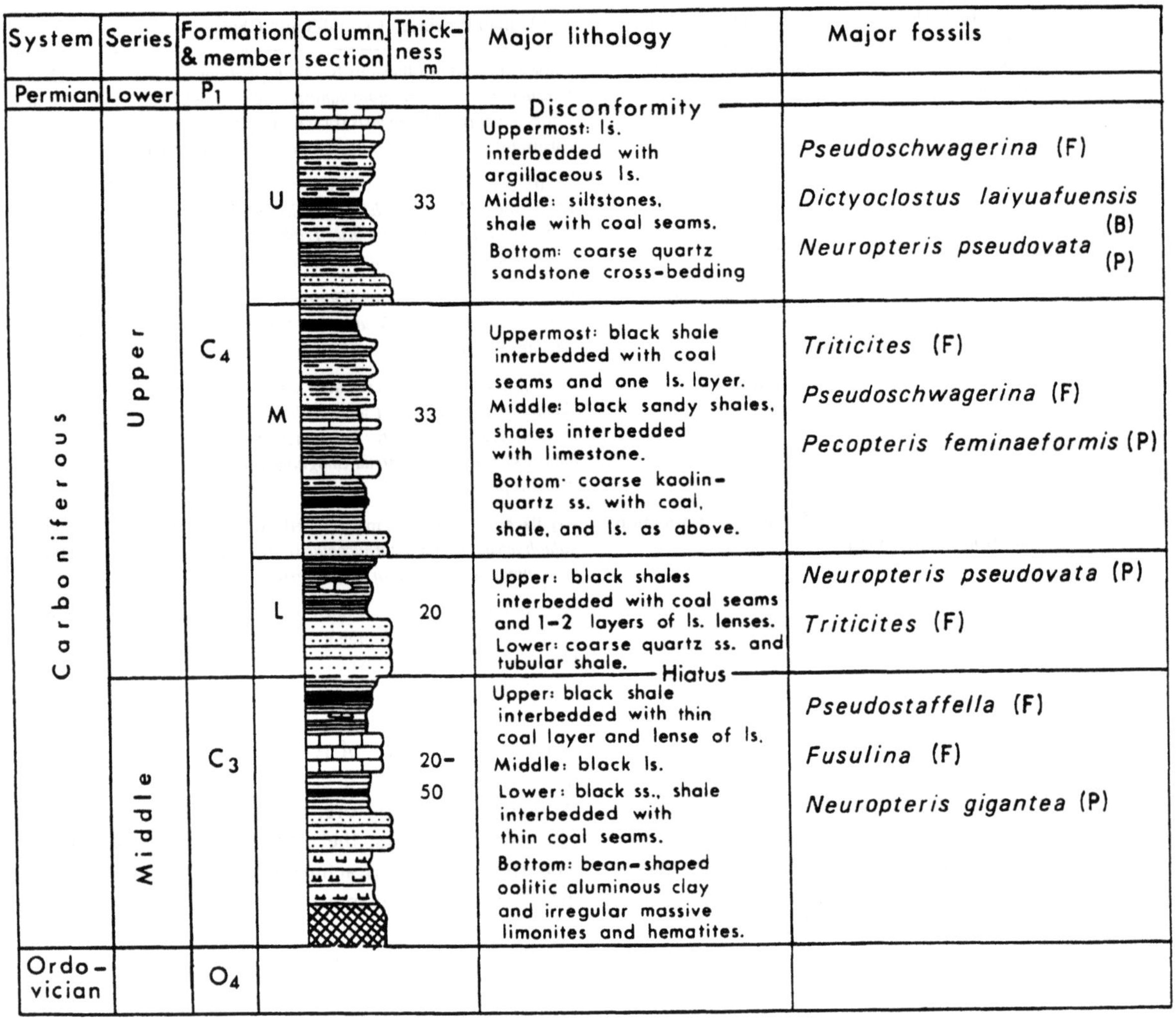

Fig. II-35. Stratigraphy of Middle and Upper Carboniferous in Taiyuan, Shanxi Province, North China (from Fang et al., 1979).

C_4 coals, three are thick and also are of great economic importance. The non-marine facies, that includes the coals of C_4, yields a flora with taxa such as *Neuropteris pseudovata*, *Lepidodendron posthumi*, *Annularia pseudostellata*, and *Pecopteris*. These signal an early stage of Cathaysian flora.

Li and Zhang (1983) presented lithologic and paleontologic detail of more than twenty sequences of the Upper Carboniferous. Of these we tabulate a sequence in Shanxi Province discussed above (Table II-19). In this tabulation we show chronostratigraphic correlations with Europe, U.S.S.R., and North America.

Li and Zhang (1983) discussed the Cathaysian flora shown in our Figure II-35 and tabulated in our Table II-19. The affinities of this flora are with the European Stephanian flora, but a definite endemic element occurs.

Table II-18. Middle Carboniferous sequence - Pan Xian, Guizhou Province - Central-South China

Middle Pennsylvanian	Moscovian	Westphalian	Dalanian	Dala Formation (upper)	Limestone; *Fusulina cylindrica, F. quasicylindrica, F. schellwieni, Fusulinella praebocki Pseudostaffella paradoxa* (fusulinids)	120 m
				(lower)	Limestone and dolomite; *Profusulinella prisca, P. ovata, P. pseudorhomboides, Eofosulina triangula, Pseudostafella* sp. (fusulinids)	
Lower Pennsylvanian	Bashkirian	(upper) Namurian (middle)	Huashibanian	Huashiban Formation	Limestone and dolomite; *Gastrioceras, Brannoceras, Reticuloceras, Retites* (ammonoids); *Pseudostafella antioua* (fusulinid); *Choristites mansuyi* (brachiopod)	540 m

Table II-19. Upper Carboniferous - Taiyuan, Shanxi Province - North China

Missourian Virgilian	Gzelian	Stephanian	Mapingian	Taiyuan Formation	Sandy shale, sandstone, limestone and coal seams; *Occidentoschwagerina texana, Rugosofusulina complicata, Dunbarinella subnathorsiti* (fusulinids); *Neuropteris ovata, Lepidodendron posthumi, Annularia pseudostellata* (plants)	110 m

Geographically, the Cathaysian flora is bordered on the north and west by the Angaran flora, and on the southwest by the Gondwana flora. The border between the Cathaysian and the Angaran floras corresponds relatively closely with the southern boundary of the Tian Shan-Da Hinggan latitudinal zone. From this boundary southward the Cathaysian flora covers virtually the whole of China. Only in the southern Xizang (Tibet) Autonomous Region does a salient of Gondwana flora enter China, on the northern slopes of the Himalaya, and this is of Late Carboniferous-Permian age (Hsu, 1976; Yin and Kuo, 1978).

Faunas of the Upper Carboniferous in China include fusulinids, corals, brachiopods, and ammonoids, obviously congeneric and some of them conspecific with faunas in the U.S.S.R., Europe, and North America. On this basis the chronostratigraphy of the Carboniferous-Permian boundary should pose no problem. Yang et al. (1962) and Sheng (1962), however, drew attention to the unconformity or disconformity between the 'Carboniferous' and the 'Permian' produced after the Yunnan movement. In China the fusulinid *Pseudoschwagerina* zone lies beneath the unconformity, thereby suggesting that at least for China this zone should be placed in the uppermost Carboniferous, and not in the Permian, as is customary elsewhere, Japan included.

Li and Zhang (1983) discussed the Carboniferous-Permian boundary in terms of a sequence nearly 1,000 m thick near Pu'an and Qinglong in Guizhou Province. The lithofacies of this sequence consists of neritic clastics and limestone. Unconformity or disconformity does not appear to enter into discussions of this sequence. At issue in the boundary question are the top strata (Baomoshan Formation). These lie above the significant ammonoid *Propopanoceras*, but geologists in China believe that there continues to be room for argument. Some consider these post-*Propopanoceras* strata Carboniferous; others Permian; and still others 'transitional.'

Paleoecology and Paleogeography

Early Carboniferous

Yang et al.(1962) wrote that the paleogeography of the Early Carboniferous was closely similar to that of the Late Devonian (Figure II-30). A clear exception to this similarity involved the submergence of the Devonian upland in the Xizang (Tibet) Autonomous Region, and the incursion of marine waters from the Indo-Pacific region by way of present-day Burma. Contemporaneous marine waters also entered and covered Central-South China from the Indo-Pacific region, but by routes east of present-day Burma. In Northwest China, the incursion of marine waters may have taken place along two routes: (1) from the west (Europe) by way of the Tian Shan geosyncline to the Qilian Shan; and (2) from the Middle East to the Himalaya, thence to the Kunlun Shan. East China, North China, and the southern part of Northeast China remained above sea level. Local uplifts at the end of the Early Carboniferous diminished the area of marine waters.

Yang Shihpu et al. (1983) divided the Early Carboniferous into a lower Aikuanian stage and an upper Datangian stage. During Aikuanian time a marine transgression 'extended over a comparatively small area, divided into southern and northern domains by an uplift' in the central area of China, which gave rise to corresponding southern and northern biogeographic provinces. During Datangian time an expansion of marine waters united not only these two biogeographical provinces but also others. Only the uplift of North China withstood this expansion of marine waters, and some of these waters may have come from the north, as shown by Siberian elements in the fauna. Coal-swamp conditions were widespread in East and Central-South China, where coal gas has been generated (Han and Yang, 1979, 1980).

Middle Carboniferous

This interval witnessed the submergence of an extensive oldland, the Sino-Korean platform of North China and the southern part of Northeast China, which had remained above sea level since the Middle Ordovician. Encroaching seas were shallow in many areas, and coastal swamps developed on their margins in a climate interpreted as hot and humid. Coal swamps formed across northern China (Han and Yang, 1979, 1980). Fang et al. (1979) used two pieces of evidence to determine the direction of transgression: (1) the Middle Carboniferous is thickest in the southern part of Northeast China, becomes thinner toward central North China, and pinches out completely in the southern and western parts of that region; and (2) the zones containing iron and aluminum ore (at and close to the Carboniferous-Ordovician contact) are diachronous. In Northeast China and in the central part of North China, these two ores lie low in the Middle Carboniferous. In the southern and western parts of North China, they lie in the basal Upper Carboniferous. Such evidence indicated to Fang et al. that marine waters transgressed from northeast to southwest during the Middle Carboniferous and that the locations of coastal zones were shifting. The ores deposited in or near such coastal zones consequently may have undergone a long period of weathering and erosion. Gao Lianda et al. (1983) added to the paleogeographic picture of the Middle Carboniferous by observing that the uplifts of Yin Shan, Huaiyang, and Kam-Yunnan (Xizang-Kangdian) remained positive areas throughout this time.

Late Carboniferous

Li and Zhang (1983) stressed the extent of transgression during the Late Carboniferous. In addition, they inferred the strength of this transgression by its invasion of land areas that had resisted submergence since the early Paleozoic. Uplifts that remained emergent were: (1) Yin Shan; (2) southern Kam-Yunnan (Xizang-Kangdian); and (3) 'Paleoyangzi' (mainly Sichuan Province). Smaller land areas were Liaodong (peninsula, Liaoning Province), and Jiaodong (Shandong Peninsula). This is the Jiaoliao oldland of the Sinian to Paleozoic interval according to Mu (1962) (see Figures II-8, II-12, II-19, etc.). Small uplifts remained along the Qin Ling (Shaanxi Province) and along the Dabie Shan (Anhui Province), a line that constituted 'a natural geological boundary between South and North China.'

On the Sino-Korean platform the transgression did not reach northward as far as that of the Middle Carboniferous. Major coal fields developed (Figure II-32) from the swamps that flourished from the Middle Carboniferous through the Early Permian, and these generated large amounts of gas which now have accumulated in Carboniferous, Permian or younger reservoirs. The northern extent of the marine transgression is close to a line that runs from the Datong coal field in Shanxi Province to Beijing, and from Beijing to the Benxi coal field in Liaoning Province. Farther west, transgression was strong, able partly to invade the local uplift of Dongsheng-Longshan in southeastern Sichuan Province, which had stood as land in the Middle Carboniferous. In this way marine waters came through southeastern Gansu Province to the present Ningxia Huizu Autonomous Region in North China.

In much of North China and Northwest China, marine waters retreated as much as they advanced. The regimes of these areas correspondingly alternated between the littoral and the deltaic. Where the paludal (marshy) phase of the deltaic cycle intervened, coal commonly was deposited during the Late Carboniferous. In Central-South China, the powerful transgression of marine waters linked this area with the marine waters of Tethys, west and northwest, and with the ancestral Pacific of the Late Carboniferous on the east (Li Siguang (J. S. Lee), 1939). A second link between Tethys and the Pacific resulted from marine waters that moved along the latitudinal zone of Tian Shan-Da Hingan Ling in northernmost China. The faunas of Xizang (Tibet) Autonomous Region, including those of the Himalaya, are related to those of central and southeastern China, Nei Monggol, the U.S.S.R., and Western Europe, with clear links to North America. However, presence of Gondwana fauna in the stratigraphic sections of the Xizang-Qinghai (Tibet) Plateau, Himalaya, and western Yunnan may suggest that there was another link between the Tethys and the southern 'cold seas' (Mu En-zhi et al., 1986).

PERMIAN SYSTEM

Permian sequences in China are well developed, and also are highly fossiliferous, two properties that make China one of the world's best areas for lithostratigraphic and chronostratigraphic studies of the Permian System. Marine strata predominate in Central-South China; continental deposits in North China (Figure II-36). Fauna shows notable variation. In Central-South China, fusulinids constitute an important fossiliferous element throughout the sequence. Corals are more prominent in the Early Permian. In the southwestern sector of Central-South China, brachiopods are characteristic of the whole Permian sequence. Throughout Central-South China, ammonoids and conodonts constitute lesser elements in the fauna. In the Himalaya, corals and fusulinids dominate. Brachiopods and a few ammonoids occur. Farther north, the fusulinids exhibit four assemblages in a fauna that contains brachiopods, corals, ammonoids, and bryozoans. In North China, plants dominate the fossil assemblage (Sheng et al., 1979).

Distribution and Facies

Sheng (1962), in a study of depositional environments and fossil assemblages, recognized five Permian regions in China (Figure II-36). We have modified his names of areas somewhat and we have added the Gondwana-related paleogeographic province of southern Xizang-Qinghai.

Central-South China and East China

The facies (on Figure II-36) of this region is mostly marine with only minor exceptions. Much of the section consists of shelf limestone deposited in shallow water. Abundant fossils include fusulinids, corals, and brachiopods. Along the Ali Shan, the metamorphosed backbone of eastern Taiwan, the low grade of metamorphism has permitted the preservation of rare fusulinids and corals. These may include Late Carboniferous, but their main indication is for Permian (Yen et al., 1951; Yen, 1957; Chang and Lin, 1984; Chen, 1984). The Permian fossil assemblages of Central-South China as a whole correlate closely with Permian fossil assemblages elsewhere in the greater Tethyan realm of Asia, as well as with those

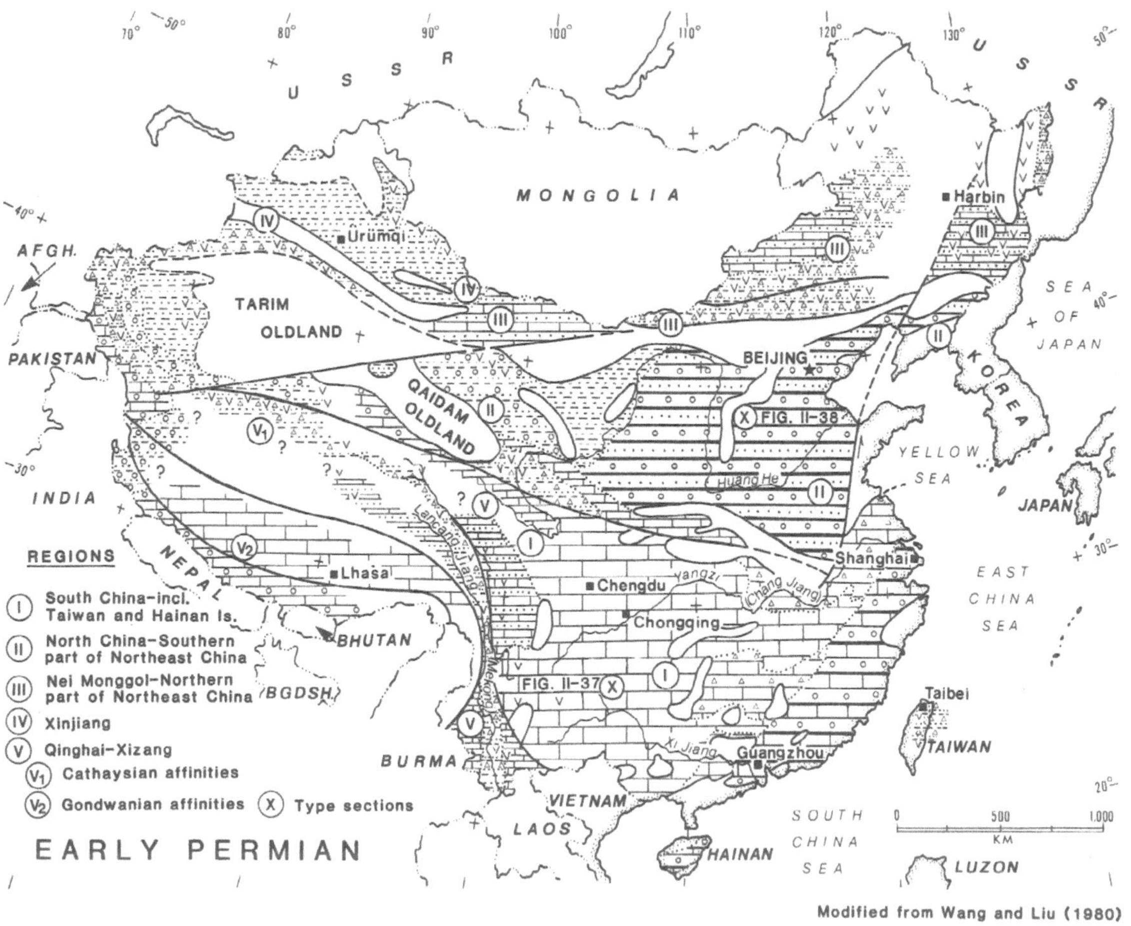

Modified from Wang and Liu (1980)

Fig. II–36. Paleogeography and lithofacies map of Early Permian, China.

in Tethyan realms beyond Asia (Sheng Jin-zhang et al., 1985).

Figure II–37 presents a typical column of Permian marine rocks. Some coals enter the sequence near the top. Although this section lies in central Guizhou Province, it is nearly identical with equivalent sections in Sichuan Province.

North China-Southern Part of Northeast China

Continental deposits dominate this region (II on Figure II–36), and the flora that they contain is typically Cathaysian. The large coal swamp and deltaic plain that developed in the Middle Carboniferous persisted into the Permian; Figure II–38 shows a typical Permian section from the Shanxi Plateau, central Shanxi Province. The section is mainly non-marine. Coal measures in the Lower Permian are known source rocks for some gas in northern China.

Nei Monggol Autonomous Region-Northern Northeast China

The Permian sequence in this latitudinal belt (III on Figure II–36) exhibits a decidedly variable lithofacies, with low-grade metamorphic and volcanic rocks in addition to sedimentary deposits. Of the latter, the lower part consists of marine deposits with fusulinids; the upper part consists mainly of non-marine deposits with plants, bryozoans, and pelecypods.

Xinjiang Uygur Autonomous Region

The Permian sequence in this large area (IV on Figure II–36) of northwestern China consists of continental deposits, scattered outcrops of marine deposits, and a few volcanic rocks. Floral assemblages, not surprisingly, relate closely to those across the border in the U.S.S.R. In the various geosynclinal belts (Tian Shan, Kunlun Shan), thick sequences of flyschoid rocks associated with volcanics are present in many areas.

Northern-Central Qinghai-Xizang (Tibet) Plateau

Strongly folded marine limestone, quartz sandstone, slaty shale, and related shallow-water to neritic-shelf deposits comprise the lower part of the Lower Permian (Tingrian stage of Wang and Mu, 1983); the upper part, or Maokouan stage, consists mainly of marble and limestone. The best-studied area lies in the western part

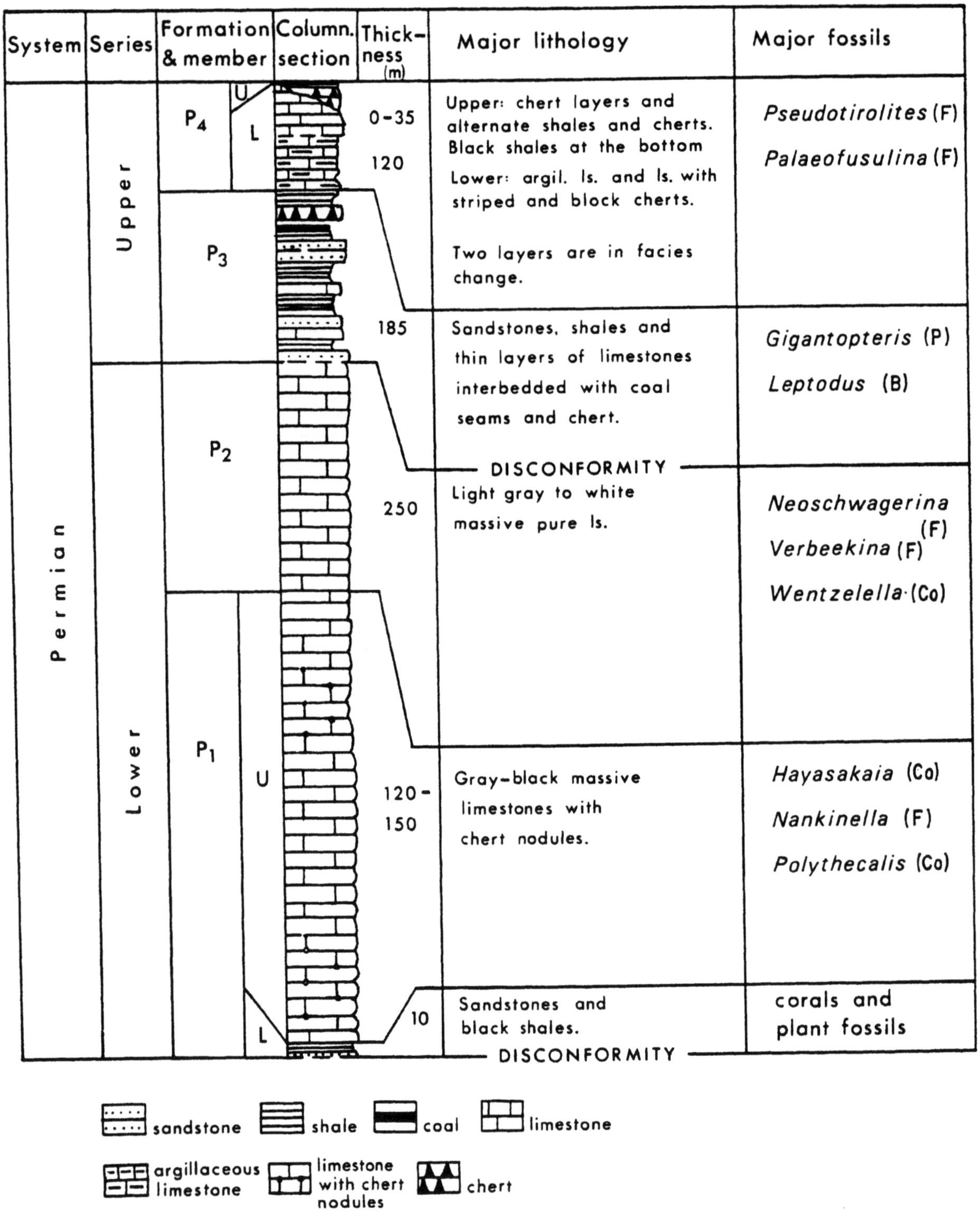

Fig. II-37. Stratigraphy of Permian in central Guizhou Province, Central-South China. Location is shown on Figure II-36 (from Fang et al., 1979).

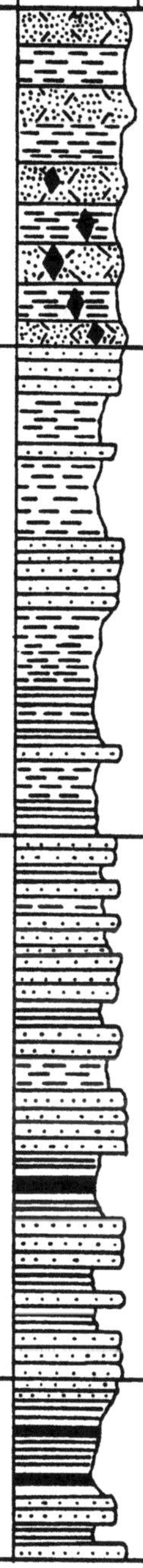

System	Series	Formation	Column section	Thickness (m)	Major lithology	Major fossils
Permian	Upper	P_4		133	Purplish-red and yellowish-white coarse quartz, sandstone and arkose, sandy mudstone, interbedded with gypsum.	*Shihtienfenia permica*
		P_3		129	Upper: purplish-red mudstone interbedded with grayish-green and yellowish-green thin layers of mudstones and sandstones. Middle: Grayish-white, and yellowish-green alternate mudstones and sandy mudstones. Lower: yellowish-green, grayish-purple and purplish-red alternate sandy shale and sandy mudstone with ss.	*Gigantopteris nicotianaefolia* *Lobolannularia ensifolius*
	Lower	P_2		74-84	Upper: yellowish-green thick layers of ss. with shale, mudstone, sandy shale and carbonaceous shales. Lower: yellowish-green and grayish-green ss. with grayish-black shale and irregular coals. Bottom: grayish-green massive coarse kaolinitic quartz sandstone with cross-bedding.	*Cathoysiopteris whitei* *Callipteris conferta* *Emplectopteris triangularis* *Taeniopteris multinervis*
		P_1		56.77	Upper: fine sandstones with siderite nodules. Middle: black shales, calcareous shales with coal seams. Lower: medium size of kaolinitic quartz ss. Black shale with coals.	*Emplectopteris triangularis* *Sphenopyllum*

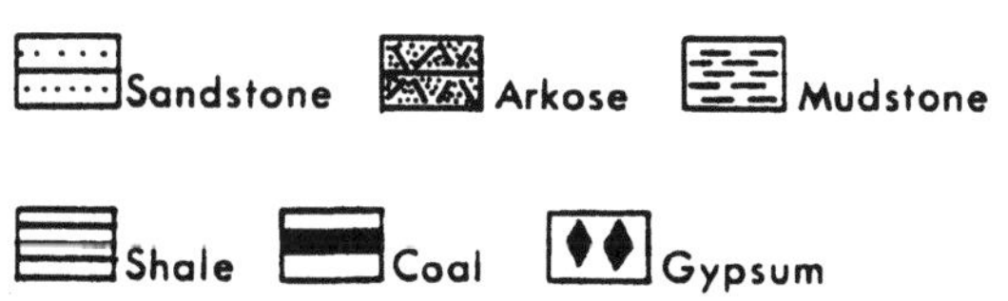

Fig. II-38. Stratigraphy of Permian in Taiyuan, Shanxi Province, North China. Location is shown on Figure II-36 (from Fang et al., 1979).

of the Qiangtang block north of Rutog (Figure II-31). Among recent studies conducted in this area are those by He and Weng (1982, 1983), Liang Dingyi et al. (1983), Liu and Cui (1983), Wang Hongzhen (1983), Wang and Mu (1983), and Hu Changming (1984).

Liang Dingyi et al. (1983) studied the section near Caggar Co (Tsaggar-tso) in northern Rutog County, Ngari (Ali) Prefecture. The Caggar Co is just south of the Xinjiang border, and just east of that part of Kashmir annexed by China, along the northern margin of the Qiangtang block. Liang Dingyi et al. reported the presence of nearly 6,000 m of Upper Carboniferous and Permian, from the basal Cameng diamictites to a position in the middle of the Permian. Most workers state that Upper Permian has not been found in Xizang, but Liang Dingyi and his colleagues place the upper 570 m of the section in the Upper Permian. At least 4,000 m (composite thickness) is Lower Permian, and possibly 4,660 m (composite thickness) is Lower Permian, if there is no repetition by faulting. The actual thickness at any one locality probably is not more than 3,500 m (Liang Dingyi et al., 1983; Hu Changming, 1984).

The Qudi Formation, near the base of the Permian (middle Sakmarian to early Artinskian), lies just above a Late Carboniferous to early Sakmarian shale that contains the coldwater *Eurydesma* fauna (discussed under Carboniferous). The Qudi includes 3,274 m of sandstone, shale, and lesser amounts of siltstone, conglomerate, and limestone. Many of the beds are turbidites with graded bedding and mixed shallow/deepwater faunas. Allochthonous faunas include the fusulinids *Pamirina*, *Rugosofusulina*, and *Pseudofusulina*; the brachiopods *Neospirifer fasciger* and *Subansiria ranganensis*; and the gastropods *Schizodus tibeticus*, *Oriocrassatella rutogensis*, and *O. intermedius*. Above lies the middle to late Artinskian Tulonggongba Formation, 1,325 m of mainly limestone, below, and shale and sandstone, above. The fusulinid zone of *Monodiexodina* occupies this sequence. Another common fusulinid is *Parafusulina*. Brachiopods include *Paraderbya*, *Jipuproductus*, *Costiferina*, and *Juresania*; gastropods include *Bellerophon*; and corals are *Lytvolasma*, *Paracanina*, *Polythecalis*, *Chusenophyllum*, and *Tachylasma*. Above the Tulonggongba lies the Longge Formation (Kungurian, Kazanian, early Tatarian) with the fusulinids *Yabeina* and *Neoschwagerina*; the gastropod *Bellerophon*; and the corals *Iranophyllum* and *Tibetophyllum*. The Longge is 761 (+) m thick. The basal part of the Permian (and the underlying Upper Carboniferous) consists of coolwater beds related to the Gondwanian realm; the Tulonggongba suggests mixed elements of Gondwana and Tethys; the Longge suggests Tethys.

The location of the Caggar Co section illustrates an important fact that has not been realized by most earth scientists interested in this area. This fact is that the Gondwanian facies does not parallel the major east-west fault zones of this region (Figure II-31). The Gondwanian facies in the Rutog area, instead, crosses from the Himalaya block near Zhongba southeast of Rutog to the Qiangtang block north of Rutog, thereby crossing the Indus-Yarlung Zangbo fault zone, the Gangdise block, the Bangong-Nu Jiang fault zone, and almost all of the Qiangtang block, northward to within less than 150 km of the Tarim basin.

Another fact of importance is that scores of taxa have been recorded from the various fossil localities, and some of them are common to both the Gondwanian faunal province and the Tethyan province (Jin Yu-gan, 1981). The separation of one faunal province from another, therefore, is based on only part of the taxa present, and not all. The existence of common taxa in both provinces suggests very close proximity, as Waterhouse (1983) has commented. Waterhouse (p. 347), indeed stated that the Permian faunas and lithologies of the Karakoram (just west of Rutog, north of the Indus-Yarlung Zangbo fault zone) and the Himalaya (south of the fault zone) are close enough to permit the statement that they 'developed in one faunal province.' The differences are rather small, and they suggest that we are dealing here with small changes in latitude (i.e. climate), a suggestion made many decades ago. Waterhouse (1983) suggests that the two areas were within 1° to 10° of each other (latitudinally). The postulate of a wide Permian Tethyan ocean cannot be sustained, as all the biogeographical evidence shows clearly that the strata everywhere developed in shallow water. For this reason, Gansser (1982), Waterhouse (1983), and Stocklin (1984) stated that the concept of a broad ocean, great water depths, and early Tertiary collision along what is now the Indus-Yarlung Zangbo fault zone must be abandoned. The only 'collision,' in fact, which can be supported in this area is the collision that now is taking place between old hypotheses and newly gathered evidence!

The Caggar Co area is not the only place on the Qingtang block where the Permian section has been studied. Wang and Mu (1983) described another section at Tulonggongba (Toinlunggomba), approximately 55 km straight north of Rutog at Dyap Co, along the road to Yecheng (southwestern Xinjiang Uygur Autonomous Region). Here only about 520 m of section is present above the Late Carboniferous diamictite and shale. The lower unit (of two), the Tulonggongba (Toinlunggomba) Formation, 300 m thick, consists of sandstone, slaty shale, and limestone, and this formation contains the *Monodiexodina* fusulinid zone. The fusulinid assemblage consists of *Monodiexodina sutschanica*, *M. wanneri*, *M. kattaensis*, *Pseudofusulina pseudosuni*, *P. houziguanica*, *Parafusulina cincta*, *P. visseri*, *P. elliptica*, *P. rothi*, and *Schwagerina hupehensis*. Above the Tulonggongba is the 200 m-thick Minzho Caka Limestone with the Neoschwagerina-zone assemblage. Common fusulinids include *Neoschwagerina cheni*, *Chusenella schwagerinaeformis*, *Minojapanella* sp., and *Schwagerina* sp. Clearly the fauna here lived in waters warmer than those in the Caggar Co sections. It is also clear, however, from the position of the Tulonggongba section, southwest of the Caggar Co area, that climate was not the only factor that influenced the biota; facies also was important in determining which organisms flourished in which locale. The possibility of large-scale tectonic transport (from north to south) in this strongly thrust-faulted area also must be kept in mind.

Sections studied on the Gangdise block, south of the Bangong-Nu Jiang fault zone (Figure II-31), throw additional light on paleogeographic relations in this area during the Early Permian. At Rutog, 60 km south of the Tulonggongba section, the cool-water Late Carboniferous Zhanjin Formation contains the *Eurydesma* fauna. This unit grades upward into the Qudi Formation, which has been reviewed here previously in some detail; this is the same unit which extends onto the western Qiangtang block (150 km northeast of Rutog at the Caggar Co). A recent study by Hu Changming (1984) has revealed the presence in this unit of turbidites and of coarse layers with mixed allochthonous and autochthonous marine faunas. Much of the 'paraflysch' described by Liang Dingyi et al. (1983) is, in fact, turbidite deposited in fairly deep water, possibly as deep as 500 to 1,000 m (Hu Changming, 1984). The autochthonous fauna is a mixture of Gondwanian and Tethyan taxa that clearly lived together. The fauna includes the gastropods *Schizodus tibeticus*, *Oriocrassatella rutogenis*, and *O. intermedius* (Liu and Cui, 1983). Although younger Permian formations were not described by these authors, the section appears to have undergone the same history as the Caggar Co section. If the various authors cited here have reported the locations of their sections accurately (their text descriptions are vague), it is evident that there, in southwestern Xizang (Tibet) Autonomous Region close to the Western Himalayan syntaxis, the lithofacies strike at a high angle (60° to 90°) to their nearly east-west trends in central and southeastern Xizang. Depositional and structural strikes thus diverge considerably, a most significant relationship not mentioned by most generalizers on the geology of this huge region.

Hu Changming's (1984) paper is important in showing that a local deep-water basin developed in southwestern Xizang (Tibet) Uygur Autonomous Region during the Carboniferous and the Early Permian. Among the many evidences he cited for the deep-water origin of the 3,500 m turbidite section are: (1) abrupt and numerous thin beds of sandstone showing graded bedding with interbeds of shale; (2) the presence of shallow-water fossils mixed with pelagic fauna and deep-water trace fossils; (3) the existence of numerous bioturbated layers; and (4) wide development of load casts.

Farther east on the Gangdise block, Wang and Mu (1983) described a Lower Permian section at Lhunzhub, 65 km north of Lhasa (Figure II-31). This, the Urulung section, is 620 to 630 m thick. The basal 40 to 50 m represents Sakmarian and Artinskian, and is called the Urulung Formation, equivalent to the Tulonggongba Formation near Rutog. The Urulung consists of interbedded black slaty shale and thin-bedded shallow-water limestone with the corals *Tachylasma minor* and *Verbeekiella* sp. In addition, there are brachiopods (*Stepanoviella flexuosa*, *Liraplecta* cf. *richthofeni*, *Asioproductus gratiosus*, *Cancrinella cancriniformis*, *Stenoscisma timorensis*) and bryozoans (*Goniocladia indica*, *Rhabdomeson consimile*, and *Girtypora occultolamina*). The corals belong to the *Lytvolasma* fauna.

The overlying Kungurian through early Tatarian section is 580 m thick (Lobadoi Formation). The lower 400 m consists of limestone with nodular chert; the upper 180 m contains limestone and limestone conglomerate interbedded with andesitic to basaltic volcanogenic rocks. The fauna of the lower 400 m is rich in fusulinids: *Neoschwagerina globularis*, *Lantschichites minimus*, *Dunbarula nana*, *Verbeekina* sp., and *Rugoschwagerina tibetica*. Associated corals are *Iranophyllum tumicatum*, *I. shirasakiensis*, *Ipciphyllum persicum*, *Praewentzelella* cf. *multiseptata*, and *Neokueichowpora haydeni*. In the upper 180 m of limestone and calcirudite section the fusulinids *Yabeina multiseptata*, *Y. shiraiwensis*, *Neoschwagerina margaritae* and coral *Iranophyllum* sp. occur. This demonstrates that the unit is equivalent to the Lasarla (Chitichun) Limestone south of the Yarlung Zangbo fault zone.

Wen (1981), He and Weng (1982), and Wang and Mu (1983) pointed out close similarities between the faunas of the Gangdise-Qiangtang blocks of Xizang (Tibet) Autonomous Region and those of southern and southeastern China. Numerous taxa include the same species. In addition, some also are found in Nei Monggol Autonomous Region.

Southern Xizang (Xizang) – the Himalaya Block

Although, in some areas, the fauna and flora of the Carboniferous and the Permian change across both the Bangong-Nu Jiang fault zone and the Indus-Yarlung Zangbo fault zone, in other areas, especially in the west, such changes do not take place across the fault zones. As more detailed field studies are completed, indeed, less and less support is being found for the concept of formerly separated blocks now joined along so-called sutures. Thus many ideas – especially those supporting the concept of facies zonation parallel with regional structural features – need major revision (e.g. compare the works of Jin (1981), Li and Yao (1981), Wen (1981), Wen et al. (1981), and Wu and Liao (1981) with later works and contrary ideas, by He and Weng (1982, 1983), Liang Dingyi et al. (1983), Liu and Cui (1983), Wang Hongzhen (1983), Wang and Mu (1983), and Hu Changming (1984)).

A moderately large number of sections now have been studied south of the Yarlung Zangbo (Zangbo = River), beginning with the section on Qomolangma Feng (Mt. Everest) described by Mu Anzi et al. (1973). Yin Jixiang et al. (1983) have summarized some of the completed studies; Wang and Mu (1983) have summarized additional studies.

The most complete section is the Lasarla section close to Zhongba and Burang, along the Yarlung Zangbo, 650 to 680 km west of Lhasa. The lower unit, called the Serlung Group, is 300 to 400 m thick, and it consists of calcareous, silty shale interbedded with argillaceous limestone. Brachiopods include *Neospirifer ravana*, *Costiferina alata*, *Stenoscisma gigantea*, *S. timorensis*, and *Spiriferella rajah*; corals are *Soshkineophyllum zhongbaensis*, *Amplexocarinia*, and *Tachylasma variable*. This Lasarla section is the equivalent of the Urulung Formation north of Lhasa. Above lies the 100 m Lasarla (Chitichun) Limestone, which contains fusulinids (*Neoschwagerina sosioensis*, *N. fusiformis*, *N. leei*, *Reichelina cribroseptata*, *Dunbarula pisilla*, *D. nana*, *Lantschichites minimus*, *Kahlerina pachytheca*, and *K. usurica*); brachiopods (*Leptodus* cf. *nobilis*, *Richthofenia lawrenciana*, *Chonetella nasuta*, *Spiriferella grandis*, *Neospirifer* cf. *kubeiensis*, *Cancrinella cancriniformis*, *Waagenoconchia abichi*, and *Notothyris simplex*); corals (*Wentzelella zhongbaensis*, *W. regularis*, *Wentzelloides* (*Multimurimus*) *minor*, and *Tachylasma* cf. *schematicum*); ammonoids (*Cyclolobus walkeri* and *C. oldhami*); and calcareous algae in abundance. All are Tethyan taxa that lived in warm water; many are found also in southeastern and southern China. The clear correspondence in fauna, age, and facies with equivalent sections north of the Yarlung Zangbo can hardly be doubted.

The Chubuk section near Tingri lies 240 km east-southeast of Zhongba, 450 km west-southwest of Lhasa, and north of Qomolangma Feng (Mt. Everest). The Lower Permian is represented only by the Sakmarian-Artinskian Serlung Group; post-Artinskian strata are not known. At the base is the 20 m Chubuk Formation with the plants *Glossopteris communis* (strongly dominant), *G. indica*, *Sphenophyllum speciosum*, *Dizeugotheca qubuensis*, and *Dichotomopteris qubuensis*. The 375 m Chubujeka Formation overlies the Chubuk, and includes the following fauna: corals (*Lytvolasma asymmetricum*, *Tachylasma paradoxicum*, *Lophophyllum* sp.), brachiopods (*Neospirifer kubeiensis*, *N. stratiformis*, *Taeniothaerus* cf. *subquadratus*, *Costiferina indica*, *C. alata*, *Cancrinella cancriniformis*, *Spiriferella rajah)*, and one ammonite (*Uraloceras xizangense*). This is the section originally published by Hsu Jen (1973, 1976) and also by Mu Anzi et al. (1973). Its Gondwanian affinities with southern China, and Tethys are clear from the contained fauna; and its partly continental nature (at the base of the section) is reflected in the lithology, with fluviatile sandstone and lacustrine shale in the Chubuk section grading upward into shallow-marine sandstone, silty shale, bioclastic limestone, and bank limestone in the Chubujeka Formation. Yet the fauna also has close affinities with contemporaneous biotas in Europe and North America.

Like the sequence at Tingri, that at Kangmar, 250 km farther east, lacks a post-Artinskian section. The basal Kangmar Formation, 100 m of speckled argillite, and the gradationally overlying Baidingpu, 200 to 250 m of limestone and marmolized limestone, are thought to be equivalent to the Chubujeka alone. Both units are marine and both contain the coral *Tachylasma* sp. as well as the brachiopods *Spiriferella* cf. *qubuensis*, *Neospirifer kubeiensis*, *Marginifera* sp., *Stenoscisma gigantea*, *Chonetella nasuta*, and *Athyris xetra*.

From the preceding brief review, we conclude, that the Lower Permian (and Upper Carboniferous) section was deposited on a notably broad shelf or epeiric sea, bordered by some land areas to the south (Himalaya block). Approximately the same lithologic units cover large areas of the Himalaya, Gangdise, and Qiangtang blocks. The lithologic and faunal changes may parallel the major fault zones in southeastern Xizang (Tibet) Autonomous Region. A very gradual transition takes place in faunal contents in the shallow-water sections, although the major part of the biota does not change. Clearly the concept that the Indus-Yarlung Zangbo and Bangong-Nu Jiang fault zones are 'sutures' between once far-distant landmasses is mistaken. Studies by Petrushevsky (1971), Saxena (1971, 1978), Meyerhoff and Meyerhoff (1972, 1974, 1978), Crawford (1979), Auden (1981), Gansser (1981), Jin (1981), Wang and Mu (1983), Waterhouse (1983), Chatterjee (1984), Stocklin (1984), Dickins (1985), Helmcke (1983), and many others, point to the virtual impossibility of drifting of continental blocks to an ultimate collision and 'suture' with Paleoasia.

Type Sections

Fang et al. (1979) described a type section of the Permian in central Guizhou Province (Figure II-37), and they discussed also the Carboniferous-Permian boundary. Li and Zhang (1983) returned to the question of the Carboniferous-Permian boundary in Guizhou Province, and, as already mentioned, reported a division of opinion among Chinese geologists. Whatever the horizon of the actual boundary may be, fauna and lithology alike are transitional across it.

Yang Zunyi et al. (1980) tabulated general features of the Permian sequence in Central-South China in terms of formations, and in terms of lithology and facies. In the table that follows, Table II-20, we place their data in one of three columns. In a second column we place the fusulinid foraminiferal zones of Sheng et al. (1979). In a third column we place stages of the Permian according to the general 'western' consensus (Harland et al., 1982; and the Changxingian stage by Furnish and Glenister, 1970). Our correlations of the three columns necessarily are tentative, but those for the zones of *Palaeofusulina*, *Codonofusiella*, and *Schwagerina* probably are reasonable.

The Permian sequence of China, and essentially the Permian marine sequence of Central-South China, differs possibly from all other Permian sequences by exhibiting a fauna that indicates a stage younger than any known in the world outside China. Kummel and Teichert (eds., 1970) considered this possibility through a study of ammonoids. Furnish and Glenister (1970), in view of this same possibility, designated the Changxingian stage with the fusulinid foraminiferal genus

Table II-20. Correlation of the generalized Permian sequence in Central-South China

		Changxingian[a]	*Palaeofusulina*	Changxing Formation	mostly marine carbonates
				Dalong Formation	siliceous strata, siliceous limestone, terrigenous clastics, all marine
	Upper			Xuanwei Formation	alternate marine and terrestrial strata
				Longtan Formation	paralic, coal-bearing
		Tatarian	*Codonofusiella*	Wujiaping Formation	mixed facies and marine carbonates
PERMIAN		Kazanian, Ufimian	*Yabeina-Neomisellina*		
		Kungurian	*Neoschwagerina*	Maokou Formation	marine carbonates
	Lower		*Cancellina-Parafusulina*		
		Artinskian	*Misellina claudiae*	Danchong Formation	marine siliceous strata, terrigenous clastics, mixed facies
		Sakmarian	*Schwagerina tschernyschewi*	Qixia Formation	marine carbonates; local coal-bearing terrigenous clastics at base (Liangshan Formation)

Sources: Harland et al. (1982); Sheng et al. (1979); Yang Zunyi et al. (1980): Changxingian[a] from Furnish and Glenister (1970).

Palaeofusulina as an index fossil (Table II-20). ('Changxingian,' in all of the texts available to us, appears as 'Changsingian,' a pre-pinyin spelling; pinyin spelling should have been used, as it already was in use in China in the 1960s.)

The lateness of the Changxingian stage suggests a transitional or gradational condition across the Permian-Triassic boundary, and this receives strong support from lithofacies. Thus, Yang et al. (1981) cited a gradational change, from mainly siliceous limestones and siliceous mudstones below, to mainly argillaceous carbonates above, in the marine sequences of Central-South China. On the basis of fauna, these same authors noted the persistence of certain brachiopods, from which they postulated uninterrupted marine transgression across the Permian-Triassic boundary. Kummel and Teichert (eds., 1970) had noted a similar persistence across their selected Permian-Triassic boundary in the Salt Range of (then) West Pakistan.

Elsewhere, Yang et al. (1981) reported disconformity across the Permo-Triassic boundary at: (1) the western margin of the Central-South China basin, near the Kang Dian oldland; (2) the eastern rim of the basin near the Cathaysian oldland; and (3) Northwest China. In terms of lithofacies, alternate non-marine and marine clastics below the disconformities change to predominantly marine terrigenous clastics above. In the northwestern corner of the South China basin, a local non-conformity is present between volcanic rocks below, and sandstones and slates interbedded with limestones above.

In North China, Fang et al. (1979) recorded a type section of the Permian sequence at Taiyuan, Shanxi Province. The sequence at this locality, as elsewhere in much of North China, is entirely non-marine. Plants constitute the indicative megafossils (Figure II-38), and the character of the floral assemblage is Cathaysian. Li (1962), noted a disconformity at the Permian-Triassic boundary in North China, but commonly a transition across the Carboniferous-Permian boundary.

Paleoecology-Paleogeography

Throughout the Permian, a marine Tethyan realm occupied the southern areas of China, and a continental Cathaysian realm occupied its northern areas, where formation of some of the great interior Mesozoic-Tertiary basins began (e.g. Junggar basin). The marine realm lay between Tethys proper farther west (i.e. west of China) and an ancestral Pacific sea on the east. From west to east the seaway broadened and, at the longitude of the present Pacific coast of China, stretched across more than 15° of latitude. Deep-sea deposits are unknown, although some pelagic deposits (turbidites) developed locally in Xizang (Tibet) Autonomous Region. Mainly shallow-water deposits of an epeiric sea characterize southern areas (Stocklin, 1984). Abundant fusulinids in marine Permian waters indicate that these commonly were little deeper than sublittoral, and perhaps rarely (if anywhere) pelagic. The only areas where pelagic conditions may have existed lie in Southwest China and Taiwan (Sheng Jin-zhang et al., 1985). The presence of paralic facies in several of the known sequences suggests to us that a significant number of islands stood up in the Permian marine waters of Central-South China, a conclusion reached also by Sheng Jin-zhang et al. (1985). A landmass (Huaxia = Cathay) occupied the present coastal zone of Central-South China and East China.

For the latest Permian, Cheng Zhengwu (1980) traced the boundary between Permian Tethys on the south and Permian Cathaysia on the north along a line from the Kunlun Shan on the west, through the Qin Ling to the Dabie Shan on the east. Available evidence suggests that this line may not have changed greatly from the beginning to the end of the Permian - also that it may have crossed the present coastline in the vicinity of Shanghai (Rui Lin et al., 1983). Thus, the line between Tethys on the south and Cathaysia on the north lies not far from the latitudinal. The variation in latitude is no more than from 36°N in westernmost China to 32°N in eastern China at the present Pacific coastline.

Tethys of Central-South, Southwest, and Northwest China

In the earliest Permian, a topography with considerable relief was present. While marine waters covered certain areas, intermontane lakes appeared in others. In shoreline areas, coastal swamps developed. In the latter part of the Early Permian, marine waters with abundant *Cancellina* came across Southwest China in an episode of maximum transgression. Widespread limestones deposited in these waters represent the fusulinid *Cancellina* zone (lowest fusulinid zone of P_2). By the Late Permian, the depositional environments of Central-South China showed a pattern of neritic environment on the west, and coastal and lagoonal environments with marine gulfs on the east. The Huaxia (Cathay) landmass remained elevated along the present coastal zone of Central-South China and East China.

At the end of the Early Permian, a moderate uplift brought a regression of marine waters, and a limited effusion of basalt (Emei Shan Basalt). The Emei Shan Basalt is widely distributed along the western Yangzi platform and in western Yunnan Province. Generally, the Emei Shan is assigned to early Late Permian. However, a recent study by Yang Zunyi et al. (1984) showed that the boundary in between Lower and Upper Permian is located in the middle part of the Emei Shan Basalt. Yang Zunyi et al. (1984) discovered limestone lenses within the basalt section that yielded the following late Early Permian fauna: *Neomisellina cf. douvillei*, *Chusenella sp.*, and *Verbeekina sp.* Coastal plains were typical of the prevailing environments, but they were subject to change through oscillatory episodes. Toward the end of the Permian, transgression took place once again. Neritic environments returned accordingly, with local developments of gulfs in which quiet water permitted deposition under conditions of low energy.

Cathaysia of North China and the Southern Part of Northeast China

Uplift of this region took place during the Chongyu movement (in Ordovician time; see Table III-1) and the resulting land stayed above sea level through the remainder of the Phanerozoic. Swampy forests and associated valley flats developed at the beginning of the Middle Carboniferous, and these provided environments for the formation of abundant coal seams with siderite nodules in sandstones. During the late part of the Early Permian, a diagnostic flora in this part of China suggests a hot and humid climate, whereas the strata themselves suggest fluvial, lacustrine, and paludal physical environments. The remains of labyrinthodonts and stegocephalian amphibian vertebrates support the concept of humidity. The enclosing beds of purplish-red and variegated, fine- to medium-grained, terrigenous-clastic sediments with abundant cross beds furnish the evidence for interpreting high temperatures. For the early part of the Late Permian, available evidence suggests the predominance of a lacustrine environment and gradual change to a less humid climate in North China. During the late part of the Late Permian, lithofacies of the corresponding deposits point to an environment of rivers and lakes. The deposits of this interval, now principally arkose, shale, and gypsum, may indicate a hot and partly dry climate, evidently inimical to the preservation of plants, of which relatively few appear.

Peninsular India

Two small-scale yet geographically far-reaching marine incursions took place during the Early Permian on peninsular India. Both are of great regional importance in large-scale paleogeographic reconstructions. One is the Manendragarh transgression, during which Asselian waters advanced some 400 km southward from the Tethys seaway on the north, or 350 km from the present southeastern coast of India. The second is the Umaria transgression from west of India during the Sakmarian. This transgression crossed 850 km of the peninsula. Following the studies of earlier authors, the details of these occurrences were summarized by Meyerhoff and Teichert (1971), Shah and Sastry (1975), and Chatterjee and Hotton (1986).

The Manendragarh fauna, in Chinese terms, is unmistakably equivalent to part of the Asselian Zhanjin Formation of the Gangdise block; and the Umaria fauna is clearly Tingrian, or Sakmarian, and equivalent to part of the Serlung Group (Wang and Mu, 1983).

These two marine incursions onto the Indian shield clearly place India in close proximity to Asia during the Early Permian.

Additional to the evidence of marine incursions, and to the evidence of their associated faunas, is the evidence of tetrapod (amphibians, reptiles) fauna. Nine taxa of tetrapods have been reported from India and Kashmir (Chatterjee and Hotton, 1986). Three taxa are known elsewhere only in Europe. All but one of the families of these tetrapods represented in India are known in Europe. Other areas with close relationships are Africa and North America. Not a single similarity was found in the tetrapod faunas of Australia, South America, and Antarctica (Chatterjee and Hotton, 1986). India thus may well have lain far distant from these three latter continents. Evidence from the tetrapods (Africa excepted) strengthens considerably the concept that India lay in close proximity to Asia during the Early Permian.

Magmatism

Evidence of 'Variscan' movement appears along major tectonic latitudinal sectors such as Kunlun Shan-Qin Ling, Tian Shan-Yin Shan, and the arc of Nei Monggol (Compilation Group, 1976). Magmatism during this movement constitutes a major episode, developed in three phases.

Phase 3

Intrusion of granite and alaskitic granite in the northern part of Northeast China, in Nei Monggol Autonomous Region, in the Qilian Shan, and in western Yunnan Province. Mafic and ultramafic rocks are found in Northeast China, in Northwest China, and in Southwest China. Isotopic ages range from 260 to 230 m.y.

Phase 2 (middle)

Mainly granitic intrusion in the Tian Shan, in the Altay, in the two Hinggans, and in the Bei Shan. Mafic and ultramafic rocks accompany the granites in North China and Northwest China. Isotopic ages are around 300 m.y.

Phase 1

Mainly mafic and ultramafic rocks in the Tian Shan, Sichuan Province, and western Yunnan Province. Along the Tian Shan, the Kunlun Shan, and the Qin Ling, associated granites yield an isotopic ages of approximately 350 m.y.

MESOZOIC

General Stratigraphy

In comparison with Paleozoic strata, Mesozoic strata show a strong development of non-marine facies, to the extent that non-marine deposits cover large areas of Northwest China, North China, and Northeast China. In the more southern regions of China, significant areas remained marine through the Triassic. In contrast, marine facies diminished considerably through the Jurassic and the Cretaceous. Distinguishing depositional features of the non-marine facies include redbeds, coals, evaporites, and marked changes in lithofacies and thickness.

General Tectonic Pattern

The Mesozoic tectonic belts of China show a strong component of the meridional (north-south) direction in their trends, considerably different from the predominantly latitudinal (east-west) direction of Paleozoic trends. The near-meridional trends of the Mesozoic (as well as those of the succeeding Cenozoic) appear mainly in East China and Central-South China. As a consequence, the eastern and western sectors of China differ entirely from one another in the characters of their structural histories (Compilation Group, 1976).

The principal Mesozoic (and succeeding Cenozoic) trends belong to the Neocathaysian system, which strikes north-northeast across eastern China. The paired structures of horst and graben that characterize this tectonic system are bounded by normal faults, high-angle reverse faults, and strike-slip faults. In total, the characteristics of the Neocathaysian system include not only a series of uplifts and depression, but also metamorphic and magmatic events. Where the north-northeast trends of this system intersect the older east-west (latitudinal) trends in Southeast China and Northeast China, a complex mosaic of structure results. One notable phenomenon observed at such intersections is the presence of large- to medium-sized tectonic basins together with an echelon, arc-shaped belts of uplifts whose apices point toward the southeast. No such intersections occur in any part of western China, because the Neocathaysian system is absent from that region. Only the east-west (latitudinal) system is present, and it is characterized by folded geosynclinal systems associated and parallel with normal, reverse, or wrench faults, and by related magmatic and metamorphic phenomena.

In Southwest China, in the Sichuan-Yunnan region, the tectonic trends are north-south, are characterized by tight folds and associated faults, and exhibit zones of tectono-magmatism together with zones of metamorphism. At its northern end, this tectonic system leads southward away from the Eastern Himalaya syntaxis, on a course through Thailand and Malaysia. In the Qinghai-Xizang (Tibet) Plateau, east-west structures parallel the Tethyan arc proper, with a moderately wide separation of tectonic lineations at the center, but with tapering at both ends of the Western and Eastern Himalaya syntaxis.

Mesozoic Floral Groups

Sze and Chou (1962) recognized six groups of Mesozoic flora in China. Figure II–39 shows these floral groups and their correlations with chronostratigraphic divisions of the Mesozoic in Europe.

The *Danaeopsis-Bernoullia* flora occurs mainly in North China and Northwest China, and it relates closely to Late Triassic floras from Kurashashayskiy and Kuraylinskiy in Kazakhstan, U.S.S.R., also to coeval floras in Europe and North America. A few species of this flora persist into the succeeding flora.

The *Dictyophyllum-Clathropteris* flora belongs mainly to southern China, with close relationships to coeval floras collected in East Greenland, Sweden and Germany. Species from the preceding flora persist into this one in some places, while late species in this flora persist into the succeeding one.

The *Coniopteris-Phoenicopsis* flora occurs widely in the southern sector of Northeast China, in the Nei Monggol Autonomous Region, in North China, and in Northwest China, in all of which it is similar to the floras of Siberia and Kazakhstan in the U.S.S.R.

The *Ruffordia-Onychiopsis* flora occurs mainly in Northeast China, and it relates closely to floras from the Ussuri River in the region of the China-U.S.S.R. frontier, also to floras in the Tetori Formation of Japan, and to floras in an equivalent Korean formation. The earlier

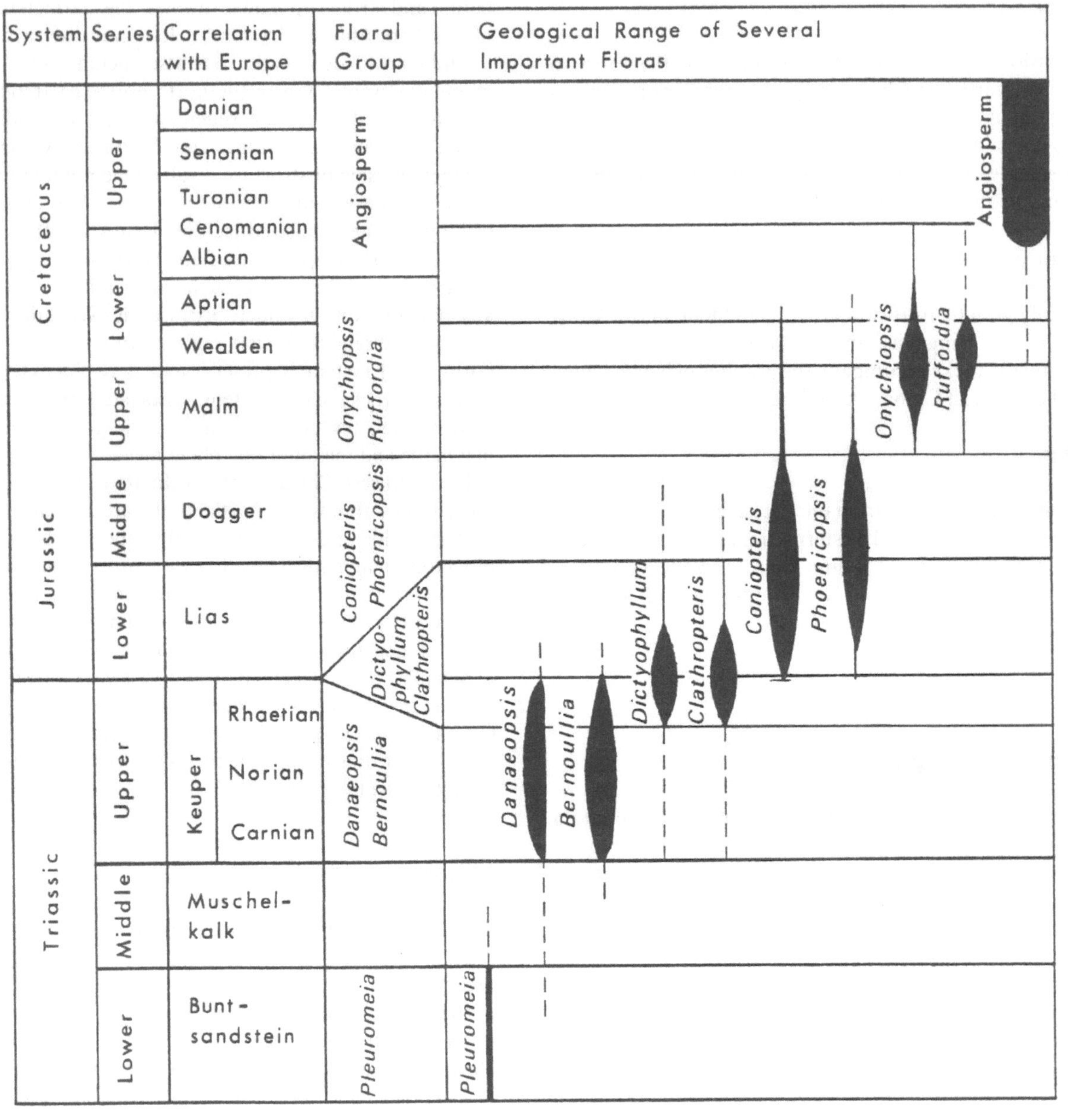

Fig. II–39. Zonations of flora in the Mesozoic of China and correlation with Europe (from Sze and Chou, 1962).

portion of the *Ruffordia-Onychiopsis* flora (Late Jurassic) contains a few species that have persisted from the preceding flora.

The angiosperms, which include *Trapa microphylla*, characterize the Upper Cretaceous formations of Northeast China.

TRIASSIC SYSTEM

Toward the end of the Permian, transgression gained strength across Central-South China, Southwest China, and parts of Southeast China (Figure II–40). The advance of marine waters into these areas, however, was not total when the Triassic began. The presence of disconformities and non-conformities indicates local instability; volcanicity associated with such instability suggests a 'Variscan' episode typical of the boundary between the Paleozoic and the Mesozoic (i.e. between the Permian and the Triassic (Sheng Jin-zhang et al., 1985)). In general, marine waters did not move north of the line that joins the Kunlun Shan, the Qilian Shan, and the Dabie Shan (Wang Yigang et al., 1981). Others have suggested that the transgression actually was more limited.

North of the Kunlun Shan-Dabie Shan east-west line, Triassic continental facies deposits stretch across the whole of China. Only a small area in northeasternmost China is an exception (Wang Yigang et al., 1981).

Triassic marine sections in eastern and northernmost

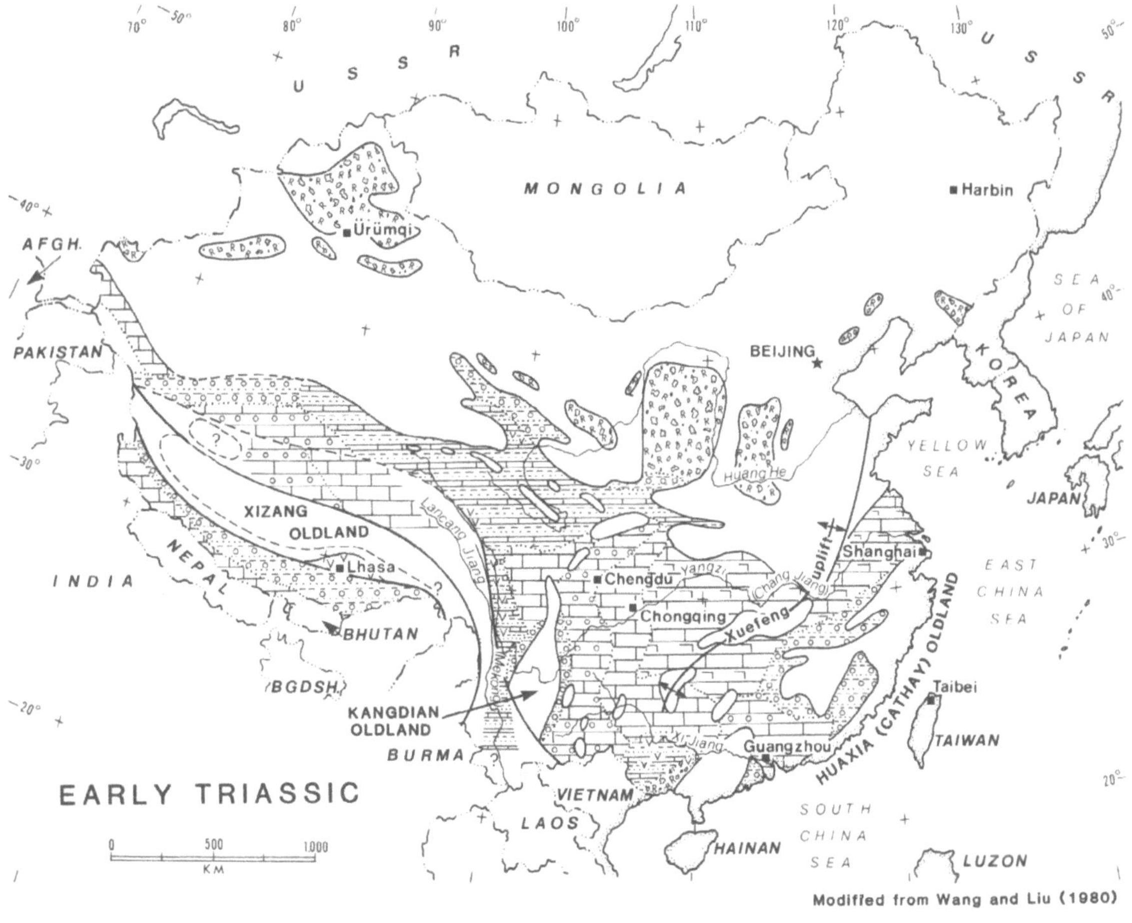

Fig. II–40. Paleogeography and lithofacies map of Early Triassic, China.

Yunnan province, eastern Xizang (Tibet) Autonomous Region, western Sichuan Province, southern Qinghai Province, and western and southwestern Guizhou Province are among the thickest and most extensive Triassic sequences in the world. Triassic deposits are abundantly fossiliferous, also largely Tethyan in their faunas, except in far northeastern and southeastern China where Pacific-realm faunas are dominant (Compilation Group, 1976). Commenting further on marine Triassic faunas, Wang Yigang et al. (1981) listed the presence of algae, foraminifera, radiolaria, corals, stromatoporoids, hydrozoans, bryozoans, brachiopods, bivalves, gastropods, nautiloids, ammonoids, belemnites, ostracods, crinoids, conodonts, and ichthyosaurs. Ammonoids, bivalves, and brachiopods may be the most important elements of the fauna for biochronology. Gastropods, conodonts, foraminifera and corals follow. Faunal successions provided by five of the fossil taxa appear in Figure II–41.

Distribution and Facies

Wang Yigang et al. (1981) noted the wide extent and the good development of Triassic deposits in China. They noted, in addition, a significant relationship between Triassic sedimentary development and tectonic lineations. In Northwest China, Triassic lithofacies parallel a system of latitudinal tectonic belts. In Southwest China, Triassic strata are roughly parallel with the Hengduan Shan, and thus are almost meridional in trend. In Southwest China and in East China, Triassic strata parallel the Cathaysian tectonic belt, consequently with a northeasterly trend.

Southern China (Central-South China and East China)

The area in which Triassic marine sediments were deposited was restricted by the positions and the trends of several major and ancient tectonic elements. These include the Xizang (Tibet)-Kangdian oldland on the west and southwest, the Huaxia (Cathay) oldland on the southeast, the Sino-Korean platform on the north, and the Tarim platform on the northwest. These areas of uplift provided: (1) debris for the marine basin; and (2) barriers to more widespread transgression. The northern boundary coincides in part with one of China's oldest east-west structures, the line of weakness that is associated with the Qin Ling and Dabie Shan (Sheng Jinzhang et al., 1985).

		AMMONOID ZONE	PELECYPOD ASSEMBLAGE		GASTROPOD ASSEMBLAGE		CONODONT ZONE	BRACHIOPOD ASSEMBLAGE
UPPER TRIASSIC	NORIAN	*Aimavatites columbianus* *Cyrtopleurites socius*	*Yunnanophorus* - *Permophorus* *Burmesia lirata*	*Entomonotis ochotica* bed	*Sisenna ninglongensis*- *Trachynerita wanneri* *weishanensis*- *Eucyclus multistriata*	*Magnicapitatus* *Omphaloptycha*	*Epigondolella abneptis*	*Himalairhynchia media-Foseptaliphora tulungensis* (Himalayan region); *Halorella dongquoensis-Septamphiclina qinghaiensis-Sacothyris sinosa* (Tethyan Northeast marginal region); *Orientospira cf. gregaria* (in Nadanhadaling region)
		Indojuvavites angulatus *Griesbachites-Gonionotites* *Nodotibetites nodosus*						
	CARNIAN	*Parahauerites acutus* *Hoplotropites* *Indonesites dieneri* bed	*Costatoria kweichowensis*- *Heminajas forulata* *Cassianella beyrichi*	*Tosapecten* *Palaeopharus* *Bakevelloides*	*Zygopleura walmstadti*- *Trypanostylus waageni*- *Palaeonarca orientalis*	*Euomphalus* *Pericarinata* *Fascivalvata*	*Neogondolella polygnathiformis*	*Neoretzia tibetensis- Oxycolpella oxycolpos- Sanqiaothyris elliptica* (Himalayan region and Tethyan North-east Marginal region)
		Beds with *Sirenites*, *Trachyceras aon*, *Trachyceras aonoides*, *Austrotrachyceras* etc.						
MIDDLE TRIASSIC	LADINIAN	*Protrachyceras deprati* *Protrachyceras primum*	*Daonella lommeli* *Halobia kui* *Asoella illyrica*		*Polygyrina cf. lommeli* - *Neritaria comensis*		*Neogondolella excelsa*	*Volirhynchia multicostata*- *Volirhynchia himalaica* (Himalayan region)
	ANISIAN	*Paraceratites trinodosus* *Paraceratites binodosus* *Nicomedites yohi* *Paracrochordiceras*, *Japonites*, *Lenotropites*	*Costatoria goldfussi mansuyi*		*Worthenia gagis* - *Zygites qingyanensis* - *Riselloides guizhouensis*		*Neogondolella constricta* *Neospathodus germanicus*- *N. kockeli* *Neogondolella regale*	*Nudirostralina griesbachi*- *Talungospirifer stracheyi* (Himalayan region) *Nudirostralina subtrinodosi*- *Diholkorhychia sinensis* (Guizhou-Sichuan) *Qilianoconcha opima-Aequispiriferina multiplicata* (in Qinghai)
LOWER TRIASSIC	SPATHIAN	*Procarnites-Ziynites* *Columbites asymmetricus*	*Eumorphotis inaequicostata* - *Pteria cf. murchisoni*		*Neritaria lendziensis*- *Trachynerita Quadrata* *Supercarinata-Natiria* *Sichuanensis*		*Neogondolella jubata* / *Pachycladina obliqua*- *Neospathodus homeri* *Platyvillosus*	
	NAMMALIAN	*Anasibirites kingianus* *Owenites costatus* *Kenickites lingyuenris* *Proptychites kwangsiensis*	*Eumorphotis multiformis*				*Neospathodus waageni* *Neospathodus pakistanensis* *Neospathodus cristagalli*	
		Gyronites psilogyrus	*Claraia aurita*				*Neospathodus dieneri*	
	GRIESBACHIAN	*Ophiceras (Lytophiceras) sakuntala* *Otoceras latilobatum*	*Claraia stachei* *Claraia wangi*		*Bellerophon asiaticus*- *Polygyrina depressa*		*Isarcicella isarcica*	*Fusichonetes pigmaea* *Paryphella triquetra* *Nouellerella pseudoutah*

Fig. II–41. Chief marine Triassic faunas of China.

Lower and Middle Triassic sequences in this region consist of neritic and lagoonal carbonates, together with red sandstone, sandstone, shale and evaporites. The evaporites are found principally in the northern part of the marine depositional area (e.g. the present Sichuan basin), whereas redbeds (continental facies) and terrigenous clastics deposited in the marine littoral zone were the main types of deposits around the positive areas (e.g. Kangdian and Huaxia-Cathay oldlands) east and west of the marine basin. Between these two uplifts (Figure II–40), the carbonate rocks are characterized by numerous facies changes (Sheng Jin-zhang et al., 1985). Ammonoids and pleecypods dominate the faunas, which are largely Tethyan. Only in the far southeastern part of China do faunal (and sedimentary) affinities change to those of the Pacific realm (Compilation Group, 1976).

During the Late Triassic, partial emergence of the Xuefeng uplift in southeastern China exerted a pronounced effect on the distribution of Upper Triassic sediments (Figure II-42). This uplift divided southeastern China into two depositional areas. The area east of the uplift (Provinces of Guangdong, Fujian, Jiangxi, western Zhejiang, and a small part of Jiangsu) contains coal-bearing formations with marine intertongues. The marine beds and the coal contain respectively a fauna and flora of the Pacific realm. In contrast, in the area west of the uplift (Provinces of eastern Yunnan, eastern Sichuan, western Guizhou, and western Guangxi Zhuang Autonomous Region) a Tethyan fauna and flora predominate. The lower part of the Upper Triassic consists of continental and littoral terrigenous-clastic strata, with shallow-water carbonates and shale. The upper part is regressive, and contains non-marine terrigenous-clastic sediments and coal.

Himalaya (Southern Xizang)

A complete Triassic sequence is present on Qomolangma Feng (Mt. Everest). This sequence consists mainly of neritic carbonates and terrigenous clastics, with little variation in lithofacies. A rich fauna contains abundant ammonoids (11 zones), bivalves, pelecypods, and ichthyosaurs (Mu Anzi et al., 1973; Yin Ji-xiang et al., 1983). Similar sequences and faunas extend southward into India, Nepal, and Bhutan (Kapoor and Tokuoka, 1985), and westward into northern Pakistan and Europe (Yin Ji-xiang et al., 1983). The fauna is distinctly of neritic facies, is Tethyan, and is related to those of equivalent strata in India, Pakistan, Europe, and even

Fig. . II–42. Paleogeography and lithofacies map of Late Triassic, China. It shows also floral domains.

parts of western North America.

The section in the Himalaya thickens northward toward the Yarlung Zangbo valley, increasing from 1,700 m to at least 3,000 m (Mu Anzi et al., 1973). The Lower and Middle Triassic are poorly exposed, and most of the observed Triassic section belongs to the Upper Triassic, which lithologically is most complex, with turbiditic flysch, mafic volcanic rocks, and radiolarian chert (Bassoullet et al., 1981; Wang Yi-gang, 1981; Yin Ji-xiang et al., 1983). The development of such deeper-water facies with mafic volcanics indicates the time of initial formation of the modern Indus-Yarlung Zangbo fault zone.

Qinghai-Xizang (Tibet) Plateau

Triassic marine deposits are widespread in eastern Xizang (Tibet) Autonomous Region, in Qinghai Province, western Sichuan Province, and western Yunnan Province, where it is developed in eugeosynclinal facies, in many places with great thicknesses. The Triassic here still has not been studied in detail (Huang and Chen, 1987).

The Kunlun Shan-Qilian Shan latitudinal trend separates this extensive marine Triassic realm from the continental deposits of Northwest China and North China. This region is separated from the shallow-water areas of southeastern China by the Kangdian uplift, and by a related uplift on the site of the present Longmen Shan, each with a north-south trend.

The Triassic consists mainly of littoral to neritic terrigenous clastics with some carbonate interbeds in central and northern Xizang (Tibet) Autonomous Region (Chang and Pan, 1981; Yin Ji-Xiang et al., 1983; Wang and Sun, 1985).

The Triassic sequence of the Tanggula Shan consists of intermediate to silicic volcanic rocks, melange-type deposits, and terrigenous-clastic flyschoid and shallow-water strata (Chang and Pan, 1981; Zheng Yan-zhong, 1981). Similar formations are present along the Bangong-Nu Jiang fault zone (Huang Jiqing, 1978; Chang and Pan, 1981; Wang and Sun, 1985). Yin Ji-xiang et al. (1983) reported the presence, in the Qamdo area of northeastern Xizang (Tibet) Autonomous Region, of 5,000 m of eugeosynclinal and flyschoid sediments associated with silicic to intermediate volcanic rocks.

In southeastern Qinghai Province and western Sichuan Province, in a triangular area known as the Songpan triangle (east of Lancang Jiang on Figures II-40 and II-42), Triassic deposits consist mostly of monotonous flysch-type sediments, in some areas interbedded with mafic-to-intermediate and silicic-to-intermediate volcanics (Compilation Group, 1976; Wang and Sun, 1985; Huang and Chen, 1987). The Upper Triassic, however, contains abundant non-marine strata intercalated with marine zones.

The faunas, primarily Tethyan, also show Pacific affinities. Brachiopods and hexacorals accompany the dominant pelecypods and ammonoids. Important pelecypod genera include *Claraia*, *Halobia*, *Daonella*, *Eumorphotis*, and *Myophoria*. The flora associated with the marine biota is of two types: (1) the Yipinglang type in the southern part of the plateau; and (2) the Yangchang (North China) type in the northern part of (and north of) the plateau. Freshwater unionid bivalves provide additional evidence for continental influence during the deposition of the Upper Triassic. Jaeger et al. (1982) reported the presence, just north of Lhasa, of a Late Triassic flora distinctly boreal or Northern Hemisphere.

Northeast and North China

A small embayment of the ancestral Pacific occupied the Wanda Shan sector in northeasternmost China (Heilongjiang Province and Jilin Province) during the Triassic (Figure II-42). Otherwise, as with the preceding Permian, these regions are outstanding for a continental Triassic sequence of wide development, of considerable thickness, and of many fossils, with inland basins (Cheng Zhengwu, 1980). Lower and Middle Triassic deposits consist mainly of red terrigenous clastics interbedded locally with marine layers and with extrusives. A flora of *Danaeopsis*, and another of *Bernoullia*, both with possible antecedents in the Permian, represent the lowest assemblage of Mesozoic plants (Yanchang type).

In the small marine embayment of northernmost China, the Upper Triassic consists chiefly of phyllite and of siliceous rocks with tuff. *Entomonotis ochotica*, a pelecypod, indicates the presence of at least one endemic taxon of the West Pacific fauna.

Northwest China

The Triassic of this region is in continental facies as it is in North China. The Triassic sequences are complete in the Junggar basin and in the Turpan basin. The Lower and Middle Triassic section consists mainly of red, brown, and purple terrigenous clastics with much coarse-grained sandstone and conglomerate. The conglomerates contain porphyritic sills, especially in the northern part of the Turpan basin. Vertebrate faunas in the Lower Triassic continental facies of the Junggar and Turpan basins contain the cosmopolitan vertebrate genera *Lystrosaurus*, *Chasmatosaurus*, and *Kannemeyeria* (or kannemeyerids) (Sun Ai-lin, 1973).

The Upper Triassic consists predominantly of gray strata, reflecting change in climatic conditions from arid (during Lower-Middle Triassic) to humid. Coal beds are sporadically present in the southern part of the Junggar basin.

In the Tarim basin, the Lower and Middle Triassic sections generally are missing. Some Middle Triassic terrigenous clastics are known from the northern part of the basin. Upper Triassic gray terrigenous clastics of lacustro-fluvial origin, preserved throughout the basin, thicken in the northern part.

Type Sections

Central-South China

The type section of Triassic for this region lies in the southwestern part of Guizhou Province. Fang et al. (1979) interpreted this section as the complete unit of a cyclothemic sequence, whose detailed lithology appears in Figure II-43. This section, although from Guizhou Province, is closely similar to the Triassic section of the Sichuan basin. The lowermost Triassic (T_1) lies disconformably on the Permian in much of this region. Its terrigenous lithofacies is clastic and variegated; its environment is coastal. The overlying Lower Triassic (T_2) exhibits a dominance of neritic cephalopods in the fauna, which indicates a transgression of marine waters. Pelecypods accompany the cephalopods. Dolomites in the lower part of the Middle Triassic (T_3) with sparse fossils suggest a highly saline environment of deposition. Fine-grained terrigenous-clastic strata in the top half of the upper part of the Middle Triassic (T_4) reflect a gradual regression of marine waters. The Upper Triassic (T_5, T_6, and T_7), with alternate non-marine deposits that contain several coal seams, reflects definitive oscillation of underlying basement (shallow crust). An emergency of some importance may be the cause of unconformity at the Triassic-Jurassic boundary.

North China

Figure II-44 presents the lithostratigraphy of Triassic deposits in a sequence within the Shaan-Gan-Ning or Eerduos basin of Shaanxi Province, Gansu Province, and Ningxia Hui Autonomous Region (Fang et al., 1979).

System	Series	Formation	Column section	Thickness (m)	Major lithology	Major fossils
Triassic	Upper	T_7		180	Quartz ss. and conglomerate interbedded with sandy mudstone.	
		T_6		680	Gray sandstone, shale, interbedded with coal seams.	*Dictyophyllum* (P) *Clathropteris* (P) *Burmesia lirata* (Pe) *Yunnanophorus* (Pe) *Myophoria napengensis* (Pe)
		T_5		420	Grayish-green ss., shale, and mudstone, interbedded with argil. ls. and thin coal seams.	*Myophoria kueichouensis* (Pe)
	Middle	T_4		788	Upper: yellowish mudstone, silty mudstone. Lower: grayish-black argillaceous limestone and limestone.	*Eumorphotis illyrica* (Pe) *Halobia comala* (Pe) *Protrachyceras deprati* (C)
		T_3		766	Upper: gray dolostones. Lower: dark gray oolitic ls., shale, interbedded dolomitic ls., argillaceous ls. and dolostone.	*Myophoria goldfussi* (Pe)
	Lower	T_2		323	Gray limestone interbedded oolitic limestone and sandy shale.	*Tirolites spinosus* (C) *Pteria* cf. *murchisoni* (Pe) *Entolium discitesmicrotis* (Pe)
		T_1		456	Variegated mudstone, siltstone, silty shale, interbedded with thin layer of limestone.	*Claraia wangi* (Pe) *Claraia aurita* (Pe) *Eumorphotis multiformis* (Pe)

Fig. II-43. Triassic stratigraphy in the southwestern part of Guizhou Province, Central-South China (from Fang et al., 1979).

System	Series	Formation	Column section	Thickness (m)	Major lithology	Major fossils
Triassic	Upper	T_{5-7}		400–750	Grayish and yellowish-green, and grayish-black ss., shale, and mudstone, often interbedded with coal seams.	*Todites shensiensis* (P) *Danaeopsis fecunda* (P)
	Middle	T_4		100–600	Grayish-white and red arkose and siltstone, interbedded with black shale or oil shale in the upper part.	*Danaeopsis* (P)
		T_3		100–800	Grayish-yellow and green, purplish-gray massive ss. and purplish-red siltstone and mudstone, calcareous nodules in mudstone.	*Shansisuchus* (R)
	Lower	T_2		100–250	Purplish-red sandy mudstone and argillaceous siltstone interbedded with purplish-red sandy conglomerate.	*Ceratodus* (F)
		T_1		160–500	Grayish-white, pink and purplish-red ss. interbedded with purplish-red sandy mudstone and argil. siltstone.	

Fig. II-44. Triassic stratigraphy in Shaanxi Province, Gansu Province and Ningxia Huizhu Autonomous Region, China (from Fang et al., 1979).

The sequence is complete, with divisions from T_1 to T_7 all present (see above), and probably is the best developed in the whole basin. A conglomerate marks the T_1–T_2 boundary; a small hiatus is present at the T_2–T_3 boundary. T_3 consists mainly of medium- to fine-grained terrigenous clastics with calcareous nodules. T_4 shows a decided contrast, with arkose and lacustrine oil shale among its deposits. The top group, T_5, T_6, and T_7 contain terrigenous clastics and several coal seams.

Numerous sequences show conformity at the Permian-Triassic boundary; and disconformity at the Triassic-Jurassic boundary. An erosional unconformity between the Triassic and Jurassic in the Eerduos basin plays an important role in oil accumulation.

Cheng Zhengwu (1980) described a sequence representative of the Eerduos basin, as above, but also included Shanxi Province on the east. His description, in addition, effects division of the Triassic by the use of specific formations rather than by the letters and numbers in Figure II-44 of Fang et al. (1979). The Table II-21 that follows showing how Cheng's lithostratigraphic detail is comparable with that in Figure II-44. In Shanxi Province, plants and sporopollen date the lower part of the Lower Triassic. Elsewhere, a dipnoan fish and a theriodontid reptile date the upper part. A variety of vertebrates marks the Middle Triassic. Plants and a plesiosaurian occur in the lower part of the Upper Triassic.

Northwest China

In Table II-22 we summarize the Triassic of this region according to Cheng Zhengwu (1980). His treatment for this region follows that which he used for North China (i.e. division by formation). The sequence is representative of a large area, that of the Xinjiang Uygur Autonomous Region, and it appears to be relatively complete. Lithofacies apparently changes little at the Permian-Triassic boundary, and the two are conformable (Cheng Zhengwu, 1980). The facies of the Lower Triassic appears to be 'lacustro-fluvial,' an interpretation that may hold also for the Middle Triassic and the Upper Triassic. The sequence itself consists almost entirely of terrigenous clastics.

The cosmopolitan *Lystrosaurus* and *Chasmatosaurus*, together with other vertebrates, characterize the lower part of the Lower Triassic (Sun Ai-lin, 1973). A kannemeyerid and several other vertebrates occur in the Middle Triassic. Fishes represent vertebrates in the lower part of the Upper Triassic, in which division flora also

Table II-21. Triassic sequence, Eerduos (Shaan-Gan-Ning) basin and Shanxi Province, North China

TRIASSIC	Upper (T_{5-7})	Yanchang Formation	Yellowish-green to greenish-gray sandstone, mudstone, and shale topped by coal beds; fining upward	N.A.
	Middle (T_3-T_4)	Tongchuan Formation	Yellowish-green to greenish-gray sandstone, mudstone, and shale topped by coal beds; fining upward; *Hybodus youngi* (plesiosaurian fish); plants	N.A.
		Ermaying Formation	Dark purplish-red mudstone and silty mudstone, with alternate grayish-green to grayish-yellow sandstone; mudstone with calcareous concretions; upper (T_4): *Neuprocolophon, Shansisuchus, Fenhosuchos, Wangisuchus, Shansiodon, Sinokannemeyeria,* and *Parakannemeyeria* (vertebrates); lower (T_3): *Paoteodon, Ordosiodon, Ordosia, Shaanbeikannemeyeria, Parakannemeyeria xingxianensis* (vertebrates)	410 to 600 m
	Lower (T_1-T_2)	Heshanggou Formation (T_2)	Brick-red, purpish-red mudstone, sandy mudstone, intercalated with purplish-gray, grayish-green arkose; *Ceratodus hesthanggouensis* (dipnoan fish), *Fungusuchus* (theriodontid vertebrate)	100 to 280 m
		Liujiagou Formation (T_1)	Purplish-red and grayish-purple intercalated with small amount of purplish-red siltstone, conglomerate, and cross-bedded sandstone; in Shanxi Province, *Pleuromeia jiaochengensis, P. rossica, Neocalamites* sp. (plants) ; subzone of *Lundblatispora-Taeniaesporites-Cycadophytes* (sporopollen)	350 to 630 m

Table II-22. Triassic sequence, Xinjiang Uygur Autonomous Region, Northwest China

TRIASSIC	Upper	Haojiagou Formation Huangshanjie Formation	Grayish-yellow, grayish-green mudstone, sandstone with marly lenses, and thin coal seams; rich flora	180 to 830 m
	Middle	Upper Kelamayi Formation	Yellowish-green, grayish-black sandstone, mudstone, and shale; *Fukangichthys, Bogdania,* and *Fukangolepis* (fishes); flora	110 to 380 m
		Lower Kelamayi Formation	Purplish-red, dark purplish-red, sandstone, argillaceous siltstone with calcareous concentrations intercalated with grayish-green sandstone; *Sinosemionotus, Parotosaurus, Turfanosaurus, Vjushkovia,* and *Parakannemeyeria* (vertebrates)	120 m
	Lower	Shaofanggou Formation	Mudstone with calcareous concretions, purplish-red massive conglomerate intercalated with grayish-green sandstone and mudstone; lacustro-fluvial	110 to 350 m
		Jiucaiyuan Formation	Purplish and dark-red mudstone, sandstones with calcareous concretions, and grayish-green sandstone; lacustro-fluvial; *Lystrosaurus, Chasmatosaurus, Prolacertoides,* and *Santaisaurus* (vertebrates)	170 to 370 m

occurs. Coal seams in the upper part of the Upper Triassic contribute to a facies with a rich flora. Recently, an Upper Triassic flora has been described from the Urumqi area in Xingjiang Uygur Autonomous Region (Zhang Lujing, 1983). This flora contains a complete mixture of Chinese, Gondwanan, and southern European species.

Cheng Zhengwu (1980) does not describe the nature of the Triassic-Jurassic boundary in Northwest China. It is known, however, to be conformable to disconformable in the Junggar basin.

Paleoecology and Paleogeography

A 'Variscan' movement or movements at the end of the Permian (close of the Paleozoic) brought on a retreat of marine waters from almost the whole of northern China, and a deposition of the Triassic continental facies described above (Fang et al., 1979; Cheng Zhengwu, 1980). The retreat did not affect an area as far north as the southernmost part of the Qilian Shan, nor the northern sector of Northeast China and East China, which thus remained part of the ancestral Pacific (West Pacific) depositional realm (Cui and Li, 1986). In general, however, marine waters transgressed broadly into China south of the line joining the Kunlun Shan on the west (36°N latitude) to the Qin Ling on the east (34°N latitude). Available evidence suggests that Triassic waters came into China from an ancestral Indian Ocean on the south and an ancestral Pacific Ocean on the east, and that contemporaneous subsidence permitted deposition in a marine geosyncline (Kunlun Shan) in western China. Geosynclinal deposition took place in parts of Southwest China for the last time. Indosinian tectogenesis (Table II-7) affected the geosynclines of Xizang (Tibet) Autonomous Region, Qinghai Province, and Yunnan Province, as well as the Kunlun Shan-Qilian Shan geosyncline.

Central-South China

A difference in degree of subsidence distinguishes the Mesozoic basins of East China and Central-South China from those of Southwest China. Thus, the marine Mesozoic basins of Southwest China show the greater subsidence; and the line of division between the two areas of subsidence is the Xuefeng Shan in western Hunan Province (Figures II-40 and II-42). There is no line of division, however, between the two areas where the trend of Mesozoic basins is concerned. Most of these basins, regardless of the area in which they lie, strike north-northeast (Sheng Jin-zhang et al., 1985).

As marine waters entered Yunnan Province during the earliest Triassic, they surrounded the Kangdian (Sichuan-Yunnan) oldland of Southwest China (Figure II-40) and advanced toward the west and the north (Chao et al., 1962). Toward the west, they moved through the region of the present Himalaya and into what is now India. Toward the North, they invaded the southeastern part of the present Qinghai Province to a latitude as far north as the Southern Qilian Shan.

Marine waters also entered the greater Yangzi area during the Early Triassic (Sheng Jin-zhang et al., 1985). Here they were contained on the west by the Kangdian oldland and on the east by the Huaxia (Cathay) oldland (Figure II-40). Many of the clastics supplied to the coastal areas of these waters were eroded from these two oldlands. By the Early Triassic, marine waters had spread over all of Central-South China and East China, which had become the site of widespread carbonate deposition.

During the Middle Triassic, parts of the present Yangzi Valley experienced a mild uplift that produced complex depositional environments. Areas above sea level suffered erosion and supplied sediments. Others became lagoonal, and received deposits of dolomitic limestone, gypsum, and halite. Still others became sites for the deposition of typically neritic limestone.

By the middle of the Late Triassic (Figure II-42), during a stage equivalent to the Norian of Europe, the Indosinian movement had begun, and then deformed the parts of China that were mentioned above - mostly in Southwest China and in what is now the Kunlun Shan-Qilian Shan foldbelt. Land emerged gradually, with some areas rising to higher than average elevation. Erosion of the lands during the Late Triassic caused deposition of alternate beds of marine and non-marine sediments, and a variety of lithofacies. These include coal, continental strata with coal seams, coastal terrigenous clastics, and shoal-to-neritic carbonates.

Volcanic activity played an important part in Indosinian tectogenesis (See Table III-1), although largely in the northern and eastern parts of the Qinghai-Xizang (Tibet) Plateau, the Qin Ling, the Nan Ling, and the Dabie Shan. Contemporaneous intrusives consist mainly of granite and diorite, locally associated with mafic, ultramafic, and alkalic intrusions that invade Triassic and older strata. Overlain mostly by Upper Triassic or Lower Jurassic sediments, these intrusives yield isotopic ages in the range of 230 to 190 ma. Recent investigations have shown that intrusive rocks of this phase are present also in the Bei Shan of northwestern Gansu Province, in Northwest China, in parts of North China, in the Da Hinggan Ling of the northeastern Nei Monggol Autonomous Region and in East China (Cui and Li, 1986). Such a wide distribution of intrusives confers on Indosinian tectogenesis the status of a Pan-China event (Compilation Group, 1976).

Northern Regions

Many Triassic inland, non-marine successor basins developed across the northern regions of China, some as continuations of Permian events, others for the first time (Figures II-40 and II-42). Basins west of the Luliang Shan of western Shanxi Province exhibit a large and deep morphology. Those farther east are smaller and shallower. In Northeast China, erosion exceeded deposition. In Northwest China, the two were reversed. For northern regions as a whole, the Triassic inland basins follow north-northeast to northeast (Cathaysian) structural trends.

Environment through the northern regions remained largely unchanged across the Permian-Triassic boundary and through the succeeding Early Triassic, as shown by the continuous deposition of purplish-red terrigenous clastics. During the Middle Triassic, a change in deposits from variegated terrigenous clastics in T_3 to black organic shale and oil shale in the overlying T_4 suggests a gradual climatic change from dry to humid. Rich organic matter in the shales of T_4 suggest further that the subaqueous environment of their deposition was quiet (low energy) and reducing. Arkose also is present in T_4, an indication of rapid deposition and rapid subsidence during the late Middle Triassic. The presence of *Glossopteris* in the Middle Triassic of North China suggests an important extension of the Gondwanian floral realm to the north (Fang et al., 1979). Late Triassic lithofacies suggest depositional environments that ranged from quiet lakes to shallow, forested swamps.

Vertebrate Fauna

The presence of a large and varied *Lystrosaurus* (Early Triassic) tetrapod association in several parts of Xinjiang Uygur Autonomous Region is important in deciphering the paleogeography of Asia during Triassic time. Sun Ai-lin (1973), Chatterjee and Hotton (1986), and Chatterjee (1984) concluded, on the basis of tetrapod studies (amphibians, reptiles), that Africa, China, and Europe were close to each other and were connected by direct land routes. A direct land route to North America also is indicated from the data. Similar conclusions were reached on the basis of even more complete data from Mongolia reported by Kalandadze and Rautuan (1983).

Chatterjee and Hotton's (1986) analysis of the complete Triassic fauna shows the following: India's closest relations were with Europe, followed closely by Africa. North America, South America, and eastern Asia were fourth, fifth, and sixth respectively. Because the Early Triassic taxa are greatly different from those that had evolved by the Late Triassic, Chatterjee and Hotton (1986) studied similarities among the continents for Late Triassic taxa. All ten families present in India are found in Europe; nine of the ten are present in North America; and six are present in eastern Asia (mainly China). However, only five are common to India and Africa, and to India and South America, whereas India shows nothing in common with Antarctica. Clearly India's land connections were with northern continents, not with southern continents.

JURASSIC SYSTEM

Indosinian tectogenesis in the Middle and Late Triassic initiated a gradual regression of marine waters in southern China, western Sichuan Province, eastern Qinghai Province, and eastern Xizang (Tibet) Autonomous Region. Paleogeography changed accordingly, not only in the above-mentioned areas, but also in China as a whole. Uplift in southern China added a large land area to the landmass already present in northern China. Jurassic marine waters thus did not move significantly beyond Xizang, southern Qinghai, and the southern sector of southern Xinjiang. Relatively brief incursions brought marine water into western Yunnan Province, Guangdong Province, and Hunan Province; one very small marine incursion took place in eastern Heilongjiang Province (Chen Piji et al., 1982; Wang and Sun, 1983, 1985).

Distribution and Facies

The Jurassic of China is divided into stratigraphic domains by Chinese geologists on the basis of lithologic characteristics (Figures II-45 and II-46). All of the several classifications differ in some degree from one another.

Gu (Ku, 1962) divided both the Jurassic and the Cretaceous into three major regions and 61 subregions across China (Figure II-47), with depositional environment as the essential criterion.

Wang and Sun (1983) based their classification mainly on stratigraphic developments, sedimentary facies, and paleotectonic setting, and divided the Jurassic System of China into three regions and 13 districts. The following comparison shows titles of the regions and number of subdivisions of Gu (1962), and Wang and Sun (1983).

Another classification was made by Chen Piji et al. (1982). They recognized four stratigraphic domains for Lower and Middle Jurassic on the basis of paleogeography and sedimentation: (1) North China domain (three provinces); (2) Tethyan domain (two provinces); (3) Yuegan Bay domain (four provinces); and (4) Wusuli domain. By taking into account the early cycles of the Yanshanian tectogenesis, they divided the Upper Jurassic into three domains and eight provinces.

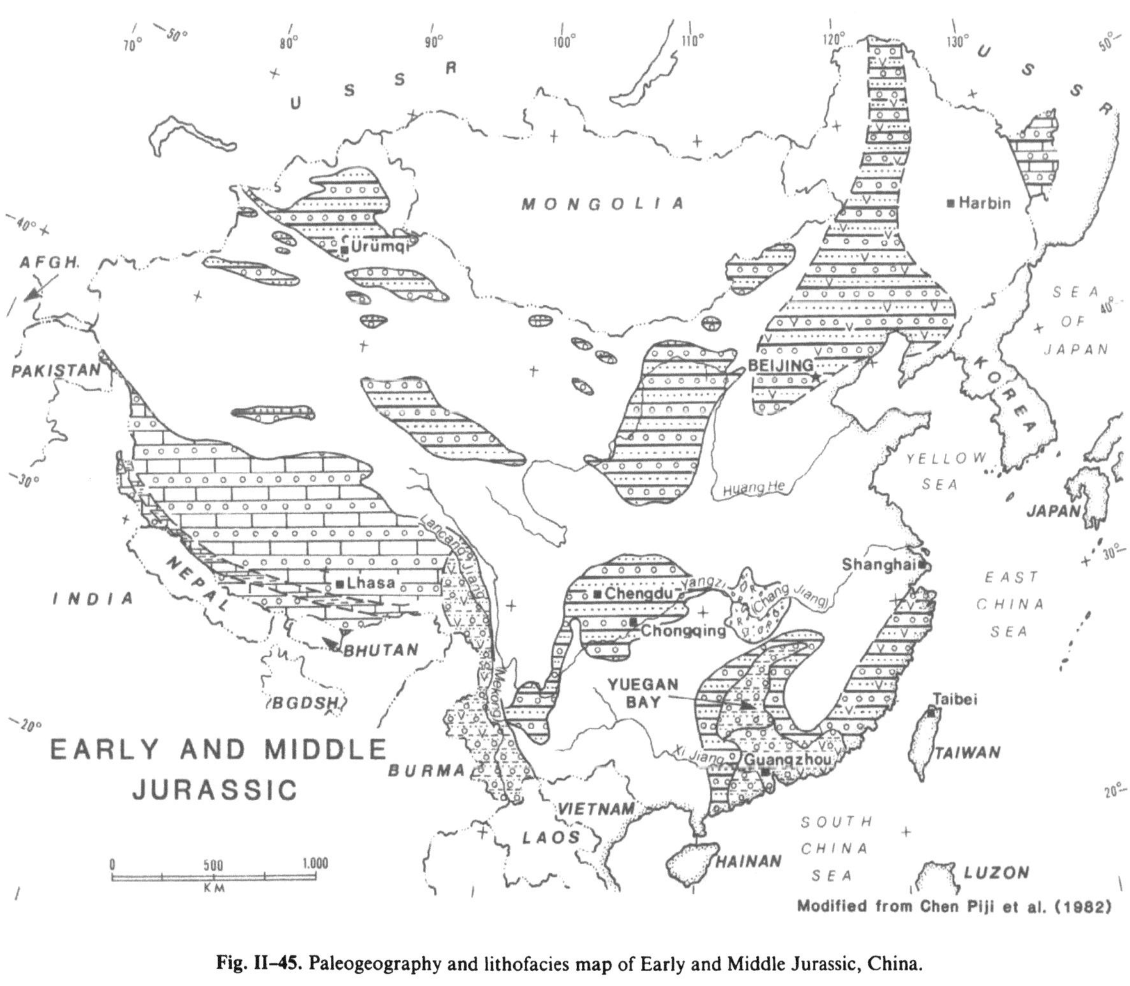

Fig. II-45. Paleogeography and lithofacies map of Early and Middle Jurassic, China.

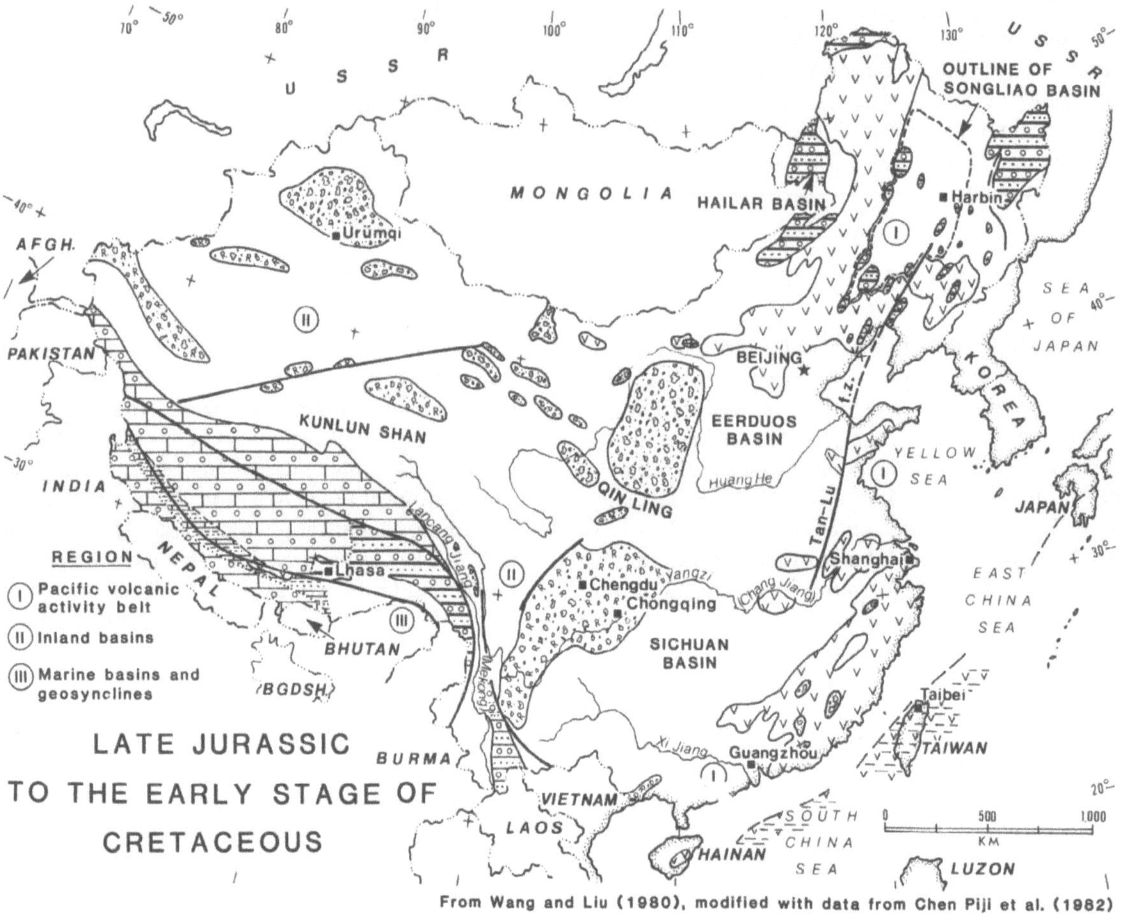

Fig. II–46. Paleogeography and lithofacies map of Late Jurassic to the early stage of Cretaceous.

Classification of Jurassic domains

Gu (Ku, 1962)	*Wang and Sun (1983)*
Marine Basins and Geosynclines in Southwest China (seven subregions), (III on Figures II–46 and II–47).	Xizang (Tibet)-Qinghai Region (seven districts).
Pacific Volcanic Belt (thirty subregions), (I on Figures II–46 and II–47).	Coastal Volcanic Region of East China (four districts).
Inland Basin Region in Central and Northwest China (twenty-four subregions), (II on Figures II–46 and II–47).	Stable Giant Basins of West China (two districts).

We adopt Gu's (Ku, 1962) system here. A description of each of his three major Jurassic regions follows.

Pacific Volcanic Belt

This belt lies east of a series of mountain ranges that strike north-south and in places northeast, rejuvenated repeatedly by Yanshanian and, most recently, Himalayan movements. These ranges include the Da Hinggan Ling, the Taihang Shan, the Luliang Shan, the Wuling Shan, and the Xuefeng Shan which extends from northeastern Nei Monggol Autonomous Region and Shanxi Province to Hunan Province. Li Siguang (J. S. Lee, 1939) called this the Da Hinggan-Shanxi-Guizhou anticlinorium. More recently, Huang Jiqing referred to it as the Da Hinggan-Taihang-Wuling trend. It coincides in places with a major fault zone dating probably as far back as the late Proterozoic or earlier. There is a sharp rise from west to east of the Mohorovicic discontinuity, and a steep gravity gradient. East of this trend lies the Neocathaysian system of horsts, grabens, half-grabens, normal faults, and strike-slip faults that strike north-northeast and northeast. The great features that strike east-west and the large intracontinental basins of western China lie west of the Da Hinggan-Taihang-Wuling trend.

Lower and Middle Jurassic strata consist mainly of fluvial and lacustrine deposits intercalated with coal layers and with intermediate to mafic volcanic rocks. These accumulated in north-northeast-trending grabens, of which more than 60 existed by Late Jurassic time (Li Sitian et al., 1982). Upper Jurassic strata consist chiefly of intermediate-to-mafic and intermediate-to-silicic volcanic rocks interbedded with lacustrine deposits. In the northern sector of Northeast China, Upper Jurassic strata are associated also with coal measures.

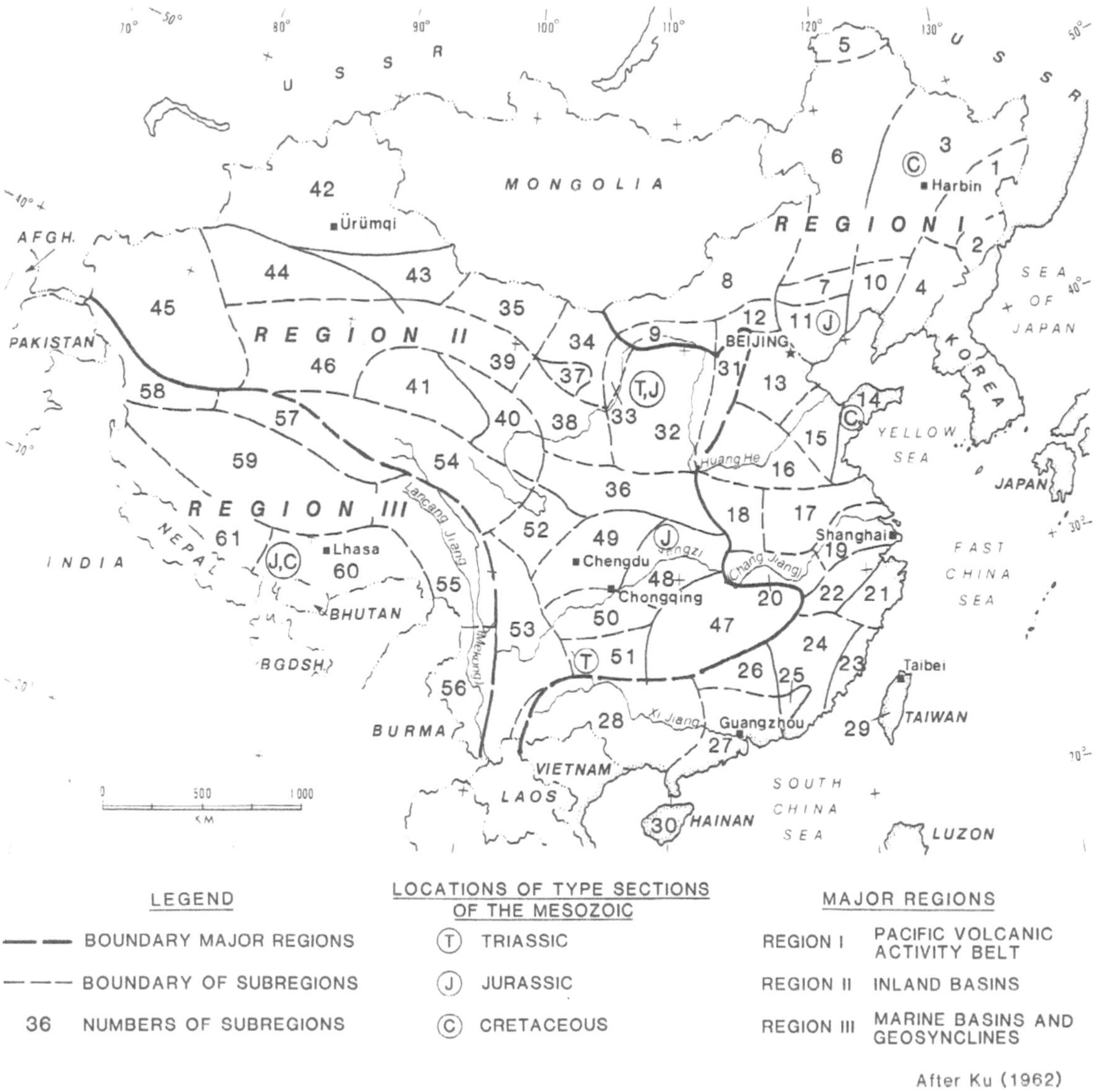

Fig. II–47. Stratigraphic regions map of Jurassic and Cretaceous in China.

In addition, they form part of a total Jurassic sequence rich in estheriids, plants, and fish. In Southwest China (subregion 26, Figure II–47), marine and non-marine deposits alternate. Central-South China (subregion 27), and Northeast China (subregion 1), contain Hettangian and Sinemurian marine deposits (early Early Jurassic). Ammonoids found in the subsurface of Taiwan once were believed to be Late Jurassic, but now are known to be Early Cretaceous (Meng, 1970; T. C. Huang, 1978). On Hainan (subregion 30), geologists have not yet found any Jurassic sequences. The paleontology of the marine sequences is summarized by Wang and Sun (1983). Ore deposits of the Pacific Volcanic belt obviously represent a segment of the Circum-Pacific metallogenic domain.

Inland Basin Region of Central China and Northwestern China

The Jurassic deposits of this region (II on Figures II–46 and II–47) accumulated in several basins west of and including the Hailar, Eerduos, and Sichuan basins. The best known of these are the Eerduos (Shaan-Gan-Ning), Sichuan, Tarim, Junggar, Turpan, and Qaidam basins. All lie north and east of the geosyclinal realm of Southwest China, which includes the Kunlun Shan to Yunnan ranges and the Qinghai-Xizang Plateau. During the Jurassic, fluvio-lacustrine deposits accumulated. Today these form the variegated redbeds in many areas. Coal seams and an abundant flora are present, especially in the Lower and Middle Jurassic.

Climatic zonation is evident from a study of the sections in each basin. South of a line joining the Kunlun Shan with the Qin Ling, in the Sichuan basin and in the basins of Yunnan Province, redbeds with abundant naiadid plants are present; coal is found only in the Lower Jurassic of northwestern Sichuan Province; no coal is found in the Upper Jurassic section. The absence of coal seems to indicate the onset of semiarid or arid conditions.

Marine deposits are severely limited, almost nonexistent in Gu's region II (Gu, 1962; Figure II–47). Gu did report some marine Jurassic, possibly Early Jurassic, in a limited area of Sichuan Province. In Qinghai Province (subregion 40, Figure II–47), some volcanic rocks are present in the Lower Jurassic sequence.

Marine Basins and Geosynclines in Southwest China

In this region (III on Figures II–46 and II–47), the main deposition of marine Jurassic occurred, and contributed to the depositional sector called 'East Tethys.' In contrast, the sedimentary component of the Pacific Volcanic belt, described above, belongs to the West Pacific depositional domain. East Tethys includes Qinghai-Xizang (Tibet) and the Himalaya of Southwest China.

In southern Xizang (the Himalaya), the Jurassic sequence consists of limestone, shelly limestone, marl, shale, and quartz sandstone with abundant fossils (Mu Anzi et al., 1973; Yin Ji-xiang et al., 1983; Wang and Sun, 1983). The upper part of the shallow-water facies grades northward (Indus-Yarlung Zangbo zone) into deeper-water siliceous-argillaceous carbonates and flyschoid strata, with monotonous alternation of thick-bedded calcareous mudstone and sandstone.

Farther north, the Jurassic sediments along the Bangong Co-Nu Jiang deep fracture zone are composed of deep-water clastic carbonates, flysch-like terrigenous clastics interbedded with volcanics, pelite, and chert (Chang and Pan, 1981; Wang and Sun, 1983; Girardeau et al., 1984, 1985).

Oscillations of sea level took place during the Jurassic along the northern edge of this region, as evidenced by littoral and neritic clastic carbonates that alternate with continental deposits.

Yin and Fang (1973) described a Middle and Upper Jurassic section from the Baoshan area of western Yunnan Province. This consists of purplish-red sandy shale with interbeds of marine dolomite and oolitic limestone and lenses of gypsum (with basalt in the upper part of the formation). This sequence lies unconformably on the Triassic. Upward, it grades into shelly, argillaceous, brachiopod-rich limestone, yellowish-gray pelecypod-bearing siltstone and shale. Lying unconformably above is an unfossiliferous unit of reddish to variegated sandstone and conglomerate.

Chen Piji et al. (1982) tabulated groups of fossil biota from both the non-marine and marine environments of the Jurassic in China. In the non-marine environment they recorded paleontological data for conchostracans (branchiopods), ostracods, charophytes, and pelecypods. In the marine environment they dealt with ammonoids, foraminifera, brachiopods and pelecypods. We tabulate their classifications in two tables that follow (Tables II–23 and II–24).

Type Sections

Northeast China

The type section of Northeast China lies in the area of northern Hebei Province and western Liaoning Province (subregion 11, Figure II–47). The sequence consists of alternate beds of continental volcanics interlayered with non-marine deposits (Figure II–48). The volcanics occur in four layers; one in the Lower Jurassic, one in the Middle Jurassic, and two in the Upper Jurassic (Fang et al., 1979).

The Lower Jurassic interval consists of fluvial clastics near the base; andesite, basalt and volcaniclastics in the

Table II–23. Biostratigraphic sequence of the nonmarine Jurassic in China

			CONCHOSTRACANS	OSTRACODS		CHAROPHYTES	PELECYPODS
				SOUTH CHINA	NORTH CHINA		
JURASSIC	LATE	TITHONIAN	*EOESTHERIA* FAUNA	*DARWINULA-DAMONELLA-DJUNGARICA* ASSEMBLAGE	*CYPRIDEA-LUAPINGELLA-EOPARACYPRIS* ASSEMBLAGE	*POROCHARA* CF. *HILDENSIS* *EUACLISTOCHARA* CF. *NUQUISHANENSIS*	*PEREGRINOCONCHA-KOREANAIA-DANLENGICONCHA* ASSEMBLAGE
		KIMMERIDGIAN OXFORDIAN	*NESTORIA-KERATESTHERIA* FAUNA	*DARWINULA-CETACELLA-DJUNGARICA* ASSEMBLAGE			*FERGANOCONCHA-MENGYINAIA* ASSEMBLAGE
	MIDDLE	CALLOVIAN BATHONIAN	*PSEUDOGRAPTA* FAUNA	*DARWINULA* FAUNA		*EUACLISTOCHARA* FAUNA	*EOLAMPROTULA-PSILUNIO* ASSEMBLAGE
		BAJOCIAN	*EUESTHERIA ZILIUJINGENSIS* FAUNA				*YANANOCONCHA-FERGANOCONCHA-SINOMARGARIFERA* ASSEMBLAGE
	EARLY	TOARCIAN DOMERIAN	*EOSOLIMNADIOPSIS* FAUNA	*GOMPOCYTHERE-DARWINULA* FAUNA		*STELLATOCHARA XIANGQIENSIS*, *STENOCHARA CHAIMENENSIS*	*QIYANGIA-APSEUDOCARDINIA* ASSEMBLAGE
		PLIENSBACHIAN SINEMURIAN HETTANGIAN	*PALAEOLIMNADIIA BAITIANBAENSIS* FAUNA				

From Chen Peiji et al. (1982)

Table II-24. Biostratigraphic sequence of the marine Jurassic in China

			AMMONOIDS	FORAMINIFERA	BRACHIOPODS	PELECYPODS	
JURASSIC	LATE	TITHONIAN	*HAPLOPHYLLOCERAS PINQUE, VIRGATOSPHINCTES DENSEPLICATUS*		*RHYNCHONELLA* AFF. *PAUCICOSTA*	*BUCHIA-ASTARTOIDES* ASSEMBLAGE	
		KIMMERIDGIAN			*RUTORHYNCHIA-MONTICLARELLA* ASSEMBLAGE		
		OXFORDIAN					
	MIDDLE	CALLOVIAN	*MACROCEPHALITES* FAUNA		*THURMANELLA ROTUNDA-KUTCHITHYRIS DENGQENENSIS* ASSEMBLAGE	*CAMPTONECTES LENS-LIOSTREA BIRMANICA* ASSEMBLAGE	
		BATHONIAN			*BURMIRHYNCHIA-HOLCOTHYRIS* ASSEMBLAGE		
		BAJOCIAN	*SONNINIA, DORSETENSIA, WITCHELLIA, FROGDENITES*		*NYALAMURHYNCHIA MIRIFICA-RHACTORHYNCHIA LANTA* ASSEMBLAGE		
	EARLY	TOARCIAN	*NYALAMOCERAS NYALAMENSE*		*CIRPA HIMALAICA-HOMOEORHYNCHIA BOLINENSIS* ASSEMBLAGE	XIZANG (TIBET)	XIANG-YUE-GAN
		PLIENSBACHIAN				*WEYLA AMBONGOENSIS-ENTOLIUM NIENIE-XIONGLAENSIS* ASSEMBLAGE	*LILINGELLA* BED *XINYUELLA* BED
		SINEMURIAN	*SULCIFERITES, GLEVICERAS, HONGKONGITES*	*ORBITOPSELLA* FAUNA			*PARAINOCERAMUS-TEINONUCULANA* ASSEMBLAGE
		HETTANGIAN	*SCHLOTHEIMIA*				

From Chen Peiji et al. (1982)

middle; and coal-swamp deposits near the top. Abundant fossil plants throughout the Lower Jurassic indicate comparatively stable tectonic conditions. The Middle Jurassic contains lacustrine and coal-swamp deposits in the lower part, andesite and basalt in the middle part, and rapidly deposited sediments laid down by fluvial currents in the upper part. The Upper Jurassic consists of: (1) a lower layer in which andesite, tuff, agglomerate, and volcanic breccia are interbedded with thick lacustrine deposits in the middle of this lower layer; and (2) an upper layer of richly organic lacustrine deposits. This succession of Upper Jurassic lithofacies indicates a period of extensive and deep lacustrine waters between two periods of strong volcanic activity, together with a quiescent marsh environment that established itself after periods of erosion.

In easternmost Jilin Province, marine strata alternate with the non-marine, and the marine strata contain in places a rich ammonoid and bivalve fauna (Wang and Sun, 1983). The marine beds range in age from Bathonian through Tithonian. Marine intercalations prove that the famous Jehol or Witin freshwater fauna is Middle to Late Jurassic and does not range into the Cretaceous (Gu Zhiwei et al., 1984).

North China and Central-South China

Fang et al. (1979) presented two type sections for the inland basins of this part of interior China. The first occurs in eastern Sichuan Province (Figure II-49) and the second in the Eerduos (Shaan-Gan-Ning) basin of Shaanxi Province, Gansu Province, and Ningxia Huizu Autonomous Region (Figure II-50).

In the first type section (Figure II-49), the Lower Jurassic (J_1), disconformable on Upper Triassic, consists of dark-colored terrigenous clastics with coal seams, plant fossils, and siderite. Lithology, lithofacies, and fossil taxa indicate the presence of lacustrine and marshy environments in a hot humid climate. The Middle Jurassic (J_2 to J_4) consists of gray variegated sandstone with freshwater pelecypods, conchostracan branchiopods, and plants, a combination that reflects deeper lacustrine waters and a less humid climate than that of J_1. The Upper Jurassic (J_5 to J_7), overlain disconformably by the Cretaceous, includes mainly red sandstone, interbedded with arkose and quartz sandstone. Fossils include vertebrates, ostracods, and conchostracan branchiopods. For the Upper Jurassic, lithology, lithofacies, and fossil taxa indicate a fluvio-lacustrine environment under a dry climate.

In the second type section (Figure II-50), the Lower Jurassic (J_1) also is disconformable on Upper Triassic. It consists of lacustrine mudstone with plants. The facies of the overlying Middle Jurassic (J_2 to J_4) includes rather coarse-grained and relatively massive sandstone in addition to mudstone, shale, and oil shale. The presence of plants and of pelecypods indicates a fluvio-lacustrine facies, which at times changed into a quiet-water, lacustrine environment, as for example during deposition of argillaceous limestone in the upper part of J_4. The Upper Jurassic (J_5 to J_7) is absent in the Eerduos (Shan-Gan-Ning) basin because of uplift.

Xizang (Tibet)

The type section of the marine Jurassic occurs on Qomolangma Feng (Mt. Everest) in the Himalaya (Mu Anzi et al., 1973; Figure II-51). Both the Lower and

System	Series	Formation		Column section	Thickness (m)	Major lithology	Major fossils
		K_1					
Jurassic	Upper	J_7			84-1,400	Yellowish-green sandy shale, congl. and coal.	*Coniopteris* (P) *Nipponanaia* (Pe)
		J_6	U		300-1,700	Grayish-black shale, oil shale, coal, sandstone.	*Coniopteris* *Lycoptera* (F) *Ferganoconcha* (Pe)
			L		200-3,500	Red-gray-purple andesitic trachyte, andesitic breccia and tuff with sedimentary rocks.	
		J_5	U		10-2,000	Grayish-white tuffaceous shale, sandstone, conglomerate.	*Lycoptera* *Ephemeropsis* *Nakamuranaia* (Pe)
			L		200-2,500	Pyroxene andesite, volcanic breccia, tuff, and agglomerate with sandstone and shale.	
	Middle	J_4			200-2,000	Red, yellow conglomerate, sandy conglomerate, sandstone, siltstone.	
		J_3			230-970	Black andesitic basalt, andesitic breccia, yellow argil. congl.	*Coniopteris* *Podozamites* (P) *Cladophlebis* (P) *Ferganocencha*
		J_2			284-543	Yellow congl., ss., siltstone.	
	Lower	J_1	U		627-1,313	Yellowish-brown sandstone, grayish-black shale and sandstone with coals.	*Cladophlebis* (P) *Dictyophyllum* (P) *Podozamites* (P)
			M		130-900	Upper: dark andesitic, basalt, volcanic clastics. Lower: tuffaceous sandy shale.	*Neocalamites* (P) *Equisetites* (P) *Cladophlebis* (P)
			L		20-280	Yellow ss., congl. with coal.	*Coniopteris*

Fig. II-48. Jurassic stratigraphy in northern Hebei Province and western Liaoning Province, North China (compiled from Fang et al., 1979).

System	Series	Formation	Column section	Thickness (m)	Major lithology	Major fossils
Jurassic	Upper	J_7		200-1,700	Alternate purplish-red mudstone and grayish-white quartz sandstone	*Darwinula* (O)
		J_6		200-500	Brownish-red mudstone interbedded with siltstone.	*Ceratodus szechuanensis* (F)
		J_5		450-1,500	Purplish-red mudstone interbedded with green arkose.	*Mamenchisaurus* (D)
	Middle				Purplish mudstone, yellow conchostracan shale in the uppermost.	*Eoestheria* (Con) *Psilunio* (Pe) *Ferganocancha* (Pe)
		J_3-J_4		100-250	Alternate grayish sandstone (or argillaceous sandstone) and gray mudstone.	*Pseudocardinia* (Pe) *Tutuella* (Pe) *Coniopteris* (P)
		J_2-J_3		100-250		
	Lower	J_1		100-464	Upper: grayish mudstone interbedded with yellow ss. and purplish-red mudstone. Lower: dark gray mudstone interbedded with quartz ss., and thin coal layer.	*Coniopteris hymenophylloides* (P) *Hsiaugchiphyllum trinerve* (P)

Fig. II-49. Jurassic stratigraphy in eastern Sichuan Province, Central-South China (from Fang et al., 1979).

Middle Jurassic (J_1 to J_4) consist of thick-bedded limestone and quartz sandstone interbedded with lesser amounts of shale. Ammonoids, foraminifera, and palecypods are abundant in this interval. The Upper Jurassic (J_5 to J_7) consists of black shale and sandy gray shale with belemnites and ammonoids. Farther north in the Yarlung Zangbo fault zone, the latest Jurassic (Tithonian) consists of red radiolarite, presumably the oldest unit in the deeper water facies associated with the fracture system. The pre-Tithonian consists of a monotonous sequence of thick-bedded calcareous mudstones that alternate with sandstones, which shows graded bedding (Mu Anzi et al., 1973).

Paleoecology, Paleogeography, and Magmatism

Whereas marine environments in southern China and non-marine environments in northern China persisted through the whole of the Permian and much of the Triassic, the Indosinian movement of the middle Late Triassic brought about a significant change in the regional pattern of environments. During the Jurassic many inland basins formed throughout China (Figures II–45 and II–46). The largest of these, as noted by Wang and Sun (1983), lay north and east of a line joining the Kunlun Shan and the Hengduan Shan. West and south of this line, the marine Jurassic environments of Tethys and of greater Tethys dominated.

Three major types of environment characterize the Jurassic: (1) inland basins with volcanism along the present Pacific Ocean; (2) inland basins not affected by volcanism, situated in central China and Northwest China - a rough boundary between (1) and (2) formed by the Luliang Shan and the Xuefeng Shan; and (3) a marine (Tethyan and greater Tethyan) domain in Southwest China. In Central-South China, whereas most basins developed typically soon after the Indosinian movement, the development of a few did not begin until the Middle Jurassic.

Of the large inland basins, two notable ones are the Eerduos (Shaan-Gan-Ning) basin, north of the Qin Ling, and the Sichuan basin, south of the Qin Ling (Figure II–46). In a comparison of these two basins, Fang et al. (1979) noted the common characteristics of a quiet lacustrine environment and internal fluvial drainage. Other characteristics in common are: (1) a depocenter

System	Series	Formation	Column section	Thickness (m)	Major lithology	Major fossils
Jurassic	Middle	J_4		50–100	Upper: red, grayish-yellow argillaceous limestones. Middle and lower: grayish-black shale and oil shale, and yellowish-green ss.	*Balaiichtys antingensis* *Darwinula*
		J_3		250	Yellowish-green and yellowish-gray ss., bluish-gray and grayish-purple mudstone with very thin coal layer. Bottom: massive sandstone.	*Coniopteris hymenophylloides*
		J_2		200–300	Upper: alternate grayish-green sandstone and grayish-black shale and oil shale. Lower: grayish-white and grayish-yellow massive sandstone.	*Todites shensiensis* *Coniopteris hymenophylloides* *Ferganoconcha* *Tuiuella*
	Lower	J_1		100	Red, purple, green mudstone.	*Coniopteris hymenophylloides* *Selaginellites*

Fig. II–50. Jurassic stratigraphy in Shaanxi Province, Gansu Province and Ningxia Huizhu Autonomous Region, Northwest China (from Fang et al., 1979).

System	Series	Formation	Column section	Thickness (m)	Major lithology	Major fossils
Jurassic	Upper	J_5–J_7		360–710	Grayish-black shale with sandy shale.	*Haplophylloceras* *Pterolytoceras*
	Lower and Middle	J_1–J_4		2,060	Limestones interlayered with sandstone and shale.	*Macrocephalites* *Kamptokephalites* *Witchella* *Dorsetensia* *Schlothemia* *Orbitopsella*

Fig. II–51. Jurassic stratigraphy on Qomolangma Feng (Mt. Everest) of the Himalaya, Xizang (Tibet) Autonomous Region, Southwest China.

on the western side of the basin; and (2) north-northeast structural trends. Jurassic climate, however, shows differences. In the Sichuan basin, hot and humid conditions in the Early Jurassic changed to dry conditions during the Late Jurassic. In the Eerduos (Shaan-Gan-Ning) basin, hot and humid conditions persisted through the entire Jurassic.

The Indosinian movement of the middle Late Triassic was the first important tectogenesis of the Mesozoic in China. The second tectogenesis, the Yanshanian (Ningjinian) movement, began during the Bajocian and the Bathonian stages of the Middle Jurassic and continued until the Late Cretaceous (Gu (Ku), 1962). The Yanshanian movement affected the whole of China, but was most intense in East China and Central-South China (Compilation Group, 1976). Folding, rifting, strong intrusive magmatism, and volcanism contributed to the formation of a tectonic pattern that basically has persisted across China until the present time. Magmatism associated with the Yanshanian movement produced intrusions throughout China, particulary in the Circum-Pacific segment represented by East China and Central-South China (Hu Huanguang et al., 1986; Zhu Ming, 1989). In those two regions Jurassic and Cretaceous Yanshanian intrusions relate closely to the mineral deposits of many endogenic metallic ores.

The intrusive magmatism of the Yanshanian movement includes Cretaceous phases as well as a Jurassic phase. In the Jurassic, the main episodes occurred in East China and Central-South China: in the middle and lower Yangzi Valley; in parts of North China and Northeast China; in Xizang (Tibet) Autonomous Region; and in western Yunnan Province. We describe the Cretaceous phases of intrusive magmatism in the next section.

The movement of Jurassic marine waters was associated with the Tethyan domain in the west and with the Pacific domain in the east. During the Early and Middle Jurassic, marine Tethyan waters advanced into Southwest China, but during the Late Jurassic these waters retreated. In the Pacific domain, a deep marine basin formed in the eastern part of Jilin Province (Northeast China) during the Early Jurassic. The waters of this basin came undoubtedly from the ancestral Pacific.

Evidence worthy of special note is the discovery of several Gondwanian genera and species in the Middle Jurassic marine section of eastern Jilin Province. These are bivalve (benthic) species, and their discovery is the first for the Middle Jurassic of this part of the world (Gu Zhiwei et al., 1984). Gondwanian genera and species of other ages (e.g. Permian) have been found and described from the adjacent area of the U.S.S.R. (Zimina, 1967; Sanylina and Yefimova, 1968). These include, in addition to bivalves, the floral genera *Glossopteris* and *Dicroidium*, which also are associated with some genera previously known only from Europe, as well as from Gondwana.

Peninsular India

Chatterjee and Hotton (1986) found that the Jurassic reptiles of India are related to forms of the Northern Hemisphere. These authors also noted that the absence of pre-Eocene mammals in India has been used to date India's 'collision' with Asia. This argument, however, no longer is a valid one. Early Jurassic mammalian remains of symmetrodont teeth now have been found. The only other known fossil symmetrodonts are from Europe, China, Mongolia, and North America. The affinities of India, faunal and paleogeographic, were northern rather than southern in the Jurassic.

CRETACEOUS SYSTEM

The Cretaceous system shows pronounced stratigraphic, faunal, floral, and paleogeographic similarity to its systemic predecessor, the Jurassic (Figure II-52). Emergence of most of the area of present mainland China continued through the whole of the Cretaceous. In the Pacific domain, a retreat of marine waters kept Cretaceous shorelines as far east as Japan on the north, and as far east as Taiwan in more southern latitudes (Hao Yichun et al., 1986). Volcanism showed a Circum-Pacific character during the Early Cretaceous, but diminished considerably during the Late Cretaceous.

Distribution and Facies

Pacific Volcanic Belt

Cretaceous volcanics and Cretaceous pyroclastics occur mainly in medium to small basins with north-northeast and northeast trends in the coastal region of Central-South China and East China (Compilation Group, 1976; Liu Xun, 1988). The section contains mainly non-marine terrigenous clastics and intermediate to silicic volcanics, locally interbedded with marine strata, or with strata that bear coal and gypsum. More than 60 interior grabens with Neocathaysian (north-northeast strike) are known (Li Sitian et al., 1982). Farther east, in the area of present marine waters, an island-arc eugeosyncline appeared possibly in the Jurassic (Figure II-46), and became prominent during the Cretaceous and later (T. C. Huang, 1978; Huang and Chu, 1979; Ho, 1982; Figure II-52).

Upper Cretaceous deposits are abundant in the onshore grabens (Figure II-53). The plant *Frenelopsis* appears commonly; and deposits of this same interval in Central-South China have gained distinction from the discovery in them of dinosaur eggs (*Oolithus*, *Rugustus*). Coal is present.

Sparse marine fossils (Cenomanian) occur in a notably thin interval in the Songliao basin of Northeast China (Subregion 3, Figure II-47). Thicker marine intertongues are present in the northeastern part of Northeast China (subregion 2, Figure II-47) and in Central-South China

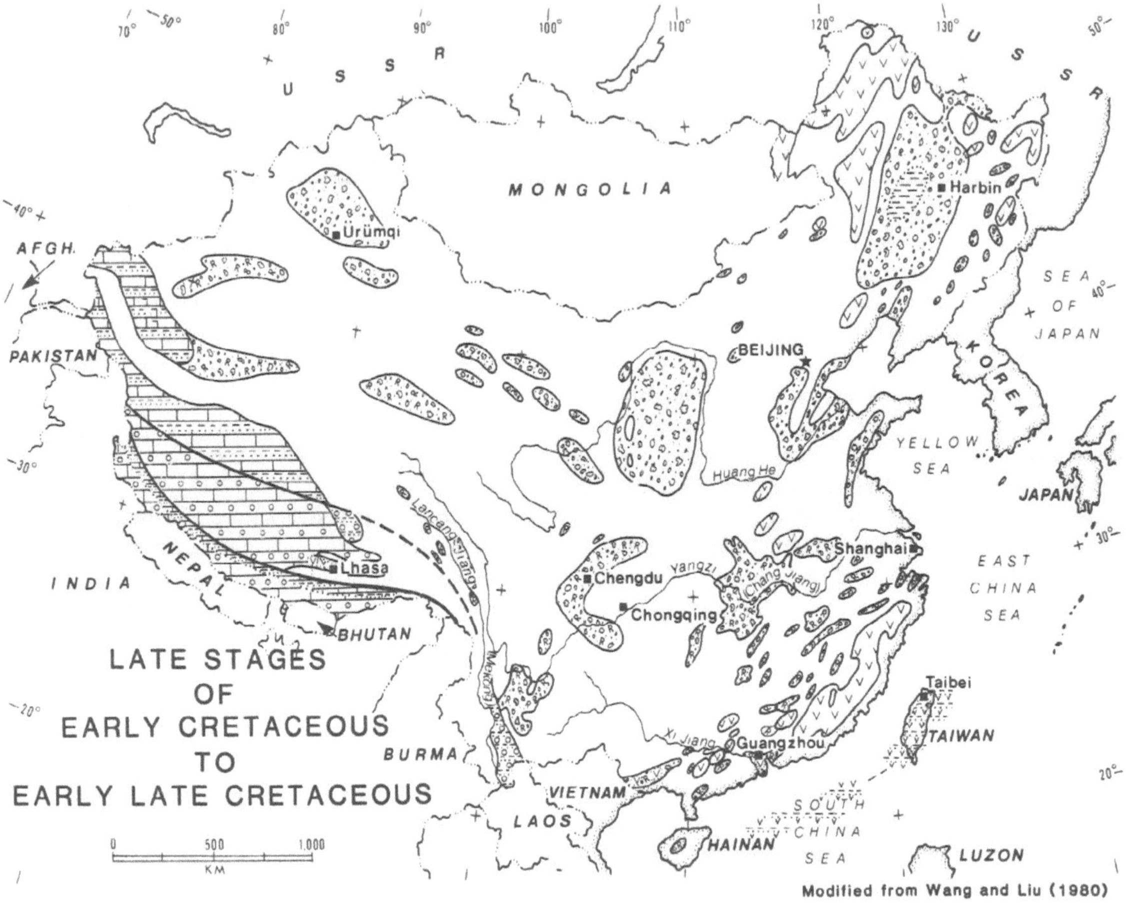

Fig. II–52. Paleogeography and lithofacies map of late stages of Early Cretaceous to early Late Cretaceous, China.

(subregion 27, Figure II–47). In the area of present marine waters of East China and Central-South China, island arc or eugeosynclinal conditions prevailed.

For the Songliao basin of Northeast China, Chen (1980) presented a comprehensive study of Cretaceous stratigraphy, Cretaceous environment, Cretaceous climate, Cretaceous hydrocarbon reservoirs, and their Cretaceous source beds, all in a non-marine setting. The non-marine Lower Cretaceous is especially well developed here, as well as in several other onshore grabens. The non-marine Lower Cretaceous is most extensive in the west; Upper Cretaceous is more extensive close to the coast as basin formation shifted eastward and southeastward (Figures II–52 and II–53). Volcanic activity was confined mostly to East China and Central-South China during the Late Cretaceous. In Northeast China, Upper Cretaceous deposits unconformably overlie Lower Cretaceous rocks, and they show evidence of uplift during this time. In the offshore and on Taiwan, an Upper Cretaceous marine eugeosynclinal island-arc system continued to be active.

Inland Basin Region

The typical representative of this region belongs mainly to a few basins of large to medium size in Southwest China, North China, and Northwest China. The Cretaceous sequences consist primarily of non-marine red terrigenous clastics interbedded with argillaceous strata, also with gypsum and salt. Marine deposits occur, however, in subregions 45 and 46 (Figure II–47) of Northwest China, the Tarim basin. Volcanics of intermediate composition are present in subregion 45 (Figure II–47). Non-marine faunas and floras are abundant. The western portion of the Tarim basin, however, yields rich Cretaceous marine faunas.

Marine Basins and Geosynclines

Marine deposits of the Cretaceous occur mainly in the western Kunlun Shan, in the Himalaya, in the Xizang (Tibet) Autonomous Region, and on Taiwan. These marine deposits consist mainly of neritic to paralic terrigenous clastics interbedded with limestone, with a fauna of ammonites, pelecypods, and foraminifera. Along the Yarlung Zangbo River of Xizang (Tibet) Autonomous Region (subregion 60, Figure II–47), there occurs a thick sequence of flyschoid or turbiditic sandy shale, black shale, conglomerate and radiolarian cherts with pelagic ammonoids. Ophiolitic clastics composed

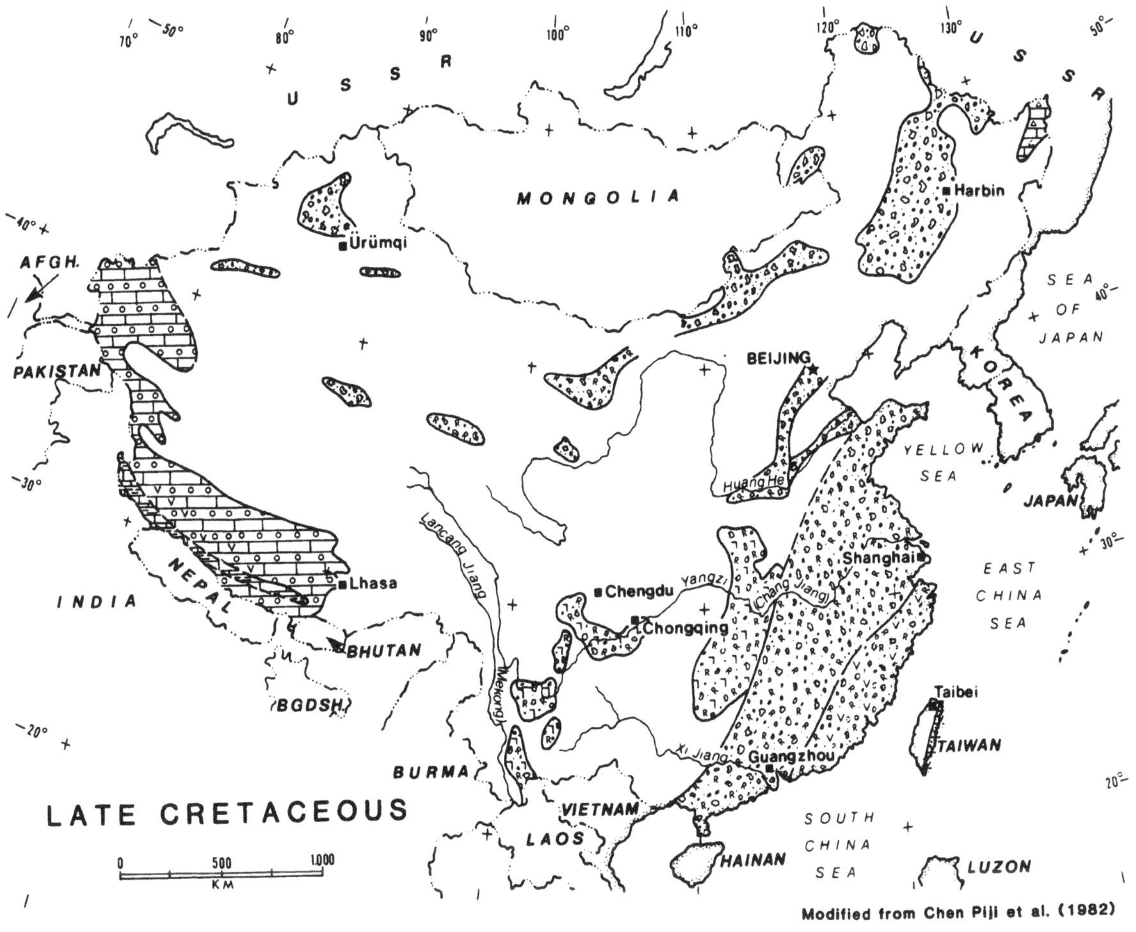

Fig. II-53. Paleogeography and lithofacies map of Late Cretaceous, China.

of cyclothems of sandstone, siltstone, and shale also crop out. In the basal cyclic deposits, siliceous shale and chert occur, with mafic lava and tuff present (Yin Ji-xiang et al., 1983; Girardeau et al., 1984). South of this belt in the Himalaya, the Cretaceous section, composed of carbonates and terrigenous clastics, indicates a shallowing depositional environment southward (Mu Anzi et al., 1973). North of the volcanic Gangdise block, Cretaceous sections of the Qinghai-Xizang Plateau consist mainly of shallow-water carbonate and terrigenous-clastic rocks, but there are local areas of continental plant-bearing redbeds and volcanics. A typically Tethyan Aptian-Albian foraminiferal fauna with *Orbitolina* has been found in several localities in the Gangdise block (Jaeger et al., 1982; Girardeau et al., 1985).

Albian (?) to Cenomanian marine waters transgressed into the western part of the Tarim basin, the southern Tian Shan, and the western Kunlun Shan. Shallow-water argillite, limestone, and mudstone with intercalations of gypsum were formed, with a rich fauna. The Tarim fauna, although Tethyan, is different in certain respects (extra-Tethyan) from the truly Tethyan fauna of the Qinghai-Xizang Plateau. It contains many elements in common with equivalent faunas from correlative rocks in Central Asia (Tang Tianfu et al., 1984). In subregions 55 and 61 (Figure II-47), volcanic rocks present in the Upper Cretaceous were extruded in a submarine environment, and their compositions range from intermediate to silicic.

On Taiwan, the Lower Cretaceous faunas consist of benthos and of numerous pelagic forms, including ammonoids of Aptian-Albian age (T. C. Huang, 1978). Some are in place but others are reworked into Paleocene shale (C. Y. Meng, 1970; T. C. Huang, 1978). Upper Cretaceous has not been found in western Taiwan beneath the coastal plain. Proprietary seismic data from the offshore north of Taiwan, however, suggest that a thick Upper Cretaceous section is widespread, but that, where absent, it is absent because of erosion.

Chen Piji et al. (1982) tabulated groups of fossil biota from both the non-marine and marine environments of the Cretaceous in China. In the non-marine environment, they recorded paleontological data for conchostracans (branchiopods), for ostracods, for charophytes, and for pelecypods. In the marine environment they dealt with ammonoids, foraminifera, brachiopods, and pelecypods. We tabulate their classifications in two tables that follow (Tables II-25 and II-26).

Table II-25. Biostratigraphic sequence of the nonmarine Cretaceous in China

Cretaceous	Stage	Conchostracans: Southwest China	Conchostracans: Southeast China	Conchostracans: Northeast China	Ostracods: South China	Ostracods: North China	Charophytes	Pelecypods	Sporo-Pollen: North China	Sporo-Pollen: South China
Late	Maastrichtian Campanian	?	?	D Zone M Zone	*Talicypridea-Cypridea-Candona* Assemblage		*Lotochara cylindrica-Charites* Assemblage	*Pseudohyria-Sphaerium shandongense* Assemblage	*Schizaeoisporites-Aquilapollenites* Assemblage	*Pterisisporites-Rugubivesiculites-Aquilapollen* Assemblage
Late	Santonian Coniacian Turonian Cenomanian	*Aglestheria* Fauna	*Tenuestheria* Fauna	*Euestherites* Fauna	*Eucypris-Quadracypris-Cypridea* Assemblage		*Porochara-anluensis* Assemblage *Atopochara restricta-A. trivolvis qidongensis* Assemblage	*Acclino Plicatounio-Plicato Trigonoides* Assemblage	*Belmeisporites Tricolpopollenites Tricolporopollenites* Assemblage	*Schizaeoisporites Pterisisporites Poroses* Assemblage
Early	Albian Aptian	*Orthestheria* Fauna	*Cratostracus* Fauna		*Cypridea (Morinia)-C. (Bisulcocypridea)-Darwinula* Assemblage	*Cypridea-"Lycopterocypris"-Candona* Assemblage	*Atopochara trivolvis trivolvis-Flabellochara hangzhouensis* Assemblage *A. trivolvis triquetra-F. hebeiensis* Assemblage	*Trigonioides Plicatounio Nipponaia* TPN Fauna	*Clavatipollenites-Tricolpopollenites* Assemblage *Cicatricosisporites-Clavatipollenites* Assemblage	*Schizaeoisporites* Assemblage *Cicatricosisporites* Assemblage
Early	Barremian Hauterivian Valanginian Berriasian		*Yanjiestheria* Fauna		*Jinguella-Pinnocypridea-Darwinula* Assemblage	*Cypridea-Yumenia-Mongolianella* Assemblage	*Perimneste ancora-Clypeator jinquanensis* Assemblage *F. xiangyunensis-C. zongjiangensis* Assemblage	*Peregrinoconcha-Koreanaia-Danglengiconcha* Assemblage	*Disacciatrileti-Cicatricosisporites* Assemblage	*Classopolis-Cicatricosisporites* Assemblage

Table II–26. Biostratigraphic sequence of the marine Cretaceous in China

			AMMONOIDS	FORAMINIFERA	BRACHIOPODS	PELECYPODS
CRETACEOUS	LATE	MAASTRICHTIAN		*ORBITOIDES, OMPHALOCYCLUS, GLOBOTRUNCANA* FAUNA	*XENOTHYRIS TUILAENSIS* ASSEMBLAGE	
		CAMPANIAN				
		SANTONIAN				
		CONIACIAN			*RECTOTHYRIS SINKIANGENSIS-ORNATOTHYRIS-CARNEITHYRIS* ASSEMBLAGE	*BOURNONIA* FAUNA
		TURONIAN	*MAMMITES, PLACENTICERAS, THOMASITES*			
		CENOMANIAN	*ACANTHOCERAS, CALYCOCERAS*	*ORBITOLINA* FAUNA	*ORBIRHYNCHIA HENPOLAICA-ALITHYRIS SHIQUANHEENSIS* ASSEMBLAGE	
	EARLY	ALBIAN	*DIPLOCERAS, OXYTROPIDOCERAS, TURRILITES*			
		APTIAN				
		BARREMIAN			*SELLITHYRIS MAYUENSIS-PLATYTHYRIS XANZAENSIS* ASSEMBLAGE	
		HAUTERIVIAN				
		VALANGINIAN	*NEOCOMITES, "NEOHOPLOCERAS"*		*PEREGRINELLA* ASSEMBLAGE	*BUCHIA SHUOMOENSIS-B. MANKANANENSIS* ASSEMBLAGE
		BERRIASIAN	*BERRIASELLA OPPELI*			

Type Sections

Northeast China

In the Songliao basin, the type section of Northeast China consists of a thick non-marine sequence typical of the inland basin (Chen, 1980; Figure II–54). In general, lithofacies and thicknesses change from fine-grained clastics, 6,000 to 7,000 m thick, in the central part of the basin, to coarser grained terrigenous clastics only 200 to 500 m thick at the margins. Coarse-grained deposits of the lowest Cretaceous units belong possibly to the category of molasse, and these deposits overlie Jurassic formations with disconformity. Of the sections that lie within the Songliao basin, the Cretaceous section is the best developed. Main components of this section include thick lacustrine deposits in possible deltaic facies, thin layers of richly coquinoid limestone, and locally occurring littoral to marine sediments with glauconite and phosphate. Sedimentation during the Cretaceous tended in general toward cyclicity. Present are: (a) normal cycles fining upward; (b) reverse cycles; and (c) a few less-definite cycles. Throughout the Songliao basin, an unconformity separates the Lower Cretaceous from the Upper Cretaceous, except in the cusp of the trough at the basin's center. The top of the Cretaceous lies unconformably below Tertiary formations. Within the Songliao basin, an unconformity thus defines the Mesozoic-Cenozoic boundary. Volcanic rocks in the Songliao basin are not present in the Cretaceous sequence, only in the underlying Upper Jurassic.

North China

The type section of this region lies in eastern Shandong Province (Figure II–55), along the present Pacific coast. The Lower Cretaceous, disconformable on Jurassic units, consists of lacustrine terrigenous clastics and of volcanics intermediate in composition. The Upper Cretaceous, unconformable below the Cenozoic, consists of red fluviatile and lacustrine deposits with no volcanics.

Southwest China

A marine Cretaceous sequence crops out on Qomolangma Feng (Mt. Everest) in Xizang (Tibet) Autonomous Region (Figure II–56). The Lower Cretaceous, whose fossils include neritic ammonoids, consists of grayish-black shale interbedded with argillaceous limestone and sandstone. In the Upper Cretaceous, the lower part consists of limestone with shale; the upper part of terrigenous clastics with a fauna of brachiopods, pelecypods, and foraminifera.

Paleoecology, Paleogeography, and Magmatism

Northern China and the Songliao Basin

As already stated, the emergence of much of mainland China continued across the Jurassic-Cretaceous boundary, and thereafter persisted through the whole of the Cretaceous. In addition, several major inland basins continued to develop (Figures II–52 and II–53). The Eerduos (Shaan-Gan-Ning) basin, and others which formed previously, lie west of the north-south Da Hinggan-Taihang-Wuling trend (Huang Jiqing, 1978). In contrast, the North China basins complex and the Jianghan basin of East China, newly formed by downwarping during the first part of the Yanshanian movement of the Middle to Late Jurassic lie east of the Da Hinggan-Taihang-Wuling trend. These eastern basins exhibit a Neocathaysian trend, that is, north-northeast

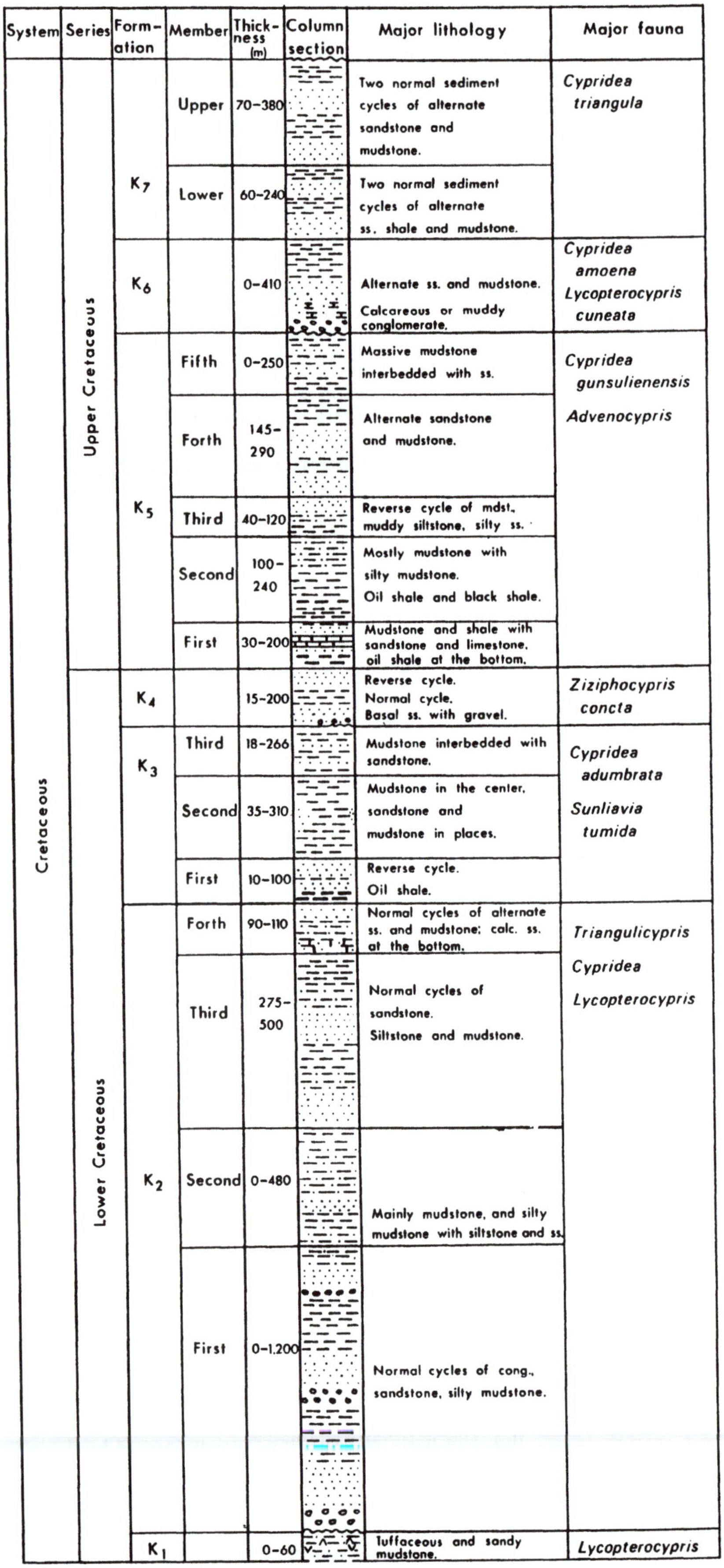

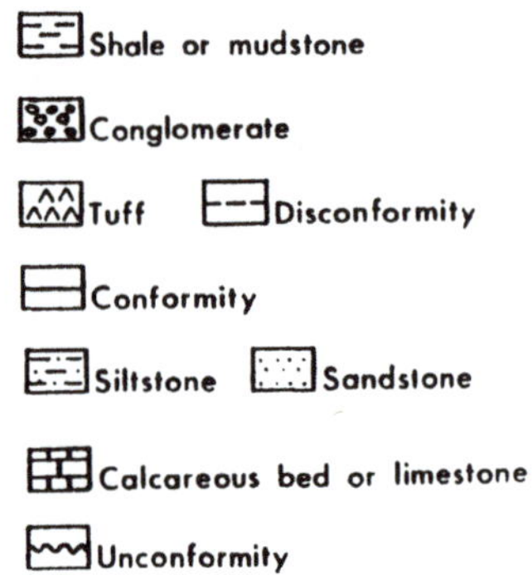

Fig. II-54. Cretaceous stratigraphy of the Songliao basin, Heilongjiang Province, Northeast China (from Chen, 1980).

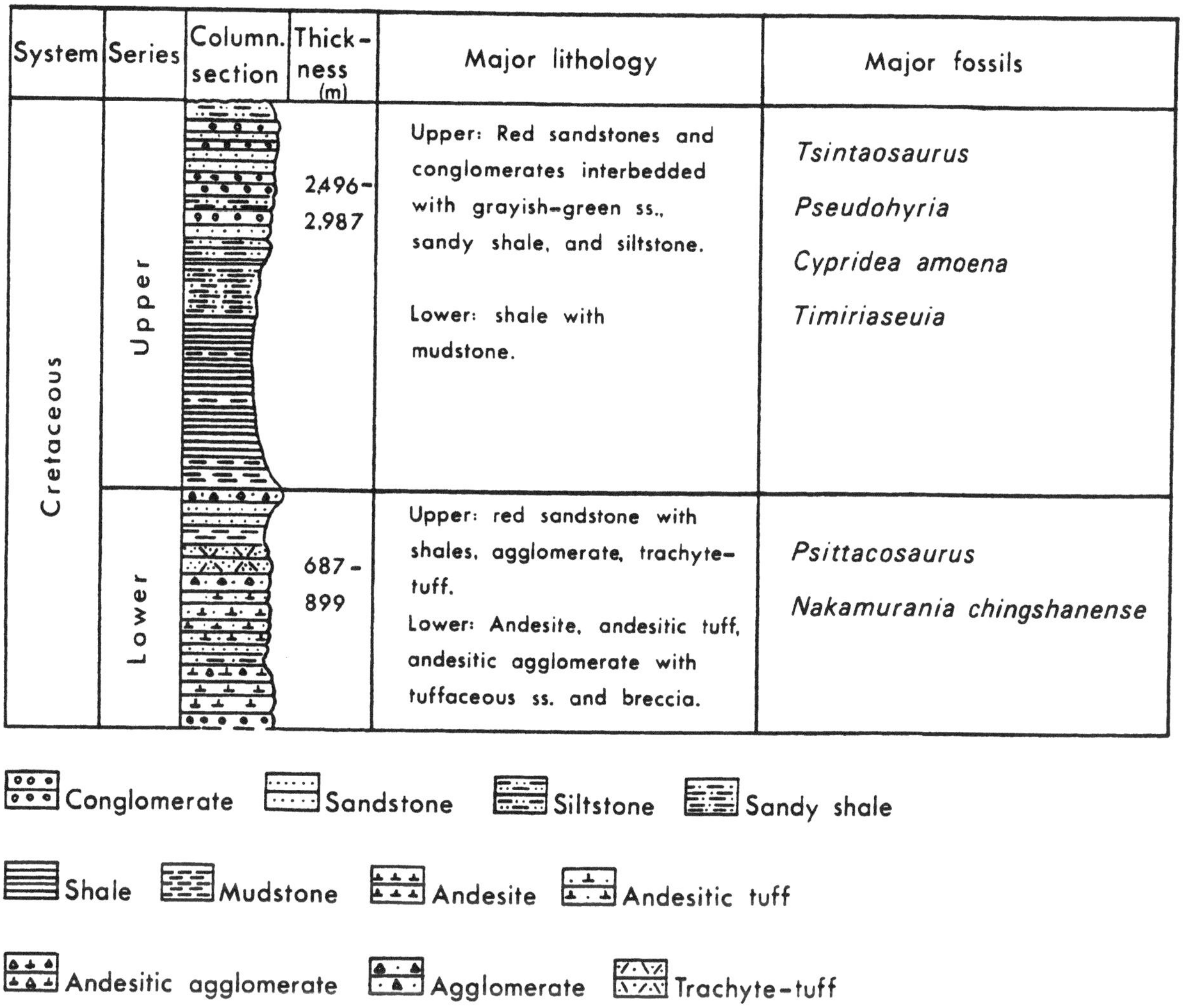

Fig. II–55. Cretaceous stratigraphy in eastern Shandong Province, North China (from Fang et al., 1979).

to south-southwest, and locally, northeast to southwest. Small basins characterized by strong volcanism formed along the Pacific margin, mainly during the Early Cretaceous. In Northwest China, north of the Kunlun Shan, a few basins, untouched by volcanism, trend north-northwest south-southeast, to west-east.

Chen (1980) synthesized the geology of the Songliao basin from the Chinese literature (Hao et al., 1974; Ye et al., 1976; Gao and Zhao, 1976) as well as from the English-language literature (Meyerhoff, 1970, 1975; Meyerhoff and Willums, 1976, 1980). The Songliao basin formed in the Middle to Late Jurassic during the initial stage of the Yanshanian movement. Structure in this basin results from complex horst-and-graben tectonics along a Neocathaysian trend. An initial phase included a small shallow basin with paludal, lacustrine, and fluvial environments, and accompanying volcanism, all during the Middle Jurassic and the Late Jurassic. The volcanism of this phase died out by the Early Cretaceous. As subsidence continued, the basin broadened, to reach a maximum in the early Late Cretaceous. Around the margins of the mature and extensive Cretaceous lake, deltas formed, particularly from the north. Only once in the long continental history did an episode of littoral to marine environment occur, quite probably synchronous with the worldwide Albian to Cenomanian transgression. Toward the end of the Cretaceous the Songliao basin decreased in size and ultimately disappeared.

Three alternate dry and wet cycles characterize the climatic history of the Songliao basin. During dry cycles, when water in the lake was relatively low, several isolated basins formed. Under these conditions, organic matter was not abundant, and biota decreased accordingly. At the same time, the grain of deposited clastics coarsened. During wet cycles, the lake rose and inundated the whole basin. Organic matter than became prolific, biota flourished, diversity of species increased, and deposition resulted in black mudstone and shale, both rich in organic content.

The sketch in Figure II–57 (Fang et al., 1979) indicates a generalized paleogeography of the Songliao basin for Cretaceous time. The environments, governed by depth, range from shallow, through semi-deep, to deep. Shallow waters occupied by far the largest area of the basin.

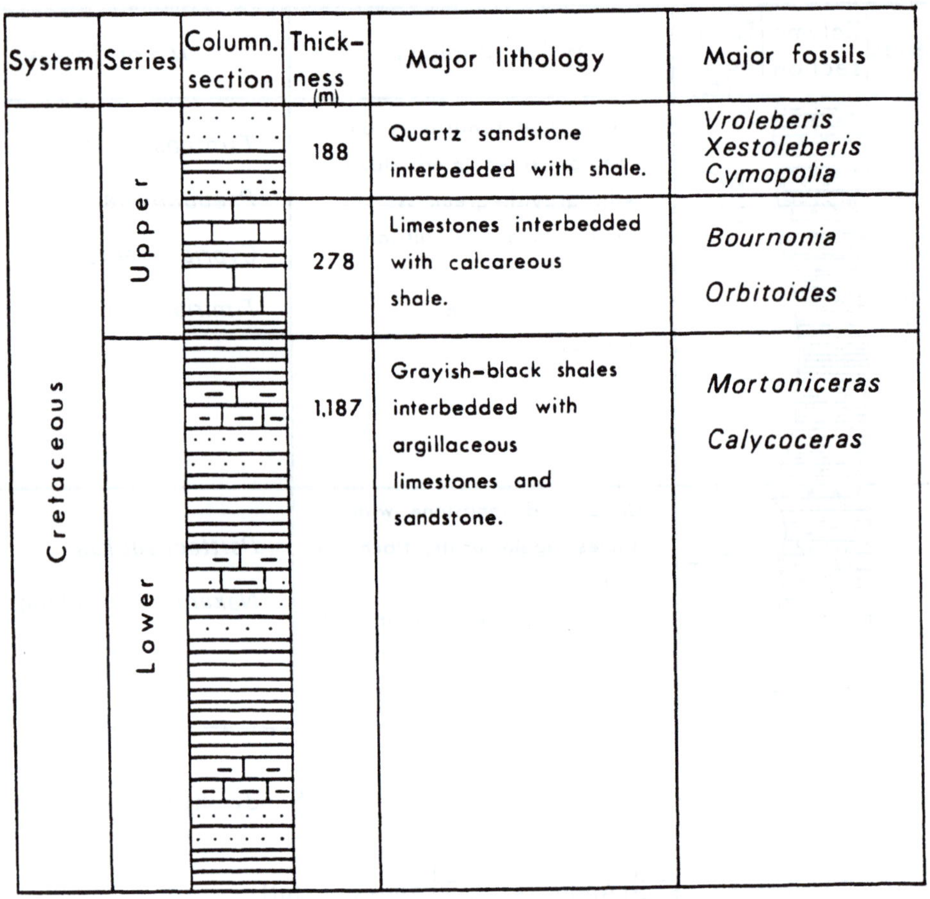

Fig. II-56. Cretaceous stratigraphy on Qomolangma Feng (Mt. Everest) of the Himalaya, Xizang (Tibet) Autonomous Region, Southwest China (compiled from Fang et al., 1979).

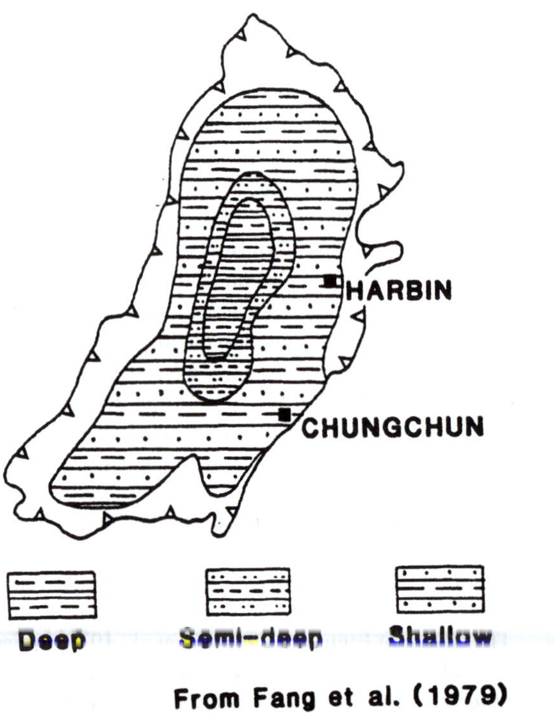

Fig. II-57. Cretaceous paleogeography of the Songliao basin, Heilongjiang Province (from Fang et al., 1979).

The alternate dry and wet cycles documented by Fang et al. (1979) for the Songliao basin illustrated effectively the possible variations of climate on a comparatively local scale. Song Zhichen et al. (1983), however, used palynoflora to deduce a pattern of climate that they believed governed major regions. For the Early Cretaceous they proposed two palynofloristic regions: (1) North China, under humid subtropical conditions; and (2) Central-South China, under arid subtropical to tropical conditions. For the Late Cretaceous they proposed another two palynofloristic regions: (1) Northeast China, under semi-humid subtropical to temperate conditions; and (2) Central-South China and western China, both under arid subtropical conditions.

Southwest China

Even though the marine waters of Tethys proper maintained at least part of their position in Southwest China during the Early Cretaceous, much of extra-Tethys had disappeared, especially from easternmost India and northern Burma. In Southwest China, Chen Piji et al. (1982) showed little more than a marine channel between Lhasa and the Laotian border with Southwest China (Figure II-58). A second, parallel marine connection, however, probably passed through what is now northe-

Fig. II–58. Early Cretaceous: Tethyan and ancestral Pacific marine waters in East Asia.

astern India and Burma. In the Pacific domain, shorelines of the Early Cretaceous sea everywhere lay east of China's present coastlines, except for Taiwan which was covered by ancestral Pacific marine waters. Farther north, parts of Japan were inundated, and the seaway of the Sikhote-Alin geosyncline of the Soviet Maritime Provinces still persisted (Meyerhoff, 1981; Figure II–58). In the Late Cretaceous, the tenuous Tethyan channel of Southwest China may have disappeared completely, although Bannert and Helmcke (1981) and Bender (1983) presented some evidence to show that a part of the seaway may have survived in Burma, especially in what is now the eastern margin of the Indo-Burma ranges (see next paragraph; Figure II–59). Tethys and extra-Tethys lay west and northwest of Lhasa, along the present coastline of southeastern India (especially in the Cauvery basin), and possibly in central Burma. In the Pacific domain, shorelines of the Late Cretaceous did not differ appreciably from those of their Early Cretaceous predecessors. Chen Piji et al. (1982) indicated a slight eastward shift that brought the Late Cretaceous shoreline through the middle of Taiwan, but the absence of Late Cretaceous in deep wells of western Taiwan is probably the result of post-Late Cretaceous erosion, as interpreted from proprietary seismic data.

Although Tethys was reduced in size during the Cretaceous, its marine waters encroached on Central Asia from the south during the Aptian, the Albian, and the Cenomanian to cover the western Tarim basin (Tang Tianfu et al., 1984; Figure II–59). This transgression probably was related to Yanshanian movements in the area of the modern Pamir and Hindu Kush. On the Gangdise block of modern Xizang (Tibet) Autonomous Region, between the Indus-Yarlung Zangbo fault zone and the Tarim block, Aptian-Albian limestones contain abundant *Orbitolina*, a foraminiferal genus known only from northern Tethys, and totally unknown from Gondwana (Jaeger et al., 1982; Yin Ji-xiang et al., 1983). The *Orbitolina*-bearing neritic strata are closely similar to those found along the eastern margin of the Indo-Burma ranges in Burma (Bender, 1983). North of Lhasa, plants are abundant locally, and tree trunks of the Lower Cretaceous are of Boreal species, with well-developed annual tree rings of a presumably temperate climate. South of the Indus-Yarlung Zangbo zone, as well as along it, a narrow trough with flysch is lithologically and faunally similar to the flysch zone of equivalent age in the central Indo-Burma ranges (Bender, 1983). Bannert and Helmcke (1981) suggested that the Cretaceous-Tertiary flysch of the Indo-Burma ranges is coextensive with that in the Yarlung Zangbo Valley, a suggestion seconded by Bender (1983). By inference, the *Orbitolina*-bearing zones north of the Yarlung Zangbo and east of the belt of Indo-Burma ranges with flysch

Fig. II–59. Late Cretaceous: Tethyan and ancestral Pacific marine waters in East Asia.

also are coextensive. Chen Piji et al.'s (1982) work does not support this suggestion, and an intelligent conclusion must await the publication of more detailed stratigraphic information from Burma, northeastern India, and the Xizang (Tibet) Autonomous Region of southwestern China.

Flysch, especially Cretaceous flysch, has been studied carefully in recent years. The Cretaceous flyschoid deposits of the Yarlung Zangbo Valley appear together with extrusives of intermediate to mafic, and ultramafic, composition. The zone of flysch, volcanics, and ultramafics of the Yarlung Zangbo Valley has been described in numerous publications, which include important contributions by Deng Wan-ming (1981), Jin Cheng-wei (1981), Wang and Wang (1981), Wu Hao-ruo (1981), Wang Hongzhen (1983), Allegre et al. (1984), Chang and Pan (1984), Girardeau et al. (1984, 1985), and Gopel et al. (1984). Most of these studies, however, rank as no more than preliminary, and much additional information remains to be gleaned from the area.

Marine Mesozoic of Peninsular India

The geology of equivalent marine strata in southern India bears strongly on any interpretation of southern and south-western China, but the relationship has received scant attention in the published literature. Cretaceous faunas have been described from the Krishna-Godavari-Cauvery basins. These lie along India's southeastern coast, and extend to offshore Sri Lanka (Ceylon). Raju (1968) described a rich Lower and Upper planktonic foraminiferal assemblage that is typically Tethyan, and which flourished in a warm-water environment. This fauna overlies a late Late Jurassic transgressive section, also deposited in shallow water, and it occurs in strata associated with bioherms and an evaporitic lagoon (Ramanthan, 1986; Sastri and Raiverman, 1968). The subsurface geology of this same area recently was reviewed by Kumar (1983), and the complete biotas by Sastri et al. (1977). It is somewhat difficult to envisage a Tethyan marine fauna flourishing off the southern Indian coast at a time when India supposedly still was attached to Antarctica or, at the very least, beginning its 'northward flight to India' (Dewey and Burke, 1973; Johnson et al., 1976, 1978; Tapponier and Molnar, 1977; Molnar et al., 1981). Similarly, Khudoley and Prozorovskaya (1985) showed, on the basis of ammonoid distribution, that India was a faunally part of Asia throughout Mesozoic time.

Faunal studies of marine Paleozoic, Jurassic, and Cretaceous of the Indus Valley of Pakistan (Voskresenskiy et al., 1970) and the western 'slope' of the Indian shield (e.g. Rajasthan; Das Gupta, 1975) show unequivocally that the same problem exists there. The faunas are of warm-water type and are Tethyan (also see Arkell, 1956, p. 606–615). In popular plate-tectonic 'models,' the western coast of India should have been just north of Antarctica and adjacent to some parts of southern Africa and Madagascar.

Vertebrate Fauna

Studies of the vertebrate faunas are even more damaging to currently accepted plate-tectonic 'models.' The fauna is not abundant, is largely fragmentary, and is of Late Cretaceous age (mainly Turonian). On the basis of identified taxa, however, India is most closely related to North America (all families in India are present in North America). Other clear relationships, in decreasing order of importance, are with eastern Asia, South America, Europe, and Africa. India has yielded no families of vertebrates in common with Antarctica and Australia.

Yanshanian Orogenic-Magmatic Activity

In the progression of regional tectogenesis, the second and third stages of the Yanshanian movement took place during the Middle Cretaceous and the Late Cretaceous, and these two stages are approximately equivalent to the early stages of the Laramide movement of North America (Fang et al., 1979). In China, the strength of the Cretaceous Yanshanian stages and of their accompanying magmatism was less than those of the first (Jurassic) stage. Yanshanian intrusives of the Cretaceous occur mainly in East China, but they extend westward as far as the Xizang (Tibet) Autonomous Region and western Yunnan Province. They intrude the entire pre-Cretaceous section; they are overlain either by Upper Cretaceous or Tertiary; and they register isotopic ages of 130 to 80 m.y. (Compilation Group, 1976; Zhu Ming, 1989). In East China and Central-South China, intrusives include a variety of types. Among them are biotite granite, granodiorite, monzonite granite, and several hypabyssal and alkalic rocks, generally in small to medium-sized bodies. Elsewhere, the habit of Cretaceous intrusives differs. On the Liaodong Peninsula of eastern Shandong Province, in the eastern Qin Ling, and in eastern Xizang (Tibet) Autonomous Region, as well as in the Karakorum, intrusive bodies consist mainly of granite and granodiorite. These bodies, of large size, occur in some places together with small intrusive bodies of hypabyssal and alkalic rocks.

CENOZOIC

The Cenozoic (Tertiary and Quaternary) in China contains a considerable number of rock types. These, in many areas, form a cover of non-marine deposits whose facies include varieties such as fluvial, piedmont, loess, laterite, cavern, glacial, intermontane, and volcanic. In parts of Xizang (Tibet) Autonomous Region, Xinjiang Uygur Autonomous Region, and Qinghai Province, on the west, and on Taiwan, Hainan Island, and certain coastal segments on the east, the facies of Cenozoic deposits is marine, or with marine tongues.

Tectonic belts of the Cenozoic, similar to those of the Mesozoic, show meridional trends. The Himalayan movement, the latest to affect China, lasted through the middle and late Tertiary. Effects are most prominent in the region of the Himalaya themselves, but as far east as the Pacific domain the Himalayan movement affected also Taiwan. Widespread uplift of the Himalayan movement brought on a withdrawal of marine waters almost completely from the present area of China, and thus laid the geomorphological foundation of existing geography (Compilation Group, 1976).

TERTIARY SYSTEM

The Tertiary System, practically synonymous with the Cenozoic in view of the fact that Tertiary time is more than 30 times longer than Quaternary time, exhibits a variety of rock types such as that already mentioned for the Cenozoic as a whole. Tectonics and paleogeography, for the same reason, need no further introductory remarks. In Central-South China and East China, deposition appears to have been continuous or seemingly continuous across the Cretaceous-Tertiary boundary. For the most part, however, a hiatus occurs at the Cretaceous-Tertiary boundary within China and below its marine waters. The Tertiary-Quaternary (Pliocene-Pleistocene) boundary also contains a hiatus in some regions. In Table II–27, drawn for the Cenozoic as a whole, the Tertiary epochs, the elements of their faunas and floras, and their equivalent European stages, all appear (Fang et al., 1979).

Distribution

Pei et al. (1963), the Compilation Group (1976), and Fang et al. (1979), recognized seven Tertiary regions in China, each defined by its depositional environment and its geological structure. We describe these briefly, together with additional Tertiary provinces or subprovinces that we have studied mainly from subsurface data.

East China Coastal Plain

In the eastern coastal plains, the four large lacustrine or fluvio-lacustrine basins are: (1) Songliao; (2) North China; (3) Jianghan; and (4) Jiangsu (I, Figure II–60). All exhibit a complex pattern of fault blocks, some overlain by Tertiary sequences that consist mainly of fluvial and lacustrine terrigenous clastics with gypsum, oil shale, and/or basalt. The predominant fossils are freshwater forms, with brackish and/or marine forms

Table II-27. Cenozoic divisions in China and correlation with Europe (compiled from Fang et al., 1979)

Erathem	System	Series	Major fossils	European stages
Cenozoic	Quaternary	Holocene	*Homo sapiens*	
		Pleistocene	*Homo erectus pekinensis* (V)	Wurmian Riss-Wurmian Rissian Mindel-Rissian Mindelian Gunz-Mindelian Gunzian
	Tertiary	Pliocene	*Ilyocypris* (O) *Candoniella* (O) *Cyprinotus* (O) *Candona* (O)	Astian Plaisancian Pontian
		Miocene	*Alnus* (P) *Salix* (P) *Secuoia* (P) *Stephanocemas* (V)	Sarmatian Vindobonian Burdigalian Aquitanian
		Oligocene	*Dongyingia* (O) *Huabeinia* (O) *Cyprinotus* (O)	Chattian Stampian Sannoisian
		Eocene	*Liminocythere* (O) *Cypris* (O) *Eucypris* (O)	Ludian Bartonian Auversian Lutetian Cuisian Sparnacian
		Paleocene	*Prionessus* (V) *Sphenopsalis* (V) *Palaeostylops* (V) *Bemalambda* (V)	Thanetian Montian

V= Vertebrates O=Ostracods P= Plants

correspondingly sparse. In three of the above four basins, Tertiary sequences are better developed than those of their Mesozoic predecessors. In the Songliao basin, on the contrary, Tertiary thickness is much less than Cretaceous thickness (Figure II-54), due to a decrease of basinal area that followed gradual uplift after the latest stage of the Yanshanian movement. In addition to the large basins, several medium to small structural basins formed in East China. Their Tertiary sequences, mainly red fluviolacustrine deposits, in general are incomplete. In addition, their Tertiary sequences show great variations in lithology and thickness.

Nei Monggol Peneplain

Several basins of medium to small size developed in the peneplain of central and western Nei Monggol Autonomous Region (II, Figure II-60). The Tertiary section of these basins consists of thin layers of grayish-green sandstone and mudstone with abundant vertebrate fossils. The basins in this region are structural, for the most part, but some are erosional, as a result of episodic

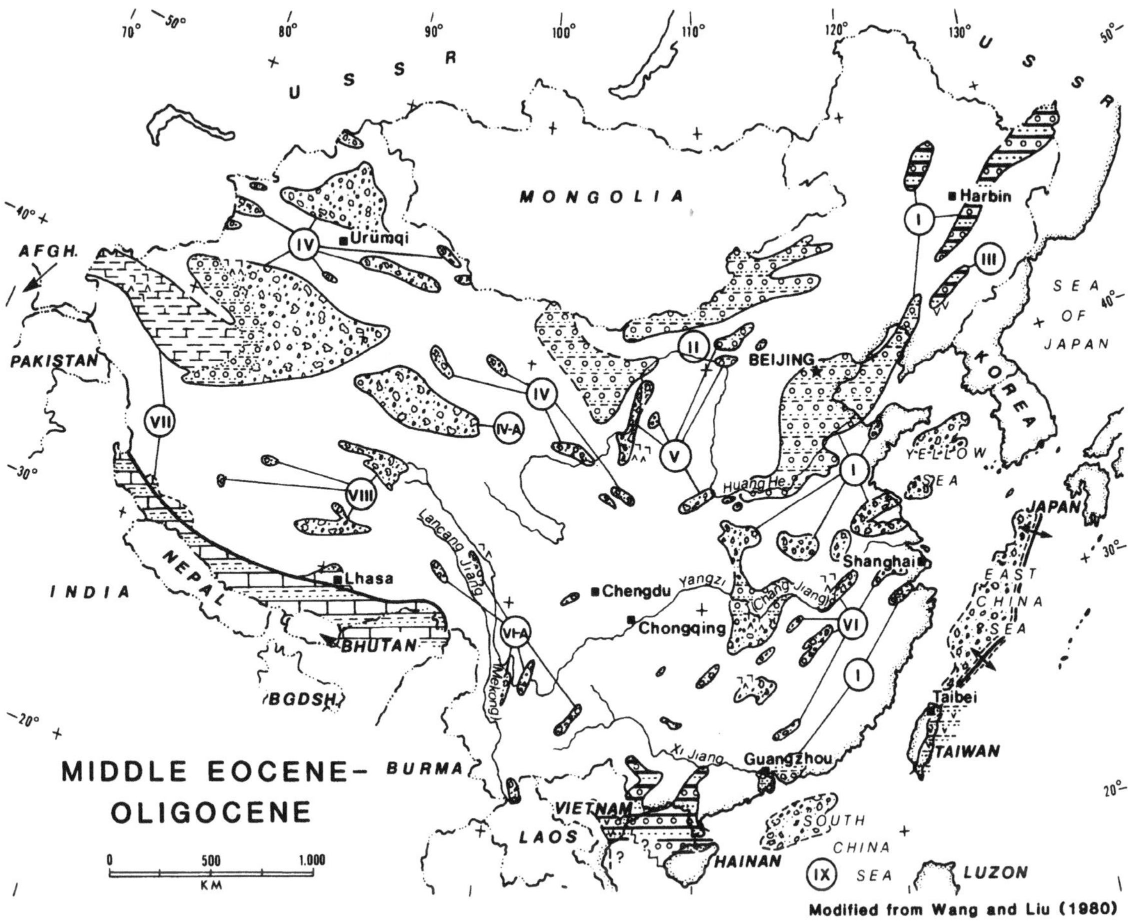

Fig. II-60. Paleogeography and lithofacies map of middle Eocene-Oligocene, China.

movement during different epochs of the Tertiary. The well-known Hannobar basalt, widely distributed across the eastern part of Nei Monggol Autonomous Region and the western part of Liaoning Province, is interlayered with sandy shale, oil shale, and marl. Miocene vertebrates include *Trilophodon* sp. and *Stephanocemas thompsoni*. The age of the Hannobar basalt may range from Paleogene to early Neogene (Compilation Group, 1976).

Eastern Liaoning Fault Basins

Several small faulted basins lie within the upwarped zone of eastern Liaoning Province (III, Figure II-60). The Tertiary (largely Eocene) section within these small basins consists of basalt, coal, oil shale, and mudstone, with great lateral variations in both lithology and thickness. The Fushun basin, with its commercial oil-shale deposits and its high-grade lignite, is the most famous of these (Han and Lang, 1980).

Northwest China Piedmont

The Tertiary deposits of this region consist primarily of piedmont talus and of lacustrine redbeds within large downwarps such as the Junggar, Qaidam, and Tarim basins of Northwest China (IV, Figure II-60). Initial accumulations in these basins took place within down dropped fault blocks of the piedmont. Around these basins, mountains rose during the Yanshanian movement (Pei et al., 1963) and these contributed detritus to the piedmonts within their periphery. As quieter conditions prevailed, inland lakes filled the basins, and sediments of finer grain succeeded the detritus that now forms the base of Tertiary sections. Intermittent uplifts of local extent provided fanglomerates and coarse-grained stream deposits that spread laterally as they came down the margins into the basins.

The Qaidam basin (IV-A, Figure II-60) is unique, however, among the continental basins of Mangnai depression in the northwestern part of the basin. More than 6,000 m was drilled in one well without reaching the base of the Miocene. Massive salt deposits were precipitated during two periods of time. The so-called 'First Salt Epoch' spanned the Pliocene and the early Pleistocene (Li Jijung et al., 1979; Song and Liao, 1982). These deposits, 2,800 m below the surface in the center of the Mangnai depression (Song and Liao, 1986), comprise one of the largest salt deposits in the world outside of the Cambrian of Siberia and the Devonian of western Canada (>60 billion metric tons). Salts include gypsum, halite, mirabilite, blodite, epsomite, potash, bischofite, celestite, and borates (Du Bomin et al., 1984).

On the Qinghai-Xizang Plateau, several sizeable Pliocene lake basins formed. The *Hipparion* vertebrate fauna and the lush tropical to subtropical plants suggest that the average elevation of the plateau was below 1,000 m (Li Jijun et al., 1979).

North-Central China Intermontane Basins

Many intermontane lacustrine basins, medium to small, now exist in north-central China, together with a lesser number in western China and southern China (V, Figure II–60). These basins formed with between folded ranges, or in areas surrounded by folded ranges, during the Yanshanian movement. Most of these basins strike northeast-southwest (Cathaysian direction). They include the so-called Shanxi (Fenhe-Weihe) and Hetao-Yinchuan graben systems (Zhang Buchun et al., 1985). The Tertiary section within such basins consists mainly of red lacustrine terrigenous clastics in the Eocene, with local deposits of gypsum at the centers of basins. Lateral gradation of sections into fluvial facies, or into piedmont facies, occurs around margins. Similar changes of facies take place in post-Eocene deposits.

Central-South and Southern Southwest China

These basins are Neocathaysian, and they contain some Upper Jurassic, Cretaceous, and Paleogene continuous sequences (VI, Figure II–60). Paleontological data are of a quality sufficient to determine the Cretaceous-Tertiary (Mesozoic-Cenozoic) boundary. The Tertiary sequence, formerly known as 'The New Red Beds' (Compilation Group, 1976), grades upward from basal red breccia to feldspathic and micaceous sandstones, cross-bedded in their middle, then upward from these to oil shale, marl, or to gypsum and other salts near the top. The whole represents a normal cycle of Tertiary deposition for this area. Tertiary sequences, uniformly thick, formed in environments as varied as fluvial, lacustrine, intermontane, and piedmont.

In southern Southwest China, specifically in Yunnan Province, several narrow elongated basins (VI–A, Figure II–60) lie parallel to the north-south structural grain of Himalayan folds and faults. Most deposits are fluviatile and lacustrine, but the presence of thick Neogene coal measures indicate intermittent marshy environments.

Southwest China (Xizang-Himalaya)-Northwest China (Tarim)

Early Tertiary marine waters of Himalayan Tethys extended onto the Gangdise block, north of the Indus-Yarlung Zangbo fracture zone (VII, Figure II–60). The waters could not reach the present Tarim basin from the south, because the northern part of the Qinghai-Xizang Plateau lay above sea level. Invading waters consequently entered the western Tarim basin from the west, and they also occupied a large area of the western Kunlun Shan (VII, Figure II–60).

In southern Xizang (Tibet) Autonomous Region, the section consists of nummulitic limestone, sandstone, shale, and occasional submarine volcanics. In the Yarlung Zangbo Valley, flyschoid turbidites are present. In the western Tarim basin, the sequence consists of limestone, marl, and argillite, with a rich marine fauna, and also gypsum. The evaporitic interval with gypsum reaches 300 m in thickness in at least one place.

Northern Qinghai-Xizang Plateau

Several elongated east-west basins controlled by faults and parallel with regional structural and depositional strike, are present on this part of the lofty plateau (VIII, Figure II–60). These basins, such as the Lunpola basin, contain up to 4,000 m of fanglomerate, redbeds, and lacustrine shale, some rich in organic matter. The length:width ratio is high, commonly 10:1 or more.

Pacific Domain

This domain, which includes all of the continental shelf east of Bohai Bay and south of Korea, extends to Vietnam, and includes Taiwan (IX, Figure II–60). Sufficient evidence has accumulated from offshore drilling, from seismic studies, and from field work on Taiwan to indicate the presence of distinctive subprovinces. Data are insufficient, however, to define the precise area covered by each.

The Tertiary of the entire Pacific domain is divisible into a lower sequence and an upper sequence separated by an unconformity. The unconformity occurs close to the base of the Aquitanian in the South China Sea-Taiwan area, and is loosely referred to as the pre-Miocene unconformity. In coastal regions and across much of the shelf, the unconformity is a gentle one, with an angular divergence generally less than 2° to 3° where lower Tertiary rocks are present. In eastern Taiwan, and along the continuation of the eastern Taiwan foldbelt to Japan (the Diaoyu Dao, or Taiwan-Sinzi, foldbelt), the discordance is much more pronounced, locally with high angles of divergence.

The Paleocene-Oligocene sequence, in general, is restricted to north-northeast-trending (Neocathaysian trend) grabens (IX, Figure II–60) that were especially active during the early part of the Himalayan orogenic cycle. These grabens, some of which had formed in the late Mesozoic, began to subside rapidly in the Eocene. Subsidence slowed in the later Oligocene, and graben formation soon ceased. Up to 4,800 to 6,500 m of largely non-marine fanglomerate, deltaic, and fluviatile sediments were deposited with redbeds, lacustrine shale rich in organic matter, and, in some places, gypsum and trona (Li Desheng, 1984). In the basal part, these sediments contain many mafic to intermediate continental volcanics.

The equivalent rocks of eastern Taiwan and the

Diaoyu Dao foldbelt consist of rocks related to island arcs, with turbidite, graywacke, volcanics, and effusive mafic volcanics, largely submarine, many spilitic, and some andesitic to trachytic. It is not known whether this sequence of volcanic-arc type formed beneath the present continental shelf and upper slope south of Taiwan, as it did north of Taiwan. An extension of the island-arc, however, did reach the Philippines.

Intergraben highs were the principal sources for debris deposited in the grabens. Relief on these highs was generally subdued, inasmuch as one or two notably thin marine intervals (identified in the Oligocene sequences of the Bohai Bay and Subei basins) extend onshore more than 100 km up the Huang He (Yellow River) Valley. During strong epeirogenic pulses, however, some of the grabens sank to quite considerable depths, perhaps 1,000 m or more, and turbidites of freshwater origin were deposited as deep, freshwater cones or fans. At times, the intervening horsts, uplifted sharply, gave rise to sudden local influxes of coarse unsorted debris.

Except for the volcanic-arc sequence of Taiwan and the Diaoyu Dao foldbelt, marine influences are scarce in the lower Tertiary sequence of northeastern coastal China. The presence of important thicknesses of marine Eocene and Oligocene in the South China Sea area is revealed by drilling. Seismic data show that Cathaysian structural trends extend into the deeper parts of the South China Sea, and it is not possible now to state just where the edge of the continent lay off southeastern China during the early Tertiary. At least one minor, yet extensive, Oligocene transgression has been recorded in the San Shui basin, upstream from Guangzhou (Canton). Conditions that prevailed in northeastern China, therefore, seem also to have existed in southeastern China.

The upper Tertiary section everywhere covers the pre-Miocene grabens and the intervening horsts, including the western Taiwan coastal plain. In that area, the Miocene (with coal measures in the west, marine strata in the east) directly overlies folded early Eocene, Paleocene, and older eugeosynclinal rocks, or rocks of island-arc type. In the South China Sea, the upper Tertiary cover overlies granites that range downward in age from 51 Ma to 185 Ma (early Eocene to early Middle Jurassic). Some of these granite bodies give ages of 90 to 95 Ma (Middle Cretaceous, Cenomanian). In the Beibuwan at the north end of the Gulf of Tonkin, the upper Tertiary overlies Early Carboniferous carbonates and other pre-Tertiary sedimentary and metamorphic rocks.

Upper Tertiary and Quaternary thickness in general ranges from 1,500 to 2,500 m, and it is greatest above the pre-Miocene grabens and their intervening horsts. East of the western Taiwan coastal plain, estimated total thickness of the Eocene through Plio-Pleistocene section is as much as 10,000 m, but strong deformation of the rocks prevents a truly satisfactory estimate. East of the Diaoyu Dao foldbelt in the Okinawa trough, the thickness of the Neogene and younger section is 3,000 to 5,000 m, locally more. Letouzey and Kimura (1985) reported thicknesses possibly as great as 8,000 m at the northern end of this basin in a section that is partly volcanigenic.

The Miocene and younger section is largely marine on and east of the Diaoyu Dao foldbelt and on the western Taiwan coastal plain. Some marine influence extended onto the East China Sea shelf, but did not reach farther west into the Yellow Sea, the North Yellow Sea (Korea Bay), or the Bohai basin complexes. In the South China Sea, marine influence extended progressively landward from the beginning of the Miocene until the present. In the Yinggehai East basin, south and southeast of Hainan Island, the entire Miocene and younger section is marine, some of it deposited in deep neritic waters. Onshore, the Maoming basin is filled with lower Tertiary (mostly Eocene) oil shale and related lake beds, into which basaltic sills have intruded. The oil shale is of the same age as that in the Fushun basin, far to the north. Like the Fushun shale, the oil shale of this area is mined commercially. In the San Shui basin near Guangzhou (Canton), 300 km northeast of Maoming, dark-gray to gray-brown non-marine shale, rich in organic matter, alternates with reddish and variegated mudstone, sandstone, and with conglomerate, which is angular, poorly sorted, and argillaceous. Tuffaceous beds and basaltic sills also are present.

The Song Koi (Hang Ha, Red River) fault zone, a southward extension of the Jinsha Jiang fault zone of eastern Xizang (Tibet) Autonomous Region, strikes southeast from Vietnam, from the northern side of the Song Koi deltaic plain to the South China Sea, passing just west of Hainan Island. The Tertiary history of the southwestern side of the Song Koi fault zone differed greatly from that of the Cathaysian trends of southern China.

North and east of the Song Koi fault zone, all structures strike north-northeast to north-east (Cathaysian trend); drape anticlines related to basement highs are common; rollover structures are present in post-Oligocene burial; and the post-Oligocene section is omnipresent, with a rather uniform lithology except in marine depocenters, and with fairly constant thickness, except above buried lower Tertiary grabens. Depositional 'thicks' do overlie each lower Tertiary depocenter.

In contrast, south and west of the Song Koi fault zone, the Tertiary section consists mainly of Miocene and younger strata. Structures strike north-northwest: closures, not as abundant as closures on the other side of the fault zone, display less relief; lower Tertiary is continental, volcanic, and comparatively thin, commonly a result of pre-Miocene erosion; marine conditions extend well onshore; and growth faults are present mainly in the early to middle Miocene section. The Miocene rocks are thickest in the Hanoi graben, whose northern margin is the Song Koi (Red River) fault system. Although parallel to subparallel with pre-existing fold structures of the 'Variscan' (?) tectogenesis and the Indosinian tectogenesis, the present Song Koi fault system belongs mainly to the early and middle Miocene,

System	Series	Formation	Member	Thickness (m)	Column section	Major lithology	Major fossils
Q							
Tertiary	Miocene	Minghuazhew (Te6)		600–1,000		Yellow and grayish mudstone and fine sandstone.	*Ilyocypris errabundis* *I. manasensis* *I. dunschanensis*
		Guantao (Te5)		300–900		Gray-white fine sandstone, grayish-green siltstone with mudstone. Sand and conglomerate.	*I. gibba* *Candoniella albicans*
	Oligocene	Dongying (Te4)	Third	0–300		Red mudstone with sandstone.	*Dongyingia flori nodosa*
			Second	200–300		Alternate variegate mudstone and sandstone.	
			First	150–800		Gray mudstone.	
		Shahejie (Te3)	Forth	200–400		Gray mudstone with oil shale, biogenic limestone and dolostone.	*Huabeinia* *Liratina tuozhuangensis*
			Third	100–250		Alternate variegate sandstone and mudstone.	
			Second	300–400		Gray mudstone with sandstone	
				400–600		Gray to dark gray mudstone.	
				100 150		Gray mudstone with sandstone, oil shale.	
	Upper Eocene–Lower Oligocene		First	100–150		Gray mudstone with biogenic ls., dolostone and oil shale.	
				100–300		Bluish-gray mudstone with gypsum with rock salt.	
				150–500		Red mudstone.	
	Eocene	Kongdian (Te2)	Third	300–500		Alternate red ss. and mudstone.	*Liminocythere wexianensis* *Eucypris wutuensis*
			Second	500–600		Gray mudstone with carbonaceous shale and coal.	
			First	300–500		Red ss. mudstone. Bottom: conglomerate.	

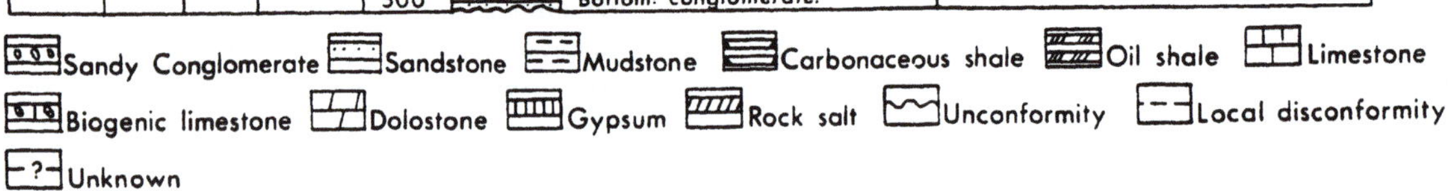

Fig. II-61. Generalized Tertiary stratigraphy in North China (from Hou et al., 1978, and Fang et al., 1979).

and it continues active. The Miocene and younger section is 5,000 to 7,000 m thick in the center of the Hanoi graben, which extends more than 300 km northwestward from the coast of the Gulf of Tonkin. The rocks are deltaic, fluvial, and coal-bearing in the central and western parts of the Song Koi Valley, but in the lower part of the valley, littoral and even shallow-water marine strata predominate except in the middle Miocene and part of the upper Miocene, which are coal-bearing (Skorduli et al., 1983). Basement below the Tertiary sequence consists of shallow-water miogeosynclinal strongly folded and thrust-faulted Silurian through early to middle Mesozoic largely marine rocks.

Type Sections

North China

A composite lithostratigraphy and chronostratigraphy of the Tertiary section in the north China basin appears in Figure II-61 (after Hou et al., 1978; Fang et al., 1979). This composite sequence contains mainly red, gray, and yellow medium-grained to fine-grained lacustrine terrigenous clastics with gypsum, rock salt, oil shale, and limestone. Paleocene formations are absent, and within the lower part of the Eocene a basal conglomerate occurs. Oligocene strata, well developed and widely distributed, contain a few thin marine layers, but because of sparse fossil content, and a lack of indicative taxa, the lowest part of the Te_3 division appears in Figure II-61 as upper Eocene-lower Oligocene. If the Eocene is present in Te_3, the Eocene-Oligocene boundary is gradational. Sparse fossils and lack of indicative taxa continue upward into Miocene and Pliocene strata. These two epochs, accordingly, remain undifferentiated.

In the coastal region of Bohai, unconformities appear to lie between: (1) the Mesozoic and the Cenozoic; (2) the lower Tertiary and the upper Tertiary; and (3) the Tertiary and the Quaternary. Taxa of the early Tertiary (or late Mesozoic) include gastropods (Yu et al., 1978), ostracods (Hou et al., 1978), foraminifera (He and Hu, 1978), charophytes (Wang et al., 1978), pollen and spores (Song et al., 1978). Most are non-marine, a few are brackish, and others are sublittoral to inner neritic.

In South China, Southwest China, Northwest China, Central China, and central Nei Monggol Autonomous Region, Tertiary basins have yielded Paleocene mammalian faunas in relatively recent discoveries. These should provide unusually valuable data for study of the origin and the evolution of mammals, as well as for reconstruction of the Cenozoic history of China.

In the Fushun area of Liaoning Province, Northeast China, a partial section of the Paleogene occurs in which deposits of oil shale and coal occur together with basalt, as follows:

		(m)	
Eocene (Fushun Group)	Green mudstone	137 to 600	Fossils include plants, insects, conchostracan, branchiopods gastropods, and fish.
	Brown oil shale	14 to 190	
	Major coal bed	1 to 115	
	Tuff	65 to 115	
	Basalt and tuff	8 to 193	
Paleocene	Basalt en coal	45 to 125	
	Basalt	35 to 223	

Taiwan

Northwestern Taiwan furnishes another type section of the Tertiary of China (Figure II-62). The sequence of this region consists principally of marine terrigenous clastics and of coastal coals interbedded with basalt, with other volcanics, with slate, and with siltstone. The Paleocene reportedly undergoes great lateral changes in lithofacies and in thickness. The Eocene and the Oligocene contain terrigenous clastics and rocks of low metamorphic grade. The Miocene shows good development and wide distribution, also three sedimentary cycles in northern Taiwan. Each cycle contains three groups of coal measures and intervening marine deposits. Chou and Lin (1974) correlated the lower cycle with the Aquitanian; the middle cycle with the Burdigalian; and the upper cycle with the Helvetian, Tortonian, Sarmatian and Pontian (we suggest that the Pontian is dubious as a stage of the Miocene). In the type area, a fault separates the Miocene from all older rocks, and several local disconformities break the continuity of the upper Miocene. Another disconformity intervenes between the Miocene and the Pliocene, the latter mainly a thick layer of shale with change of lithofacies to sandstone. The Pliocene of southern Taiwan includes facies that vary from coastal and deltaic to open marine.

In west-central Taiwan, in the Peilang area, Oligocene, Eocene, and Paleocene sections are penetrated successively by drilling beneath the Miocene (Yuan et al., 1985). The Paleocene and the Eocene are represented mainly by volcanics and volcaniclastics. The Oligocene is present locally and in marine facies. Also, from the same area Cretaceous and older rocks are reported. They are in sedimentary facies (clastics and carbonates), metamorphosed in various degrees.

Offshore from northern Taiwan, drilling has shown the oldest units penetrated to consist of arkose, sandstone, sandy limestone, claystone, siltstone, and graywacke, together with rare tuff. The lithofacies of these units appears to be mainly continental, but a few marine tongues contain Lower Cretaceous fossils.

Onshore, in angular unconformity above the Lower Cretaceous, lies middle to late Paleocene siltstone that grades upward into fine-grained sandstone, laid down largely in a neritic environment. These middle to late Paleocene deposits are for the most part lithified and fractured, but they are missing in the subsurface section

System	Series	Stage	Member	Column section	Thickness (m)	Major lithology	Major fossils F=Foraminifera Pl=Plant P=Pelecypod W=Worm
Tertiary	Pliocene				800–2,300	Mostly shale, locally grades to sandstone.	*Ostrea gigas* (P) *Globorotalia inflata* (F) *G. truncatulinoides* (F) *Pulleniatina obliquiloculata* (F) *Psendorotalia* (F)
					200–300	Dark shale, local siltstone.	*Pseudorotalia yabei*, *Eponides repandus* (F) *Bolivinita quadrilatera* (F)
	Miocene	Upper	2		500–950	Upper: sandstone. Middle: shale. Lower: ss grades to shale.	*Operculina ammonoides* (F), *Textularia* (F) *Bigenerina nodosaria* (F), *Pseudorotalia* *Globorotalia menardii tumida*
			1		500 1,200	Quartz sandstone interlayered with thin shale and coal, local basalt and volcanics.	Plant fossils, mollusks
		Middle	4		150-200	Ss. with calcareous ss.	*Operculina Ammonoides*
			3		350–400	Shales with siltstone.	*Globorotalia foshi barisanensis* (F)
			2		300–400	Subgraywacke grades to shale.	*Operculina bartschi multiseptata* (F) *Miogypsina inflata* (F)
			1		400–700	Alternate ss. and shale, and grades to shale, with 5–7 layers of coal seams.	*Ficus filiaefolia* (Pl) *Cassia* sp.
		Lower	3		300–500	Graywacke or subgraywacke and shale, local basalt, agglomerate, tuff, limestone.	*Operculina Ammonoides*, *Ditrupa* (W) *Amussiopecten* (P), *Gaudryina* (F)
			2		5–50	Alternate sandstone and shale, with 3 or 4 layers of coals, local basalt volcanics.	Plant fossils, mollusks
			1		500	Quartz sandstone, locally grades to siltstone and shale, with coal.	
				FAULT			
	Oligocene	Upper			340	Quartz sandstone with black shale and coals.	Plant fossils
		Lower			400–500	Black slate and shale with sandstone and glauconite.	
	Eocene	Upper			500	Quartz ss. interbedded with local coal, tuff, shale, and porphyrite.	
		Lower			1,000	Alternate black slate and quartz sandstone.	
				?			

C

Fig. II–62. Generalized Tertiary stratigraphy of NW Taiwan (from Chou, 1974; and Ho, 1975).

offshore. Where present they are overlain by late Paleocene to early Eocene inner to middle neritic shale and markedly fine-grained sandstone, minor limestone, and a few volcanics with flows of andesite and basalt, also ash.

Early Eocene and older sequences give way to early-to-late Eocene sandstone, siltstone, and shale, laid down in open neritic waters, all separated from younger rocks by an angular unconformity similar to that between Oligocene and Miocene in the North China basin. Data from offshore northern Taiwan suggest that late Oligocene deposits rest directly on the Eocene unconformity, and that this unconformity differs from that between the Oligocene and the Miocene in the North China basin. The case remains open, however, until regional paleontological studies can offer more definitive evidence. In the known succession of offshore northern Taiwan, the sequence of the late Oligocene and the early Miocene consists of coarse-grained to medium-grained sandstone interbedded with shale and minor carbonates, all deposited in littoral to shallow marine waters. As early Miocene deposits grade upward to late early Miocene and middle Miocene deposits their facies change from littoral to non marine. The strata become carbonaceous, and they consist of sandstone, claystone, and shale. Sandstone, coarse-grained in the basal portion of the sequence, becomes fine-grained near the top. A disconformity separates the middle Miocene from a fluvial-deltaic sequence that occupies the interval from late Miocene through Holocene, with sandstone, sand, claystone, clay shale, siltstone and coal. Most of the claystone and the shale is carbonaceous, in keeping with the environments that produced actual coal in this sequence.

Paleoecology, Paleogeography, and Magmatism

Marine waters in China continued to occupy a limited area during the Tertiary. As in the Cretaceous, a Tethyan domain dominated the west, and a Pacific domain characterized the east. In western China, marine waters entered the western Tarim basin from seaways that lay across Soviet Central Asia. In Southwest China, marine conditions persisted into the Eocene in the Himalayan realm. On the east, Taiwan, Hainan Island, and parts of coastal China experienced modest marine incursions within the Pacific domain (Pei et al., 1963). A continental volcanic belt developed on the Qiangtang block – a belt several hundred kilometers long. Although Gansser (1980) referred to this belt as being of 'subrecent age,' radiometric dates (Basu et al., 1984) are 28 Ma (Oligocene).

During the Eocene and the Oligocene, ancestral Pacific waters moved westward across Taiwan, and in the northwestern part of this island a mild movement induced submarine volcanism, also metamorphism, but, of no more than low to lowest grade. In western Taiwan, oscillatory movements on the Taiwan shelf during the Miocene gave rise to the rhythms of the three cycles of sedimentation described in the preceding section. Each cycle records a regressive phase with dominant coal measures and a transgressive phase with dominant marine terrigenous clastics. In regressive phases, the environments of northern Taiwan consisted mainly of swamp, lagoon, estuary, and tidal flat.

Through the Miocene, marine waters increased gradually in depth from north to south along Taiwan, with a marked concomitant increase in the volume of terrigenous clastics in the southern part of this island. Depths were prevalently neritic, but considerably deeper environments may have existed locally, on eastern Taiwan as well as on southern Taiwan. In zones of fracture, pyroclastics accompanied by a few lava flows appear. Volcanism, largely basaltic, was more pronounced in north-western Taiwan. The earliest cycle of the Miocene bore the most extensive volcanism, which then diminished gradually through the later cycles of this epoch. During the Pliocene, the depositional environment of terrigenous clastics varied from shallow to neritic, but depositional environments may have deepened to pelagic from time to time (He (Ho), 1975).

During the early Tertiary, on both the mainland and on the continental shelf, lacustrine basins formed as part of the development of antithetic horsts and grabens, or as intermontane depressions. Within this variety of basins Pei et al. (1963) and Fang et al. (1979) recognized several continental environments, which we list below, together with lithostratigraphy:

1. A subtropical climate, for the most part highly humid: (a) North China, (b) the southern part of Northeast China, and (c) the Nei Monggol Autonomous Region. Climatic indicators include the ostracod genus *Candona* and several floras of pollen and spores, particularly in the coastal region of Bohai (Hou et al., 1978; Song et al., 1978). All evidence points to a lower Tertiary climate much warmer than that of today. Throughout the lower Tertiary sequence, gray terrigenous clastics predominate, with carbonaceous shale and oil shale. Certain species of pollen and of spores indicate humid conditions except in the lower and middle sections of the (Upper Eocene?) first member of unit Te_3 (Figure II–61), in which red terrigenous clastics with minor gypsum and rock salt (halite) denote brief periods of aridity. The principal faunas and floras indicate fresh waters. A few brackish species and some sparse marine foraminifera suggest that the present Bohai Bay lay close to coastlines of early Tertiary ancestral Pacific waters. In the Fushun basin of the southern part of Northeast China, an abundant flora denotes strong vegetation in a subtropical forest. In Nei Monggol Autonomous Region, lacustrine lignites represent a moist subtropical environment; interspersed basalts represent episodic volcanism.
2. A subtropical semi-arid climate: (a) northern Jiangsu Province and southern Shandong Province, with evidence from increasing amounts of red terrigenous clastics and their contained evaporites; (b) Northwest China, where interspersed red terrigenous clastics suggest semi-arid to arid conditions.
3. Tropical climate with intermittent aridity: middle and lower valleys of the Yangzi River (Chang Jiang), with evidence from a section dominated by red terrigenous clastics, but also with gypsum and rock salt; interruptions of the climatic regime, with evidence from a few thin layers in the upper part of the lower Tertiary sequence; basalts enter the lower part episodically.
4. Humid tropical climate: (a) Maoming basin of Guangdong Province and certain other coastal basins, with evidence from an entire section of thick layers of oil shale in the early Tertiary; extremely rich in organic matter; (b) Shanxi Province and Shaanxi Province: Pliocene lateritic deposits formed by leaching and erosion in the uplifted plateaux of this region; (c) Southwest China, with evidence from coals for an inland swampy environment and for a warm rather than a hot humid climate. Both regions represent new depositional environments created by Himalayan movements of the middle and late Tertiary.

The Qinghai-Xizang (Tibet) Plateau did not begin to rise until the Pliocene (Xu Zhengyu, 1980; Zheng and Li, 1981). A vertebrate fauna dominated by *Hipparion* existed, and subtropical to tropical forest covered the entire area of the plateau. Study of the fauna and flora from many parts of the plateau shows conclusively that, at the beginning of the Pleistocene, the plateau's elevation was no greater than 1,000 m and that no large area of the plateau could have been higher than 1,500 m (Li Jijun et al., 1979; Liu and Ding, 1983; Liu Tungsheng et al., 1985).

Song Zhichen et al. (1983), from a comprehensive study of palynofloras, deduced climates for the Paleogene across many of the main regions of China. A summary of their conclusions follows:

Paleocene	*Northeast China*; humid, temperate to warm.
	Northwest China and East China: arid, subtropical.
Ecocene	*Northeast China*; humid, temperate to subtropical.
	Northwest China and Central-South China; arid, subtropical
	Southwest China and Central-South China; humid, tropical to subtropical.
Oligocene	*North China*; humid, temperate to warm temperate.
	Northwest China and East China; semi-arid, warm temperate to subtropical.
	Southwest China and Central-South China; humid, subtropical.

Climates during the Paleogene cooled sometimes to the temperate level but for the most part they appear to have ranged from warm to tropical, under arid or humid conditions according to time and according to region.

Intrusive and extrusive activities associated with Himalayan movements took place both in western and eastern China. In the west, most such activity took place in the Himalaya and its associated ranges of Xizang (Tibet) Autonomous Region, and in the Yunnan-Sichuan area of the southwest. Chang and Zheng (1973) and Mehta (1977) reported the presence of pegmatites and micaceous granites within the Himalaya. Radiometric dates range generally from 20 to 10 Ma. Extrusive activity took place in northern Xizang (Tibet) Autonomous Region on the Qiangtang block where subrecent dates have been suggested for an exceptionally fresh-appearing group of extrusive centers and flows (Gansser, 1980; Chang and Pan, 1984, 1986). These rocks actually yield dates from 28 to 20 Ma (Oligocene and early Miocene; Basu et al., 1984). They show evidence of considerable crustal contamination, and they include leucite basalt, tephrite, nosean phonolite, analcitite, and calcalkaline dacite and andesite. Associated intrusive rocks include two-mica granites. Tertiary igneous activity was rare outside of the area Himalaya-Qinghai-Xizang Plateau.

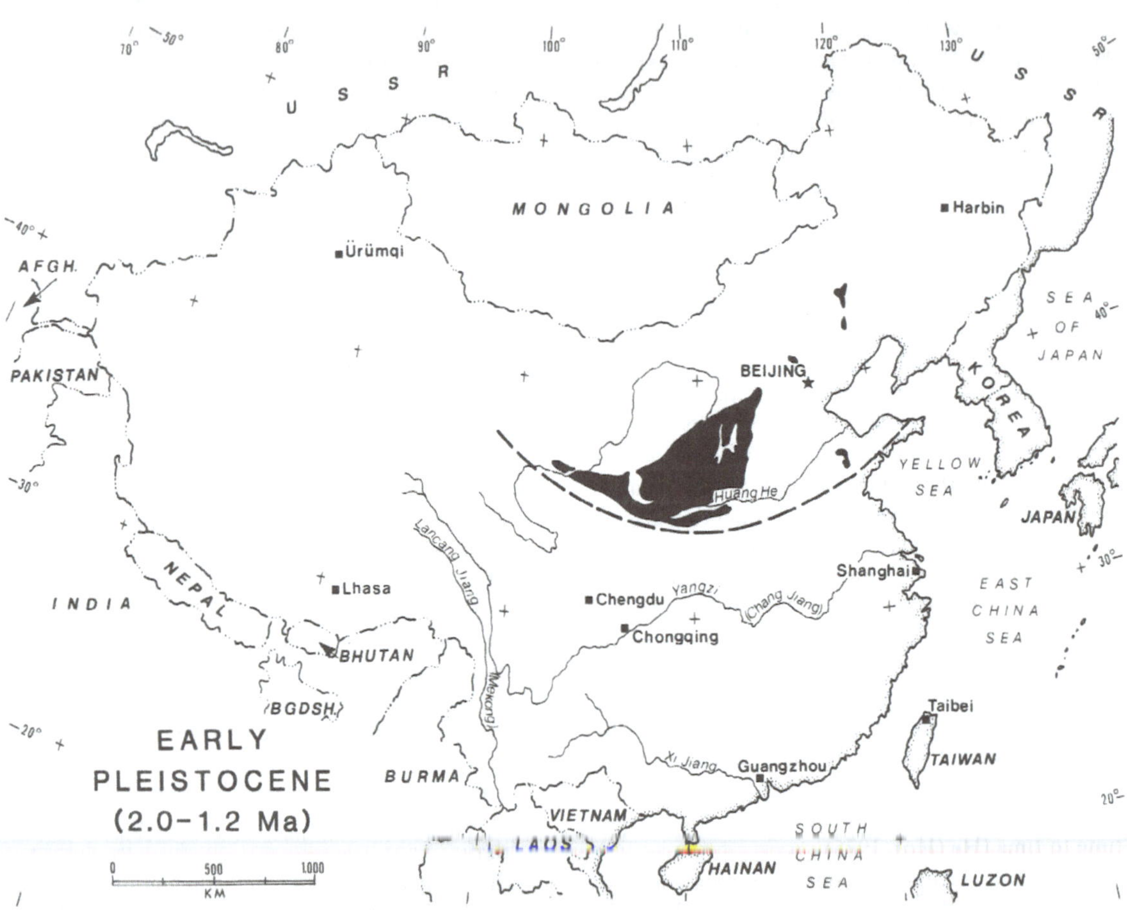

From Liu and Ding (1983)

Fig. II–63. Loess distribution in China, early Pleistocene time.

In the eastern half of China, volcanic activity is associated with great fractures, many of them, if not most, in association with the Neocathaysian structural trends. Some activity was connected also with other structural features such as, for example, the extensive Hannobar Basalt of Nei Monggol. Volcanism of island-arc type took place on Taiwan and along the Diaoyu Dao foldbelt. Small areas of flood basalts erupted in places like the Taiwan Strait (Penghu Islands). Igneous activity, largely of intermediate to mafic composition, preceded the formation of all of the grabens in the Neocathaysian system, and continued in some of these basins with Neocathaysian trends until relatively recent time. The Maoming basin, and the San Shui basin, for example, were riddled with basaltic intrusions, sills, and flows through the Oligocene and perhaps the post-Oligocene. Pliocene and younger basalts are abundant on islands from the Bohai eastward to Japan; some islands are almost entirely capped by these relatively young flows (e.g. the large Cheju Island, off Korea).

Overview of Jurassic Through Pliocene History, Eastern China

The Neocathaysian horst-and-graben and half-graben structures with associated normal faults, wrench faults, small thrust faults, and even tear faults, formed exclusively east of the Da Hinggan-Taihang-Wuling trend. They began to form with the onset of Yanshanian pulses, and they continued to form through Himalayan pulses still in progress. The first grabens appeared in the Jurassic, and they developed in the west. These structures with associated faults in general lie parallel to subparallel with the pre-Jurassic structural grain. In detail, however, there are wide divergences between Neocathaysian trends and basement structures, and parallelism to subparallelism is valid only as a generalization. What is important is the fact that these structures, which began to form in the Jurassic, continued to form through the Tertiary. The grabens and some of the faults, in addition, become younger eastward. A few major faults, however, are ancient features, as for example the Tan-Lu fault zone. The youngest basins, still in the process of forming, now lie east of Taiwan, in the Okinawa Trough, and around the Philippines. Tectonic development of these and other elements, is discussed in greater detail in the chapter on tectonics.

QUARTERNARY SYSTEM

Modern consensus assigns two million years or less to the Quaternary, which consists of a long Pleistocene

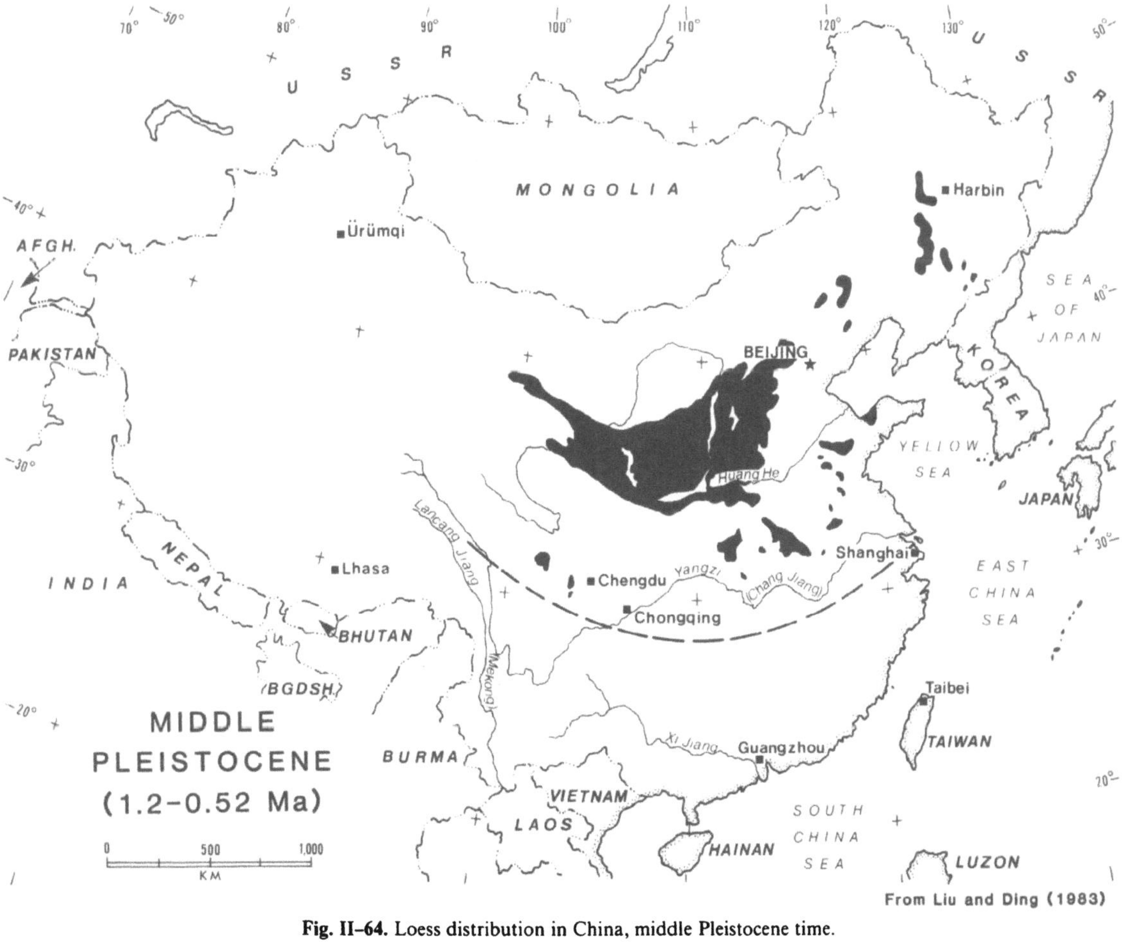

Fig. II-64. Loess distribution in China, middle Pleistocene time.

epoch and a brief succeeding Holocene or Recent epoch. In China, a relatively well-developed Quaternary is widespread, especially in basins of large and medium size north of the Kunlun Shan-Qin Ling tectonic trend (Compilation Group, 1976). Quaternary deposits show a characteristic variation of lithofacies - a reflection of changes in topography and climate that occurred both chronologically and geographically.

Distribution

North China Loess

Northern China is famous for its thick and extensive deposits of loess, the largest such deposits in the world. Figures II-63, II-64, and II-65 are maps showing the extent of the loess during the early, middle, and late Pleistocene. The increasing area of loess is related directly to the sharp rise of the Qinghai-Xizang Plateau during the Pleistocene (Liu and Ding, 1983; Liu Tungsheng et al., 1985). Figure II-63 to Figure II-65 show how close to recent the major uplift of the plateau was - middle to late Pleistocene, or within the last two million years. Winds coming from the Gobi Desert of Nei Monggol Autonomous Region and the Mongolian People's Republic could not carry the dust load farther south and were forced to deposit it north of the rising plateau.

Changbai Shan, Jilin Province

A Pleistocene to recent volcano, the Baitou Shan, rises from the Changbai Shan in eastern Jilin Province, just west of the frontier with North Korea. This volcano, 2,744 m high, possess a large crater, now occupied by a beautiful forested lake, the Tian Chi. This, the only modern volcano (even though it is extinct) in eastern China, lies close to the junction of two fracture zones, the north-northeast-trending (Neocathaysian) Dunhua-Mishan fault zone and the west-east-trending Tian Shan-Yin Shan fault zone. This volcano, by its position, serves as one illustration of the geologic importance of these intersections of fracture zones in the development of China's geology (Ogura et al., 1938).

Northwest China, Eolian Sand, and Gobi Pavement

Eolian sand and Gobi gravel pavement occur mainly along the southern margin of the Gobi Desert and the adjacent areas of southernmost Nei Monggol Autono-

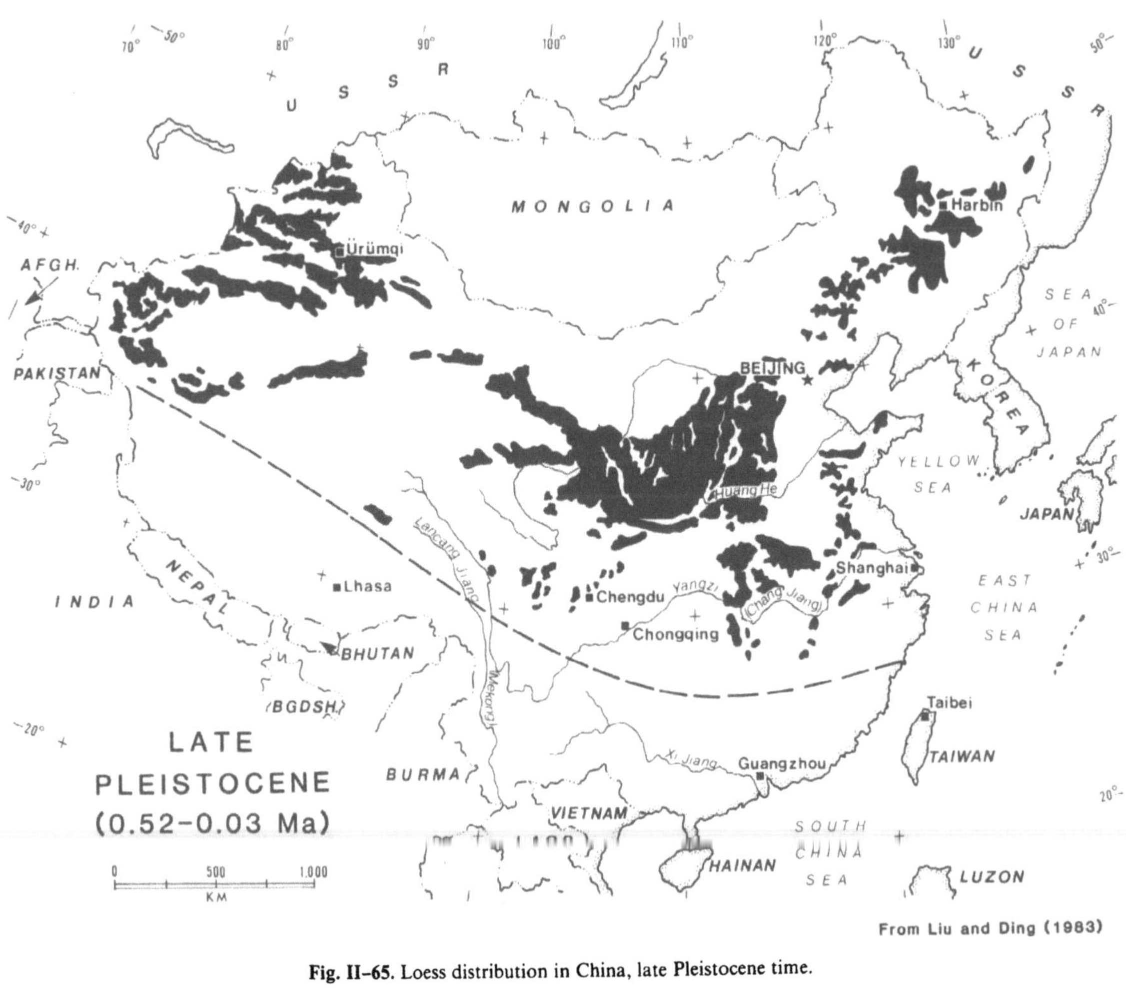

From Liu and Ding (1983)

Fig. II-65. Loess distribution in China, late Pleistocene time.

mous Region, the Tarim basin, the Junggar basin, and the Qaidam basins. Similar deposits occur in the Hexi Corridor of Gansu Province, as well as in the Helan basin of adjacent Nei Monggol. Figure II–64 shows that these basins and the Hexi Corridor also were sites for the deposition of some loess, the dust for which was derived from the Gobi Desert in the Mongolian People's Republic (Liu and Ding, 1983; Liu Tungsheng et al., 1985). In the Tarim basin, surrounded as it is by elevations of up to 6,000 m above the basin floor, large alluvial desertic fans up to 500 m thick have developed. In the center of the basin, over an area of at least 200,000 km^2, are 'live' sand dunes (barchans are numerous) and sand ridges with elevations of 300 m above the desert floor. These active and shifting ridges have presented a formidable barrier to exploration and travel within the basin (Petrov, 1967). In the Qaidam basin, below the desert floor, active playa lakes exist. The so-called 'Second Salt Epoch' of that basin began in late Pleistocene time and has continued to the present. About 30 percent of the estimated resources of various salts in the basin was precipitated during this 'epoch' (Song and Liao, 1982). The late Pleistocene to recent depocenter, as a result of the uplift of the Qinghai-Xizang Plateau, was shifted about 200 km southeast of the Mangnai depression depocenter, which lasted from the early Tertiary to the mid-Pleistocene. This shift in depocenter provides additional evidence that major uplift of the plateau was a post-middle Pleistocene phenomenon.

Western China, Southwest China, and Central China – Glacial and Glaciofluvial Deposits

These deposits are widespread in the mountainous regions of China. Pleistocene glaciation left a record of glacial deposits in the middle and lower valleys of the Yangzi (Chang Jiang), in highland areas of northern and southern China, and around the Qinghai-Xizang Plateau. For many years it was believed that the thick boulder deposits of the Lu Shan in Jiangxi Province were the products of continental glaciation, but these deposits now have been proved to be mudflow deposits. The hypothesis that they represented a period of continental glaciation consequently has been abandoned (Liu Tungsheng et al., 1985).

Pleistocene glacial history of the Qinghai-Xizang Plateau is different from what might have been anticipated in a 'normal' regime of mountain or high plateau. The oldest glaciations took place on what are now the highest peaks as glaciers never covered the plateau itself, and only a few paleoglacial centers are known, all of them small. Early Pleistocene uplift induced piedmont glaciation, but further uplift actually caused glaciation to be restricted to mountain valleys. This apparent oddity is explained by realizing that continuing and increasing uplift intensified the dry climate that developed in the lee of the rising mountains. The maximum zone of snowfall consequently was transferred to bordering mountains and to flanks of the plateau. The uplift eventually raised the plateau as high as the jet stream, one of two places in the world where this happens. The result of this raising was to cut off the interior from the Indian monsoon and, therefore, from moisture (Zheng and Li, 1981; Liu and Ding, 1983). The plateau itself, consequently, is unglaciated.

East China and North China Lacustrine, Fluvial, and Glacial Deposits

Fluviolacustrine and glaciofluvial deposits occur across the plains of China, as well as in the basins of East China and North China.

Central-South China and North China Cave Deposits

Cave deposits of several Quaternary stages occur widely across Central-South China. Early and middle Pleistocene cave deposits occur only in North China.

Taiwan, the Southern China Coral Reefs, and Hainan Island Guano Deposits

These also are Quaternary deposits, along the marine and coastal areas of China, and also on its offshore islands.

Type Sections

Beijing Region

The type section of the Quaternary in this region contains a series of alternate glacial deposits and variable interglacial deposits. These include lacustrine, fluvial, cave, and loess lithofacies (Figure II–66). The glacial deposits of this alternation consist of poorly sorted gravel and clay with sand. The fluviolacustrine deposits consist mainly of loose sand, gravel, clay, and marl with gypsum. Cave deposits characteristically contain calcareous breccia with components of various size; gravel, sand, argillaceous sand, tufa, and red clay. The lower part of the cave deposits normally is well cemented and better indurated than the upper part. The unconsolidated yellowish loess consists predominantly of silt-size sediments accompanied by material that ranges in grain size from clay to fine-grained sand, mixed with marl or loam.

An unconformity marks the Tertiary-Quaternary boundary. Within the Quaternary sequence itself, one disconformity marks the lower Pleistocene-middle Pleistocene boundary, another the middle Pleistocene-upper Pleistocene boundary.

In a cored section in the northeast suburb of Beijing marine lower Pleistocene sediments are penetrated at a depth of 428 m (Liu Dongsheng et al., 1986). The marine section includes mixed plankton and benthonic foraminefera fauna that is dominated with *Hyalina balthica* and *Globigerina bulloides*. The marine lower

System	Series	Formation	Column section	Thickness (m)	Major lithology	Major fossils V=Vertebrates M=Mollusks	Correlation with North America
Quaternary	Holo.	H_1		10	Brownish-yellow clayey soil, black soil.		Post-glacial
	Upper Pleistocene	Q_9		30	Glacial deposits: grayish-yellow soil, sand, gravel.		Wisconsinan Glacial
		Q_8		30–60	Loess and soil deposit.		Sangamonian Interglacial
		Q_7		20	Glacial deposits: yellowish-brown soil, sand, gravel.		Illinoisan Glacial
	Middle Pleistocene	Q_6		40–95	Mainly reddish clay, argillaceous sand, sand, gravel, breccia and tufa-cave deposits.	*Sinanthropus* (V) *Sinomegaceros* (V) *Bubalus* (V) *Equus* (V)	Yarmouthian Interglacial
		Q_5		10–94	Glacial deposits: brownish clay and gravel.		Kansan Glacial
	Lower Pleistocene	Q_4		75–330	Lacustrine grayish-green, brown and yellow clay.	Mammal fossils	Aftonian Interglaial
		Q_3		30–220	Glacial deposits: variegated clayey gravel.		Nebraska Glacial
		Q_2		40	Grayish sand, clay, marl, gypsum lacustrine-fluvial deposits.	*Siphneus Equus* (V) *Lamprotula* (M)	
		Q_1			Glacial deposits.		
			T				

Fig. II-66. Generalized Quaternary stratigraphy of the Beijing area, North China (compiled from Fang et al., 1979).

Pleistocene section is about 40 m thick, but the total thickness of the lower Pleistocene is 308 m.

Vertebrates are common in the interglacial deposits of North China. Remains of *Homo erectus pekingensis* were first found in 1929 in a cave at Zhoukoudian near Beijing (Peking). The Zhoukoudian cave deposits are alternating fluvial sand, silt, and clay which are interbedded with ash and limestone slabs and blocks. The Zhoukoudian section is 40 m thick and is divided into 17 layers. The Peking man fossils were found mainly in the fluvial sections together with other fossil mammals in the top ten layers. Detail investigations in recent years revealed that the Peking man at the Zhoukoudian cave lived in between 260,000 and 460,000 years B.P. Thus the Zhoukoudian deposits were assigned to middle Pleistocene (Liu Dongsheng et al., 1986).

Several new finds of human remains have been reported from Guangdong Province, Shaanxi Province, Yunnan Province, Hubei Province, and the Guangxi Zhuang Autonomous Region. In all these regions the human remains are in upper Pleistocene sediments. However, a more recent (1980) discovery of a *Homo erectus* skull was made on the northern bank of the Yangzi River at Hexian, Anhui Province (Liu Dongsheng et al., 1986). Though the Hexian man has some differences from the Peking man, both are interpreted to be the same species. Isotopic dating of the sediments containing the Hexian man remains gave an age of 290,000 to 300,000 years B.P.

Loess Plateau of North China

The second type section of the Quaternary occurs in the Loess Plateau of Shaanxi Province, where it consists of an alternate sequence of loess and nearly horizontal paleosols. Lu and An (1979) recognized three types of loess by their sedimentological characteristics. Type I, in L_1, L_2, L_5, L_6, L_9, and L_{12} (Figure II-67), exhibits the following characteristics: light grayish-yellow color, loose, porous, soft, relatively large grain size (0.02 mm) for loess, carbonate content of about 11%, secondary carbonate as cement of the mosaic texture, high content of stable minerals, thick layer with high rates of sedimentation, soil formation under weak weathering, and vertical joints that give rise to a geomorphology of cliffs.

Type II, in L_7, L_8, and L_{10} (Figure II-67), exhibits the following characteristics: light yellowish-red or reddish-yellow color, compact, less porous then Type I, relatively hard, small grain size (0.01 mm), carbonate content of about 5%, clay and secondary carbonate as cement of clay aggregates, high content of unstable weathered minerals, thin layers with low rates of sedimentation, soil formation only under severe weathering, and poorly developed vertical joints.

Type III, in L_3, L_4, L_{11}, and L_{13} (Figure II-67), exhibits characteristics between Types I and II.

Characteristics of contained clay, or carbonate content, and of illuvial process, provide a basis for recognizing several categories of paleosol. The first paleosol, the Holocene (recent) S_0 layer (Figure II-68), characteristically exhibits a dark-gray residual loamy soil with a carbonate content of about 8%. The second paleosol, in S_1, S_2, S_6, S_8, S_9, S_{11}, S_{12}, and S_{13} (Figure II-68) characteristically exhibits a brown clay layer with nodular structure, and a carbonate content of about one percent due to leaching. The third paleosol, in S_3 and S_{10} (Figure II-68), exhibits a carbonate content of about 4%. The fourth paleosol, in S_4 and S_7 (Figure II-68), exhibits a brownish-red layer with carbonate content of less than one percent, and with a well-developed nodular structure. The fifth paleosol, S_5 (Figure II-68), exhibits a dark brownish-red layer, about five meters thick, with nodular structure, and with a carbonate content reported as low as 0.08%.

Paleoecology, Paleogeography, and Magmatism

During the Quaternary, the crust remained relatively stable under many regions of China, but underwent major epeirogeny in some. The Himalaya and the Qinghai-Xizang (Tibet) Plateau rose 3,000 to 4,000 m, while the Tarim basin of Xinjiang Uygur Autonomous Region and the Qaidam basin of Qinghai Province subsided about 1,000 m. As noted above, the floor of the Qaidam basin tilted eastward. Other areas that rose include the Yunnan-Guizhou Plateau, the Eerduos (Shaan-Gan-Ning) Plateau, and the adjacent 'Loess Plateau.' Areas that subsided include the North China Plain, the East China Coastal Plain, the Jianghan Plain, and Dongting Lake. Boundaries of both the plateaux and the basins commonly consist of faults.

Glaciation, the most important geologic phenomenon of the Quaternary, occurred mainly at high altitude, as evidenced by obvious glacial topography, by associated glacial deposits, and by many rocks with glacial striations. In general, the four or five glacial stages recorded in China correlate reasonably well with those recorded in Europe and North America. The evidence for interglacial stages appears in the form of distinctly different environments. Eolian environment in the Pleistocene appears along the middle part of the Huang He (Yellow River) area as well as in Northwest China. Cave environments characterize the early and middle Pleistocene of the Beijing region and of North China, and also the Quaternary of Guangxi Zhuang Autonomous Region and Central-South China. Evidence for a fluviolacustrine environment in the Pleistocene is scattered through many localities of the country. During the early Pleistocene, intermontane environments prevailed in Northwest China. Marine environments reportedly occur along the present Pacific coast and on Taiwan. Karstification of limestone areas was intense, the most famous karst areas being in Central-South China and East China.

Cenozoic basalts are widespread in China, but mainly in its coastal segments of the Circum-Pacific belt. These segments lie in the northeast, the north, the east, and the southeast, as well as on Hainan, on Taiwan; and inland in eastern Yunnan Province, Xizang (Tibet) Autonomous Region, and the western Kunlun Shan. The majority of these basalts range in age from the Neogene to the Quaternary. We have noted previously that the supposed recent volcanics of the northern slope of the western Kunlun Shan actually belong to the Oligocene.

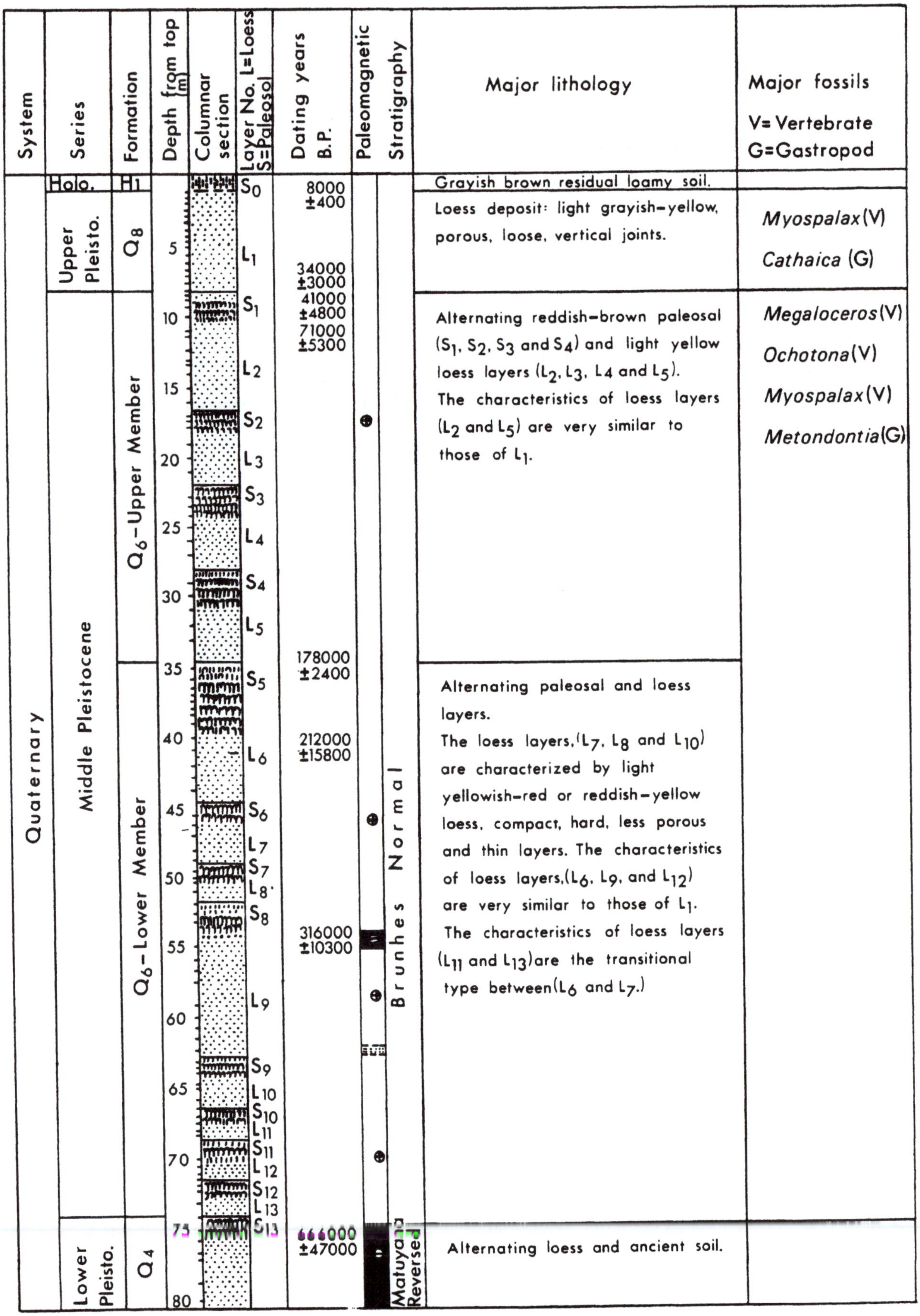

Fig. II–67. Quaternary stratigraphy in the Loess Plateau of North China (from Lu and An, 1979).

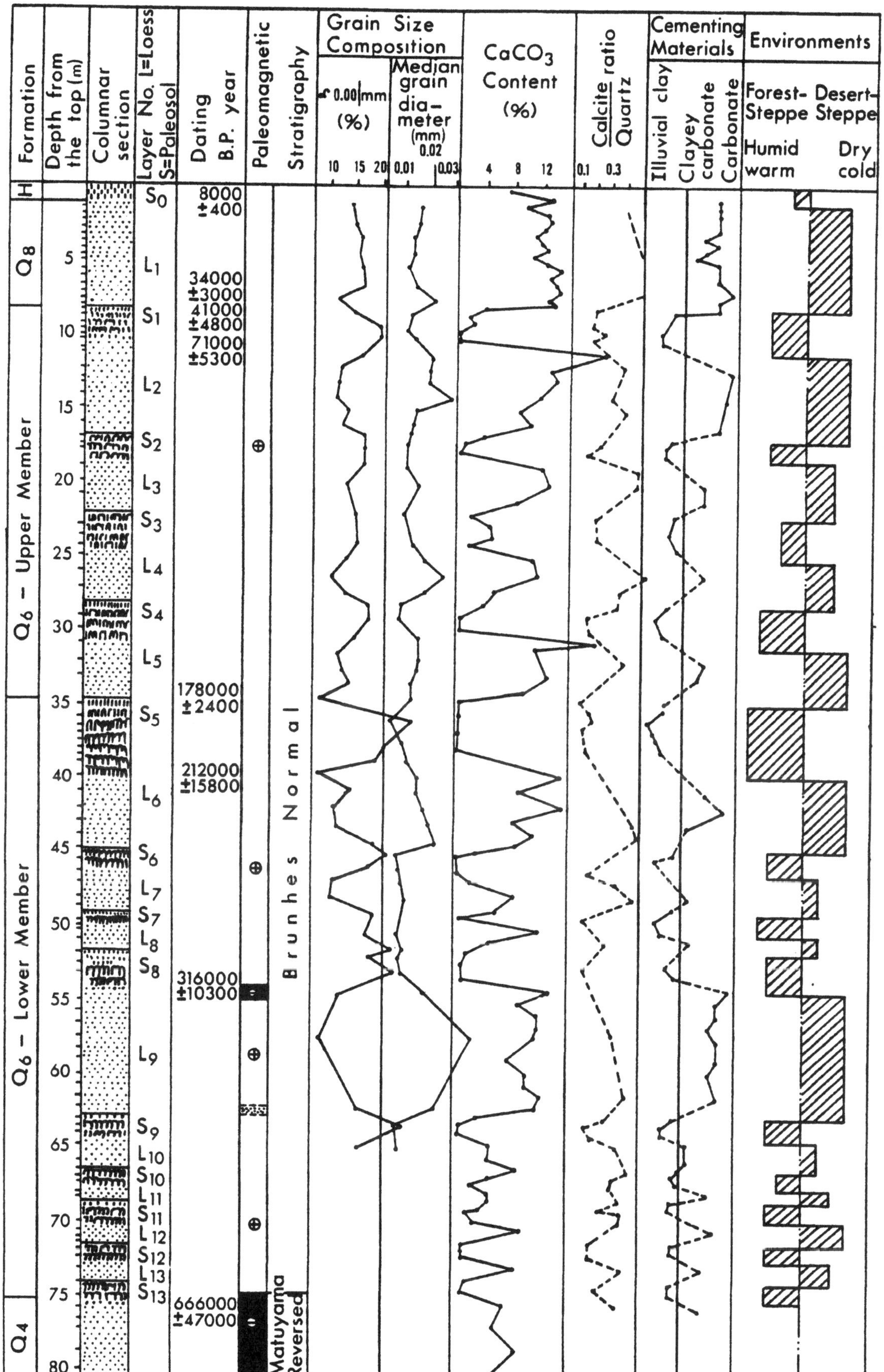

Fig. II–68. Environmental analysis of the Quaternary loess of North China (from Lu and An, 1979).

Chapter III

Tectonics and Structure of China

INTRODUCTION

Until the 1950s and 1960s, detailed geological knowledge of most of China was unavailable for the compelling reason that vast areas of the country, most of them in the central and western parts, had not been mapped. Significant geological studies of parts of China were not available until the 1970s, a statement which, even now in the late 1980s, is still true. Huge areas have been visited and sampled only in reconnaissance; some have not been studied at all. Summaries of China's structural and tectonic history, although correct in broad detail, consequently will be subject to important revisions as new data become available. We believe that our presentation, taken from many sources, presents a reasonably accurate outline of present geological knowledge.

Despite the absence of detailed knowledge of large segments of China, several tectonic and structural schools of thought have blossomed and flowered, from at least mid 1920s. Four of these appear to have the most support among Chinese geologists and geophysicists who publish on these subjects: 1) Fracture-gravity tectonic hypothesis (J. S. Lee, 1973; Chinese Academy of Geological Sciences, 1975); 2) Polycyclic geosyncline tectonic hypothesis (Huang Jiqing, 1979, 1980; Huang Jiqing et al., 1965, 1980, 1987); 3) Fault-block tectonic hypothesis (Zhang Wenyou, 1979; Zhang and Zhang, 1977, 1978; Zhang Wenyou et al., 1978, 1979, 1981; Zhang and Zhang, 1984); and 4) 'Diwa' (geodepression) tectonic hypothesis (Chen Guoda, 1989; Chen Guoda et al., 1975). Plate tectonic hypothesis also has gained supporters in early 1980 (Li Chunyu et al., 1980, 1982).

First, however, we discuss briefly our reasons for using the geosynclinal concept in China, as most Chinese geologists do, including many of the advocates of plate tectonics.

Geosynclines and Geotectonic Cycles

Following Coney (1970) and, more recently, Hsu (1982), many earth scientists have chosen to discount a geosynclinal concept built up with careful, detailed field work for more than 110 years in favor of plate-tectonic explanations for creating and tectonizing huge generally linear depressions across China. Unlike many postulates and evidences of geosynclinal theory, however, plate-tectonic concepts are new, and none of them has been demonstrated in the field to have a counterpart in reality, a point brought out extremely well by A. Gilbert Smith (1976). The same point later was emphasized and illustrated by Ilich and Meyerhoff (1980). One example of a clash between the two concepts is the demonstrable polycyclicity of most geosynclinal basins, a concept proposed first by James Dwight Dana (1873). The 'clash' between geosynclinal concepts and plate-tectonic concepts is grounded in the fundamental disagreement among plate tectonicians themselves concerning the nature of plate tectonics; do the plates move in a more-or-less steady state as all proposed mechanisms require, or do they move episodically as geosynclinal evidences require? A second example involves the difficulties that plate-tectonic hypotheses encounter where faced with the existence of several parallel but contemporaneous active geosynclinal belts, as is true across most of central and western China, where belts thousands of kilometers long lie only 400 to 1,000 km apart.

Because of these and other differences, we cannot ascribe the difficulties that some geologists have in rejecting geosynclinal theory to mental 'inflexibility,' as Schwab (1982, p. 311) has written. Schwab (p. 311) went even further, claiming that continuing adherence to geosynclinal theory means an inability to discard 'rigid mental straightjacketing.' Instead, we believe that the difficulty in rejecting long-established theory – at least in this example – is related to the fact that the geosynclinal theory, unlike plate-tectonics, is based on vast amounts of detailed, painstaking field work. We therefore think that Wang (1972, 1979), Huang Jiqing (1978, 1980, 1984a, 1984b), and Huang Jiqing et al., (1987)

Table III-1. Chronostratigraphy and tectonic stages and orogenic movements of China

ERA	SYSTEM	SUBDIVISION	RADIOMETRIC DATES (MA)	MOVEMENT	TECTONIC STAGE/CYCLE	OTHER TERMINOLOGY
CENOZOIC	QUATERNARY					
			2.0	HIMALAYAN		LATE HIMALAYAN (II)
CENOZOIC	TERTIARY	NEOGENE			HIMALAYAN	
			24.6			EARLY HIMALAYAN (I) (PULI, SHANTU, HUABEI-II)
CENOZOIC	TERTIARY	PALEOGENE			HIMALAYAN; HUABEI (WUPU)	WUBAO, HUABEI-1
			65	YANSHAN		3rd YANSHAN (ZHEMINIAN, FUJIAN-ZHEJIANG, 4th YANSHAN)
MESOZOIC	CRETACEOUS	LATE (UPPER)			YANSHANIAN	
			97.5			2nd YANSHAN (HINGGAN, XINGAN, GANDISE, 3rd YANSHAN)
MESOZOIC	CRETACEOUS	EARLY (LOWER)			YANSHANIAN	
			144			2nd YANSHAN OF SOME
MESOZOIC	JURASSIC				YANSHANIAN	1st YANSHAN (SANDU)
			213	INDOSINIAN		2nd INDOSINIAN (NANXIANG)
MESOZOIC	TRIASSIC				INDOSINIAN	1st INDOSINIAN (JINZI)
			248	"VARISCAN"		4th "HERCYNIAN" (LIWEN, QINLING)
PALEOZOIC	PERMIAN				"VARISCAN" ("HERCYNIAN")	3rd "HERCYNIAN" (DONGWU)
			286			2nd "HERCYNIAN" (TIANSHAN, QIANGU)
PALEOZOIC	CARBONIFEROUS	LATE			"VARISCAN" ("HERCYNIAN")	
			320			1st "HERCYNIAN" (HUAINAN)
PALEOZOIC	CARBONIFEROUS	EARLY			"VARISCAN" ("HERCYNIAN")	
			360			
PALEOZOIC	DEVONIAN				"VARISCAN" ("HERCYNIAN")	
			408	"CALEDONIAN"		4th "CALEDONIAN" (GUANGXI)
PALEOZOIC	SILURIAN				"CALEDONIAN"	
			438			
PALEOZOIC	ORDOVICIAN				"CALEDONIAN"	3rd "CALEDONIAN" (CHONG YU)
			505			2nd "CALEDONIAN" (YUNNAN)
PALEOZOIC	CAMBRIAN			XINGKAI (SALAIR)	"CALEDONIAN"	1st "CALEDONIAN" (LECHANGXIA, YUNGU)
			570–610			
PALEOZOIC / PROTEROZOIC (disputed)	JIDONG SEQUENCE (=SINIAN SENSU LATO OF GRABAU): SINIAN (SENSU STRICTO)	LATE (UPPER)				
			700	CHENGJIANG		YANGZI
			850	JINNING		
PROTEROZOIC	JIDONG SEQUENCE: QINGBAIKOUAN	LATE (UPPER); EOZOIC			YANGZIAN	
			1,050	SIBAO		
PROTEROZOIC	JIDONG SEQUENCE: JIXIANIAN	LATE (UPPER); EOZOIC			(UNNAMED)	
			1,400	WULING		
PROTEROZOIC	JIDONG SEQUENCE: CHANGCHENGIAN	LATE (UPPER); EOZOIC			WULINGIAN	
			1,700–1,800	ZHONGTIAO		LULIANG
PROTEROZOIC	(DISPUTED)	EARLY; PROTEROZOIC			ZHONGTIAOAN	HUTUOAN
			1,900	WUTAI		
PROTEROZOIC	WUTAIAN	EARLY; PROTEROZOIC			WUTAIAN	
			2,500–2,600	FUPING		TIEPU
ARCHEOZOIC (ARCHEAN)	FUPINGIAN	LATE			FUPINGIAN	
			2,800–3,100	QIANXI		
ARCHEOZOIC (ARCHEAN)	QIANXIAN	EARLY			QIANXIAN	
			3,800			
HADEAN						
			4,500			

SOURCES: Harland et al. (1962)
Wang and Qiao (1984)
Zhang and Zhang (1983)
Zhang Wenyou et al. (1984)
Group of Tectonics (1977)
Cheng Yuqi et al. (1982)
Sun and Lu (1985)
Huang Jiqing (1978, 1984)
Tang (1982)
Li Desheng (1984)

are almost certainly correct in their assessments of the usefulness of geosynclinal theory in China and adjacent regions. Until the premises of the geosynclinal concept are disproved, that concept continues to be a useful theory, as the earth-science community in China has demonstrated.

Tectonic and Structural History: Resume

Pre-Paleozoic

Two stages of Pre-Paleozoic tectonic and structural development are recognized by most geologists in China (Huang Jiqing et al., 1977, 1980; Huang Jiqing, 1978; Yang Weiran et al., 1984; Sun and Lu, 1985; Ma and He, 1989). The first stage encompasses the Archeozoic through early Proterozoic (pre-Sinian s.l.), and culminated in the consolidation of the Sino-Korean (North China) (para)platform during the Zhongtiaoan or Lüliangian orogeny (ca. 1,800 Ma; Table III-1). That orogeny ended a long process of accretion and growth of the Sino-Korean paraplatform during the earlier Fupingian and Wutaian orogenies (2,550 and 1,900 Ma respectively; Table III-1).* Available information indicates that the Sino-Korean massif, geologically, is the oldest region in China, and several excellent studies, as well as resumes, have been published during recent years (e.g. Cheng Yuqi et al., 1982a, 1982b; Wang and Qiao, 1984; Sun and Lu, 1985; Ma and He, 1989; and Ren and Chen, 1989; Cui Wenyuan et al., 1989).

The second stage of China's pre-Paleozoic development includes the Sinian (s.l.) Suberathem, which culminated in the Jinningian and Chengjiangian orogenies (Table III-1). During the Chengjiangian orogeny, both the Tarim platform and the Yangzi paraplatform were consolidated (Figure III-1; Huang Jiqing, 1978; Yang Weiran et al., 1984; Yang and Xie, 1984; Zhang Shuye et al., 1989). Both were then somewhat larger than they are today. Deformation during the Jinningian and Chengjiangian orogenies, as during earlier Pre-Cambrian orogenies, was directed mainly in a north-south direction, with the result that predominant tectonic trends are east-west (Figure III-1; Liu Hefu, 1984; Ren and Chen, 1989). These structural trends dominate western and central China even today, and their influence extends eastward to the margin of the present Pacific Ocean basin (Figures III-2, III-3, III-4).

Yang Weiran et al. (1984) interpreted the geotectonics of China somewhat differently from most workers, and they incorporated into their concepts (Figure III-3) the most up-to-date observations on the basement of the Qinghai-Xizang Plateau, where a major Pre-Cambrian continental massif is being discovered in notably large areas. Their interpretation, in our opinion, is the best yet published on the extent of Pre-Cambrian massifs in China.

Figure III-1, a paleotectonic map from Wang and Qiao (1984), depicts Pre-Cambrian tectonic units as interpreted from the last few decades of field, satellite, and geophysical investigations. Figure III-2 offers a popular geotectonic subdivision map of China, taken from the Group of Tectonics (1977), Huang et al. (1977, 1980). Figure III-3 is a map by Yang Weiran et al. (1984) of the geotectonic subdivisions of China. Figure III-4, modified from a map by the Institute of Geomechanics (1976), shows the main tectonic trends of contemporary China and the Mesozoic-Cenozoic basins that are superimposed on them. Figure III-5 shows the major presently active fracture zones of China.

Post-Chengjiangian orogeny has partly destroyed the original outlines of the platforms through tectonic reactivization. Such reactivization is manifested in the ensialic foldbelts generated from successor basins developed on these platforms or their margins, and in deep crustal or lithospheric fractures such as the Tan-Lu (Tancheng-Lujiang) fault (Figures III-4 and III-5). This fault formed during the late Proterozoic or earlier (Huang Jiqing, 1978; Sun Ronggui et al., 1983; Zhou Guoqing, 1989). Its subsequent active role in the tectonics of eastern China is manifest in its relation to the position of the North China (Huabei) Mesozoic-Cenozoic graben systems (Zhai and Zha, 1982; Zhao Zhongyuan, 1984) and violent earthquakes today, the most recent of which was the catastrophic 1976 Tangshan earthquake (Qiu Qun, 1976) that claimed more than 100,000 lives.

Many geologists in China believe that the Tarim, Yangzi, and Sino-Korean blocks became united with the Siberian platform as a result of the Jinningian and Chengjiangian orogenies (Figure III-1). This postulated 'superplatform' was called the 'Chinese protoplatform' by Huang Jiqing (1978, 1984a) and by other geologists. The platform was larger than it is today, as proved by the presence of Sinian (s.s.) and Lower to Middle Cambrian shallow-water marine sedimentary rocks both on the existing platform and at the base of the geosynclinal stratigraphic sections around it. This protoplatform, in Huang Jiqing's view, endured for nearly 200 million years.

There is evidence, however, to indicate that the so-called 'superplatform' may never have existed in the form envisioned by Huang Jiqing or, if existent, that is may have endured for a much shorter period. According to Cui Kexin (1984), Liu Hefu (1984), and other workers in the region, Central Asian-Mongolian marine waters invaded the zone between the Tarim and Siberian stable massifs (Figure III-2) no later than the Sinian (s.s.), and possibly during the early Sinian (s.l.) (i.e. ca. 1,700 Ma ago). Certainly an elongated trough complex had established itself in the Mongolian (including the Argun, or Ergun) region. This complex trough lay mainly north of, but locally across, the present U.S.S.R.-China-Mongolia frontiers by the pre-Sinian (s.s.). Proterozoic geosynclinal basins also began to establish themselves where the modern Kunlun Shan, Qilian Shan, Nan Shan, and Qin Ling foldbelts rise above the modern Gobi, Taklimakan (Tarim), Mu Us (Eerduos), and related deserts. These, and additional geosynclinal basins that appeared somewhat later, are dominated by eugeosynclinal facies, some of which contain ophiolites (Wang and Liu, 1976; Huang Jiqing, 1980; Cui Kexin, 1984). Liu Hefu (1984) postulated that, if the 'superplatform' did exist, it was divided quickly by invading marine waters of Central Asia-Mongolia, either through a long, depressed, synform trough or through a depression created by taphrogenesis. Liu Hefu (1984), who also espouses the concept

* From the Archean to the Quaternary, orogenic cycles or tectonic movements have become formalized largely by the use of Chinese names with standard chronostratigraphic suffices. Examples are: 'Fupingian' of the early Proterozoic, 'Xingkaian' of the Cambrian, and 'Indosinian' of the Triassic and the Early Jurassic (Table III-1). Exceptions to this usage are: 'Caledonian' for the early Paleozoic, and 'Variscan' for the late Paleozoic and the earliest Mesozoic. Both of these terms, even in their native Europe, give rise to a measure of ambiguity. Moreover, Europe is far removed from China. Yet all Chinese geologists use these two European terms widely. For these reasons we place these two non-Chinese terms in quotation marks ('Caledonian' and 'Variscan') in our own presentation of the tectonics and the structure of China. China's orogenic cycles and their relations with the geologic time scale are shown on Table III-1. The radiometric dates for the Pre-Cambrian are approximate. No two authorities agree on them.

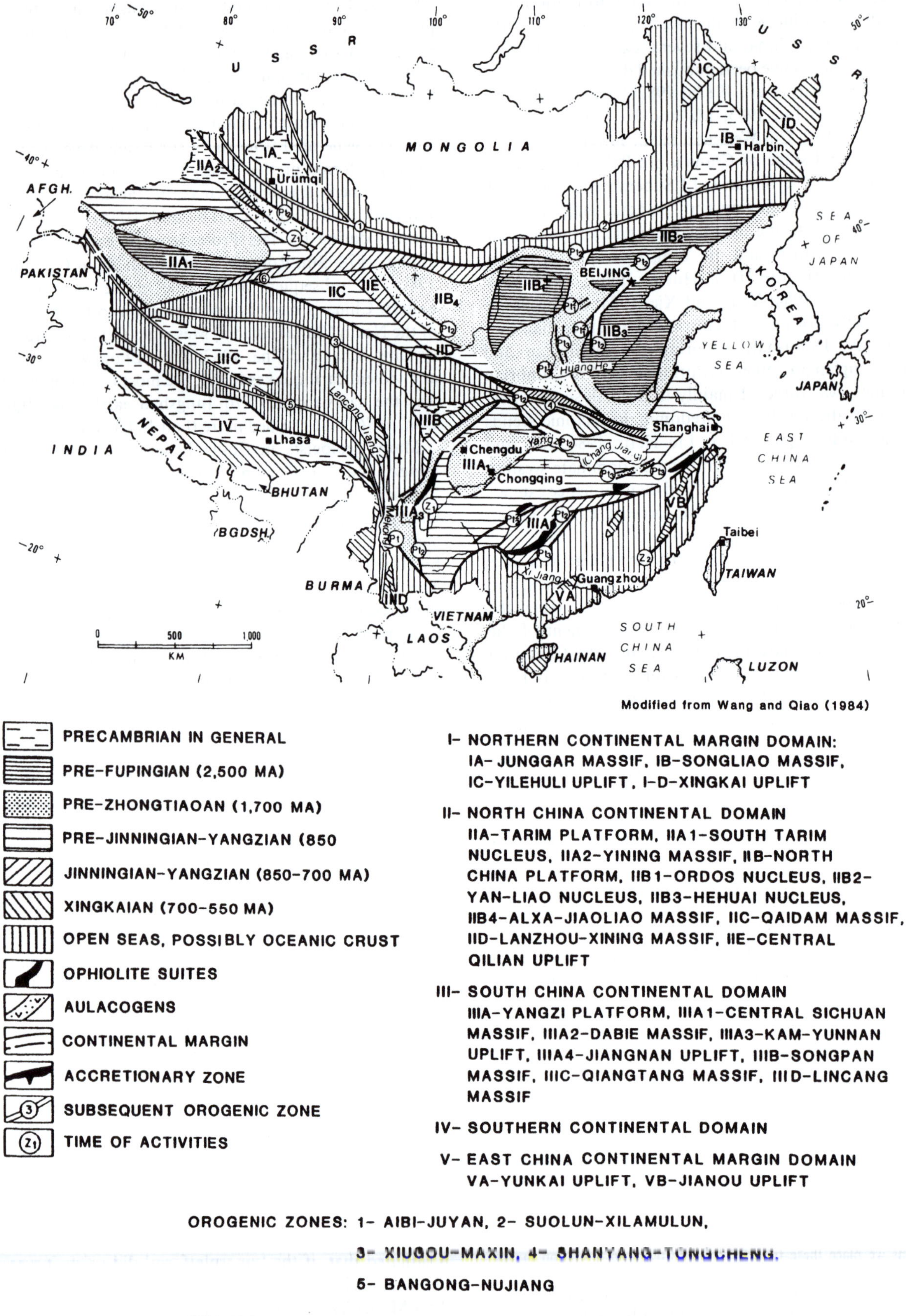

Fig. III-1. Sketch showing the interpreted structure of the Precambrian basement of China.

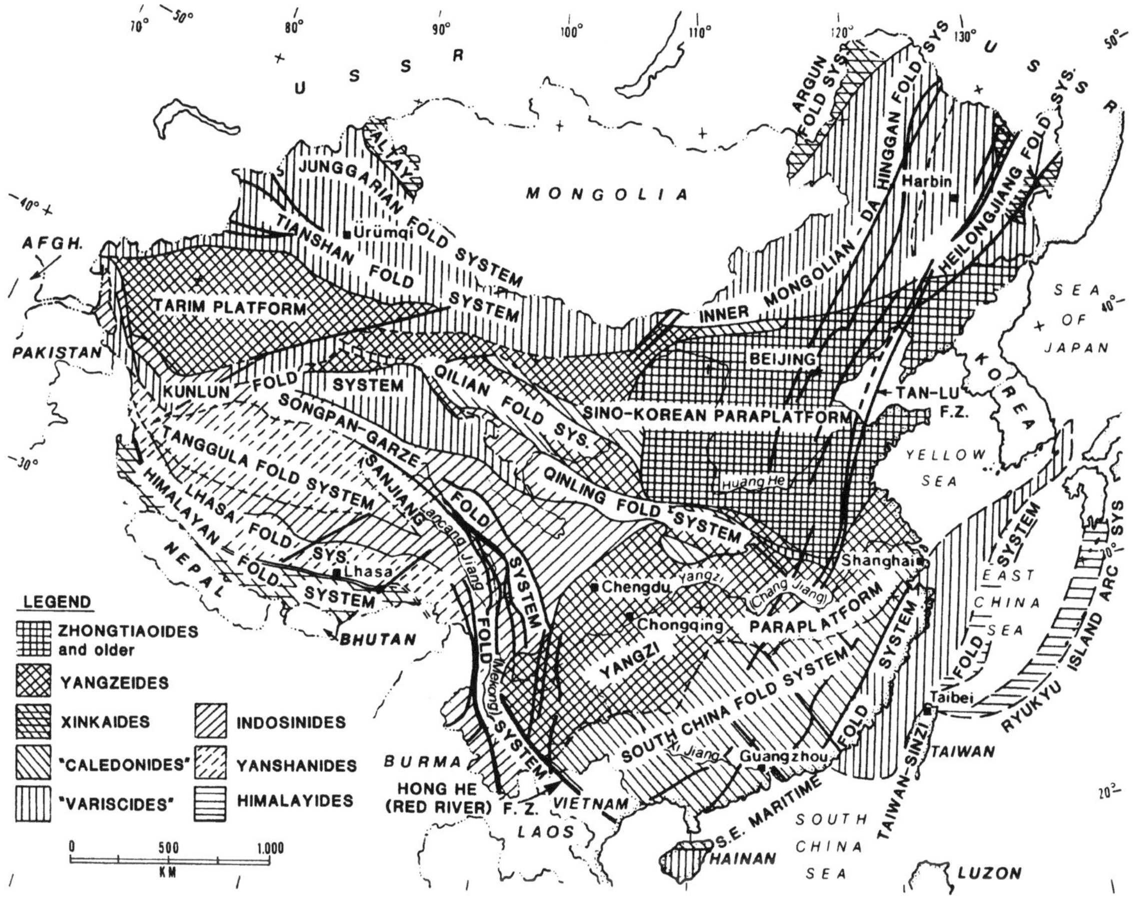

From Group of Tectonics (1977) and Huang Jiqing (1978)

Fig. III-2. Simplified map of the tectonic provinces of China.

of a Chinese protoplatform, presumably is correct, and Huang Jiqing's (1978) views, with modification, are corroborated.

In South China and adjacent parts of Southwest China and East China, the Yangzi paraplatform extends from present-day Yunnan Province to Jiangsu Province, and it includes East China Sea (Dong Hai) and part of the Yellow Sea (Huang Hai). Until somewhat recently, this paraplatform had been regarded as exceedingly old, and indeed it does contain some high-grade metamorphic rocks 2,000 Ma or older (Huang Jiqing, 1984a; Ren and Chen, 1989; Lu Liangzhao, 1989). Its area now is judged to have become a platform, however, only during the Jinningian and Chengjiangian orogenies about 850 to 700 Ma ago (Table III-1). But Lu Liangzhao (1989), in his study concluded that the Yangzi Platform has undergone at least two stages of cratonization (Early Proterozoic and Late Proterozoic) and he suggested further studies of the basement of the Yangzi Platform.

The Jiangnan orogenic belt embraces an area east and south-east of the Yangzi platform. The Jiangnan belt is made up of eugeosynclinal sequences calculated to be 10,000 m thick. These indicate the presence of Late Proterozoic orogenic cycles in this region (Figure III-1; Guo Lingzhi et al., 1984; Ma and He, 1989; Lu Liangzhao, 1989; Wang Dezi et al., 1989). Radiometric dates of 1478 to 1558 Ma, 1422 Ma, 991 Ma, 967 Ma, 932 Ma, 913 to 909 Ma, 844 Ma, 837 Ma, 800 Ma, and 760 Ma come from granitic bodies in this region. Ophiolitic suites are present along the Jiuwandashan and Wanzai-Yifeng deep faults, along which the ophiolite slabs exhibit south dip.

Elsewhere, Pre-Cambrian events equivalent to the Zhongtiaoan orogeny (1850 Ma) have been recorded from the Himalaya (Gansser, 1981a, 1981b), the Gyeanggi and Sobaek Massifs of South Korea (Lu Liangzhao, 1989), and in the Japanese arc (Sugisaki et al., 1972). Important post-Zhongtiaoan Pre-Cambrian massifs were developed in those areas (Sugisaki et al., 1972; Yang Weiran et al., 1984).

Events of the Pre-Cambrian partly determined the locations of major fracture systems such as those of the 2,400 km Tan-Lu and the 4,000 km Da Hinggan Ling-Taihang Shan-Wuling Shan (Figure III-4; Huang Jiqing, 1978; Liu Defu et al., 1983; Sun Ronggui et al., 1983; Zhu Songnian, 1984; Ren and Chen, 1989). The Tan-Lu fault zone is associated with mafic and ultramafic rocks (Huang Jiqing, 1978; Zhou Guoqing, 1989). The major fault zones commonly are 25 to 40 km wide. Those which stayed active later were loci for large ore-bearing

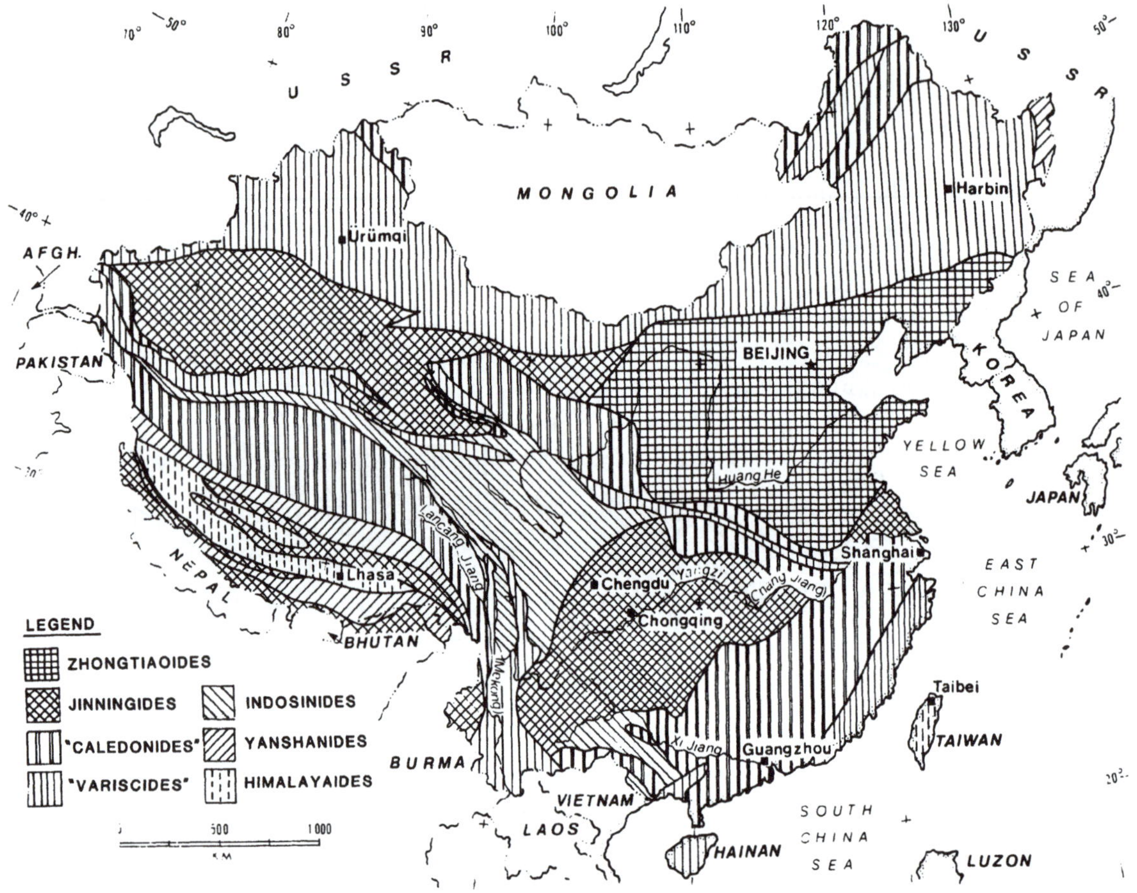

From Yang Weiran et al. (1984)

Fig. III-3. Tectonic provinces of China according to Yang Weiran and others (1984). Compare with Figure III-2. Note Precambrian massif of central Xizang (Tibet) Autonomous Region.

intermediate to silicic intrusions. Especially rich in minerals are the places where fracture zones intersect the older Pre-Cambrian east-west structural zones that extend from western and central China (Figures III-1 and III-4). The positions of the intersections between the east-west structural zones and the fault systems that intersect them appear not to have shifted more than a few – perhaps a few dozen – kilometers, which to us indicates the absence of significant amounts of post-Proterozoic strike-slip movement. Parallel with pre-existing foldbelts and structures, as well as with the present coast and continental shelf edge, the geosynclinal belt of South China formed in the latest Proterozoic to the earliest Paleozoic.

Less well known are Pre-Cambrian events that affected present Southwest China. Cui Kexin (1984) maintains that a Late Proterozoic geosyncline was well developed in present-day China, presumably in a trough parallel, or associated with, the geosyncline that occupied the northern part of the Indian shield during the Late Proterozoic. Geologists in China familiar with the Qinghai-Xizang Plateau and with northern India are nearly unanimous in stating that India, during the Proterozoic, occupied nearly the same position with respect to Asia as it does today (Chang and Zheng, 1973a, 1973b; Chang et al., 1977; Huang Jiqing, 1978, 1980, 1984; Chang and Pan, 1981, 1984; Chen Guoda, 1984b; Mu En-zhi et al., 1986). Both Crawford (1979) and Gansser (1981a) have stated that the geology of Southwest China and of northern India is not explained easily if the Indian subcontinent was far from its present position, a position demonstrated by the work of many Indian geologists (Saxena, 1971, 1978; Gupta and Janvier, 1979; Sharma, 1983; Bhat, 1984). The regional syntheses by Gansser (1981), Stocklin (1984), Kozur and Gupta (1983), and Waterhouse (1983) provide the detailed geologic data that for us effectively immobilize popular plate-tectonic models of the Indian subcontinent. Although the conclusions of these regional syntheses are at variance with widely accepted concepts, they have nevertheless the merit of being based on several decades of work on the ground and on the rocks by each of their authors. A summary of Chinese geological studies of this vast region appears in a subsequent section of this chapter.

Much Pre-Cambrian rock is present in the Qinghai-Xizang (Tibet) northern India region (Figure III-6). The largest massifs occur in the Himalaya, the tectonized northern margin of the Indian shield. K-Ar dates from the Himalaya range from 2200 to 640 Ma (Chang and Pan, 1984). Within Xizang (Tibet) Autonomous Region,

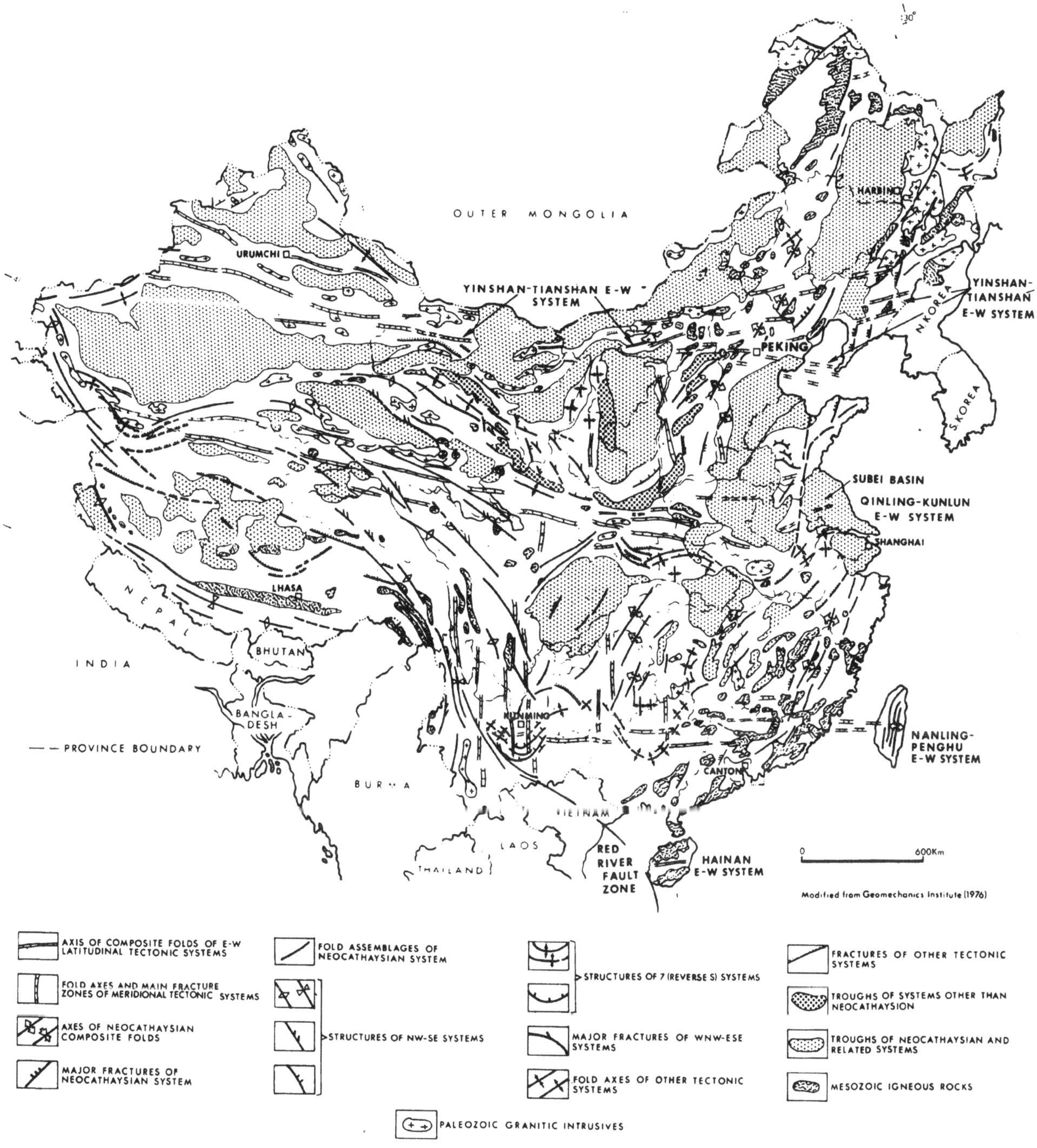

Fig. III–4. Generalized tectonic map of China. Stippled areas are Tertiary depocenters. Note east-west fracture systems.

the Lhasa block (Figure III–6) includes the Gangdise Pre-Cambrian in the west and the Nyainqentanglha Pre-Cambrian in the center and east. One Pre-Cambrian block in the latter region is approximately 200 km long and 70 km wide. Farther north in the Qiangtang block, north of the Bangong-Nu Jiang fault zone (Figure III–5), Pre-Cambrian schist is present, particularly in the western part of the block. Chen Guoda (1984b) stated that the pre-Cambrian of Xizang appears to have been metamorphosed and consolidated during the Xingkaian orogeny (Table III–1), and Yang Weiran et al. (1984) show large Pre-Cambrian continental areas within the Qinghai-Xizang geologic province (Figure III–3). In both the Lhasa and Qiangtang blocks, shallow-water lower Paleozoic marine strata blanket the Pre-Cambrian metamorphic and associated rocks (Chang et al., 1977; Chang and Pan, 1981, 1984; Wang Hongzhen, 1983; Huang Jiqing et al., 1987). In addition, the presence of Upper Proterozoic and some Middle Proterozic rocks has been established in other parts of the Qinghai-Xizang Plateau

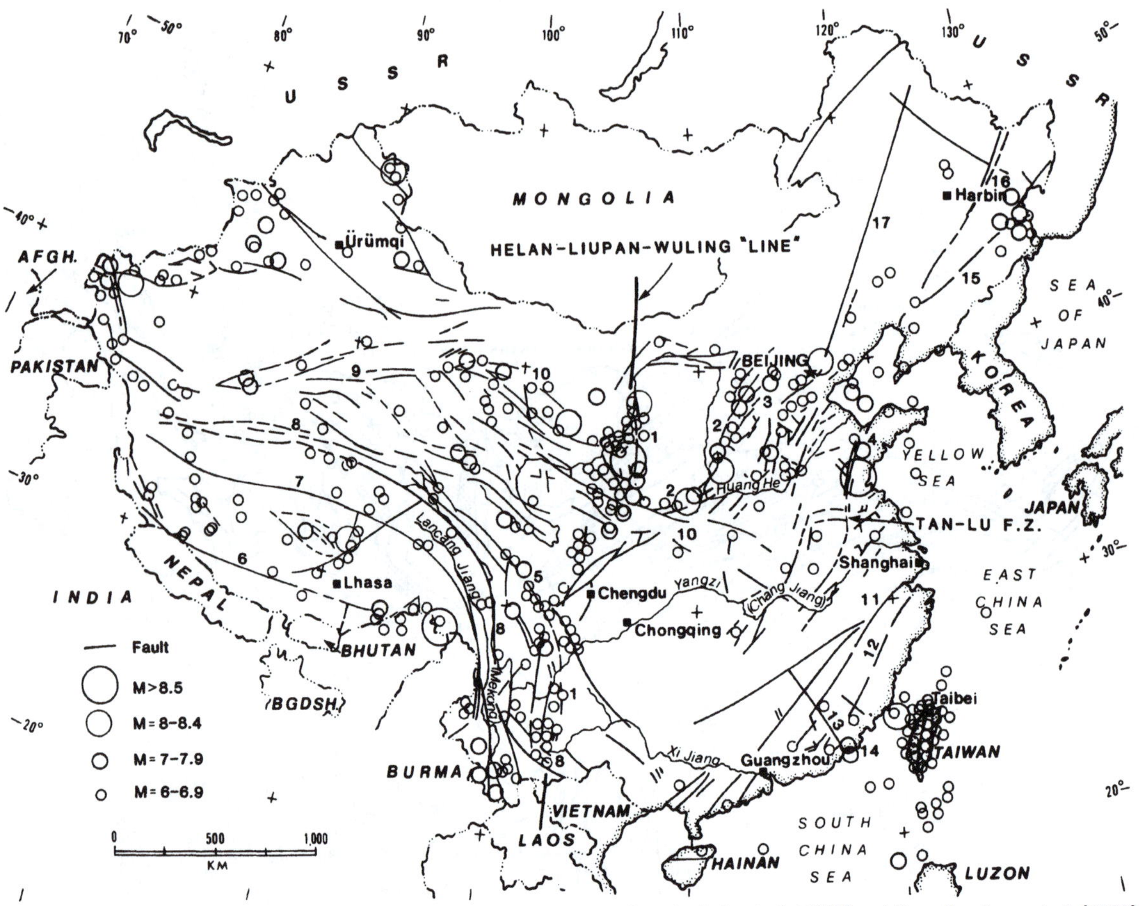

From Lu Defu et al. (1983) and Yuan Xuecheng et al. (1986)

Fig. III-5. Major, presently active fracture zones of China (with earthquake epicenters).

FAULT ZONES (f.z.)

1. Helan Shan-Liupan Shan – Longmen Shan-Hengduan Shan f.z.
2. Fen-Wei graben
3. Taihang Shan f.z.
4. Tan-Lu f.z.
5. Xianshui f.z.
6. Indus-Yarlung f.z.
7. Bangong Hu-Nujiang f.z.
8. Jinshajiang-Honghe f.z.
9. Altun f.z.
10. N. Qilian Shan-N. Huaiyang-Qin Ling f.z.
11. Xiao Shan-Zan Jiang f.z.
12. Cixi-Haifeng f.z.
13. Dongshan-Ji'an f.z.
14. Nan'ao-Changluo f.z.
15. Mishan-Dunhua f.z.
16. Yilan-Yiton f.z.
17. Da Hinggan Ling-Taihang Shan-Wuling Shan f.z.

and in western and central Yunnan Province (Liu et al., 1973; Wang and Qiao, 1984; Sun and Lu, 1985).

Preliminary reports on deformation fabrics observed in the field indicate that the Pre-Cambrian outcrops are of rocks that have been compressed tangentially toward the east, but that have not been rotated into their present positions from areas now occupied by India (Misch,

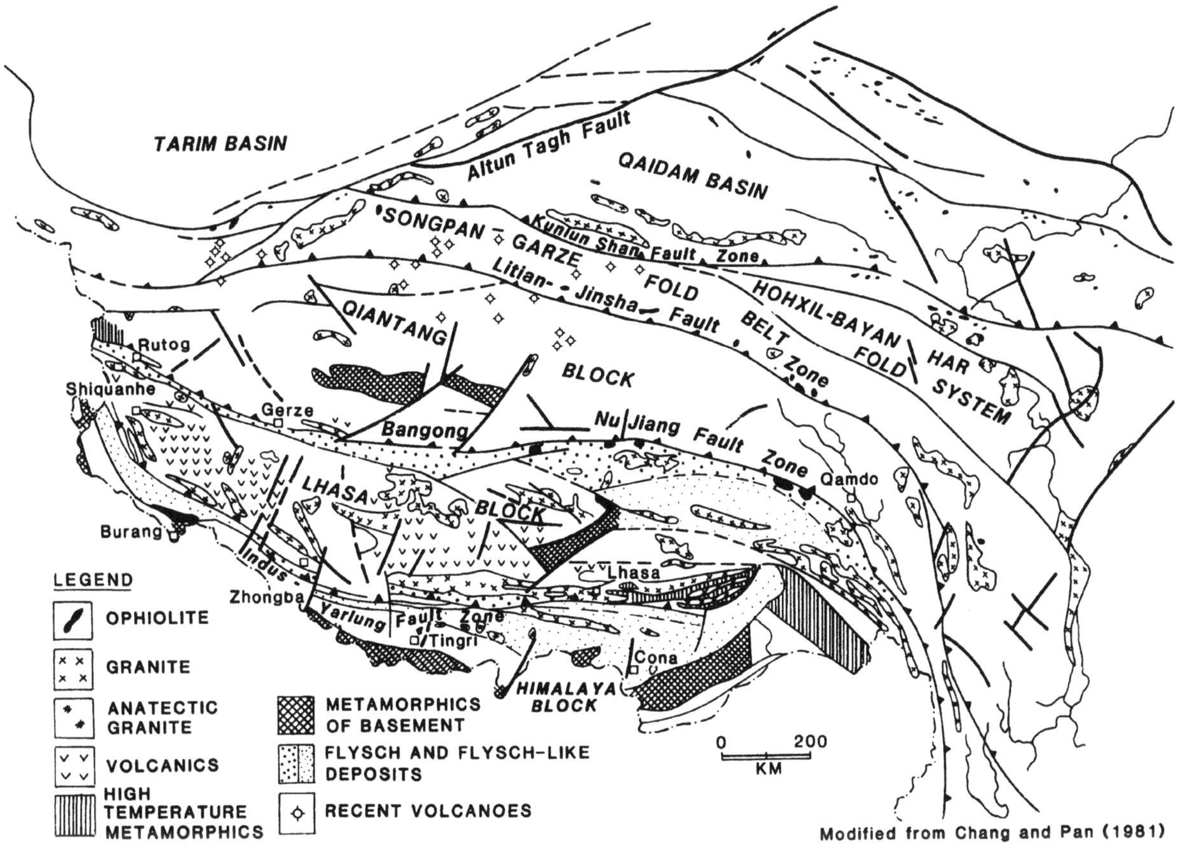

Fig. III-6. Principal. tectonic features and subdivisions of Xizang (Tibet)-Qinghai plateau region, China (modified from Chang and Pan, 1981).

1945a, 1945b, 1946). This conclusion is not based solely on a few preliminary studies of deformational fabrics measured in outcrops. It is based also on north-south alignments of facies, in Pre-Cambrian and Early Cambrian strata from Yunnan Province to Shanxi Province around the eastern margin of the 'Xizang-Yunnan oldland' of China's geological literature (Chang Da, 1959; Qian Yi, 1984). If the conclusion of non-rotation were correct, *it would have profound implications for the tectonic history of southeastern Asia*, because it would mean that the Qinghai-Xizang and Yunnan-Burma-Malaysia arcuate orogenic belts are (1) of pre-Paleozoic age and (2) owe their shapes to the *already established positions* of the Yangzi paraplatform on the east, the Tarim platform on the north, and the Indian platform on the south, and not to collision of India with Asia. No one would deny that the Qinghai-Xizang and Yunnan foldbelts have been strongly compressed and therefore may have undergone tangential shortening, but a 'collision' of India is not the only possible explanation for shortening. Future studies of tectonic fabrics within Pre-Cambrian rocks of the foldbelts thus will be of great importance.

The Pre-Cambrian did not end abruptly in China. As in many parts of the world, Pre-Cambrian events are gradational with those of the early Paleozoic. Many major structural features that affected China's Phanerozoic tectonic history were established during the Pre-Cambrian, including the large blocks of paraplatform and platform, some geosynclinal trends, and several fault zones, including the great east-west fracture zones (Figures III-1 and III-7). These have been studied by many geologists in China (see L. Siguang, 1939, 1973; Huang Jiqing, 1978, 1984a; Zha Songnian, 1984). According to Huang Jiqing (1978), the most important is the Kunlun-Qin Ling fracture zone. This broad trans-China east-west trend includes the North Nan Shan-North Qin Ling-North Huaiyang fault zone, the North Qaidam-South Qinghai Hu (Koko Nor)-North Qin Ling-North Huaiyang fault zone, and the East Kunlun Shan-Qin Ling fault zone (Figure III-7). The Kunlun-Qing Ling trend, according to Huang Jiqing (1978) and Lu Defu et al. (1983), separated (and still separates) northern China from southern China. This is a fundamental, deepseated structure whose importance cannot be overemphasized (Huang Jiqing, 1978). All east-west zones, according to Zha Songnian (1984), have been developing for at least 1,800 to 1,700 million years.

Sinian-Early Paleozoic

The distribution of Sinian and younger geosynclinal systems of island arcs foreshadowed closely the present geological architecture of China (Figures III-2 and III-

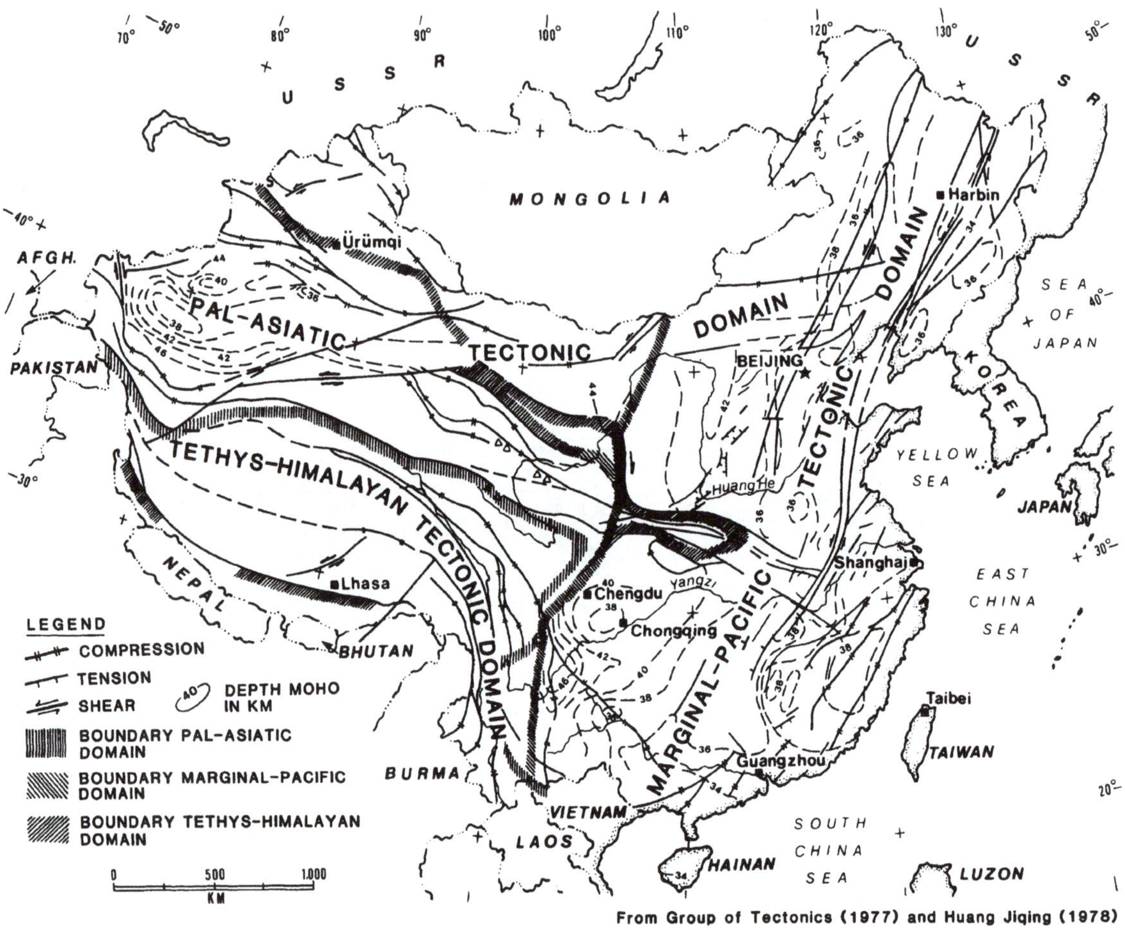

Fig. III-7. Simplified sketch map of deep fracture zones and Phanerozoic tectonic domains in China.

3). Geosynclinal or orogenic belts not only modified and reshaped the recently created Chinese protoplatform, but also presaged (a) the creation of Pal-Asia (Paleoasian or Pal-Asiatic domain of Huang Jiqing, 1978) and (b) initiation of the 'Caledonian' orogenic cycle (Table III-1). In western and central China, the geosynclinal systems trend east-west, parallel with, and in many cases nearly coincident with, older east-west trends that evolved during the Pre-Cambrian. These east-west geosynclinal belts extended farther eastward during the Pre-Cambrian than they did during the Mesozoic and Cenozoic. Huang Jiqing (1978) therefore wrote that north-south-directed stresses predominated in China during the Paleozoic, to give rise to what he termed the 'Pal-Asiatic Tectonic Domain.' The stress fields from the Pacific Ocean basin affected a smaller area than his Domain, as did the stresses that created the Himalaya, the latter stress direction being the same as that which produced the Pal-Asiatic Tectonic Domain. It was not until the Mesozoic and the Cenozoic Eras that the Pal-Asiatic stress field diminished in importance and the Pacific ('Marginal-Pacific Tectonic Domain') and Himalayan ('Tethys-Himalayan Tectonic Domain'; Figure III-7) stress fields predominated. The Phanerozoic thus may be regarded as a time when north-south stresses predominated during some 56% of its time (= Paleozoic) and when north-south and east-west stress fields were approximately equal during the remainder of its time (44% = Meso-Cenozoic).

The process of consolidation of Pal-Asia, or the Pal-Asiatic continent, was accomplished through consolidation of the various eastern Asiatic cratons during the Xingkaian, 'Caledonian,' and 'Variscan' orogenic cycles (Table III-1; Figure III-7). The predominance of north-south-directed stresses enhanced already-existing east-west trends of China (Liu Hefu, 1984). In contrast, stresses directed from the Pacific predominated only in southeastern China, the Korean Peninsula, the Japanese islands, and beneath today's continental shelves. The resulting northeast to north-northeast trends (Cathaysian to Neocathaysian) began to 'overprint' the older east-west and north-south trends.

Geosynclinal complexes began to appear in different parts of China around the margins of the Sino-Korean, Tarim, Yangzi, and Indian platforms/paraplatforms. Some of the geosynclines were folded and intruded by granitic rocks for the last time during the Xingkaian orogeny, named for the Xingkai Hu (Lake) area on the frontier between Heilongjiang Province and Primorskiy Kray (Primor'ye), U.S.S.R. The date of the Xingkaian orogeny, roughly between 600 and 500 Ma, implies Middle Cambrian, and the event was widespread in Asia, northern Europe, and parts of Australia and the Americas (Xingkaian orogeny corresponds with the Salairan orogenic phase of the 'Caledonian' orogeny in the U.S.S.R., and the Finnmarkian phase of the 'Caledonian' orogeny in Norway). Baud et al. (1982) equated Xingkaian orogeny with the terminal Pan-African event. Some of China's geologists refer to it simply as the first phase of the 'Caledonian' orogeny (see Table III-1).

Xingkaian geosynclines underlie the northern half of the area inundated by the 'Central Asia-Mongolia Sea' (Huang Jiqing, 1978), and they are characteristic of the area north of the Derbugan deep fracture zone or the Altay-North Mongolian-Argun geosynclinal fold system, which was folded, thrust, and intruded during the Xingkaian orogeny (Yang and Xie, 1954; Ren and Chen, 1989; Zhao Jian, 1989, Chang and Tang, 1989). After that orogeny, only the area south of the Derbugan fault zone persisted as a geosynclinal system – the Tian Shan-Da Hinggan Ling (Tianshan-Xinggan) geosynclinal system. Geosynclinal development in this region in fact took place from north to south along the southern margin of the Siberian platform and from south to north along the northern margin of the Sino-Korean-Tarim platform. Ophiolitic belts become younger outward from both platforms (Xiao and Wang, 1984; Figure III-8).

One truly remarkable phenomenon of the Sino-Korean platform during the Paleozoic is the total absence of Upper Ordovician, Silurian, and Devonian deposits. Such deposits are common elsewhere in China. In general, the Middle Ordovician, everywhere a shallow-water shelf deposit, is overlain by continental coal-bearing Carboniferous and Permian rocks. The contact is slightly unconformable to paraconformable.

Xingkaian and 'Caledonian' events are well developed southeast of the Jiangnan Proterozoic orogenic belt, and are closely associated with it. Products of these events

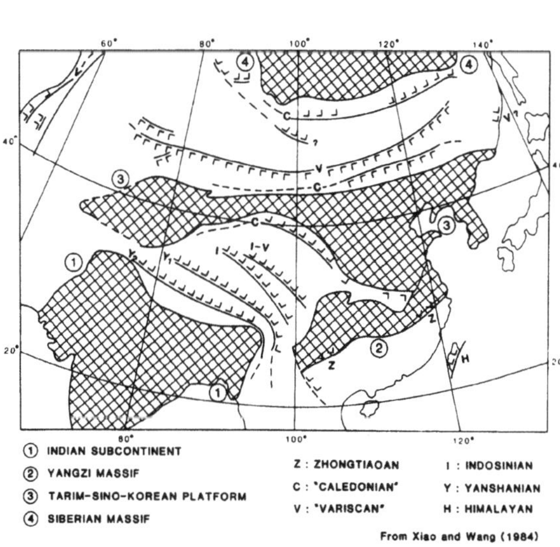

Fig. III-8. Sketch of main Phanerozoic ophiolitic suites in China. Shows accretionary growth pattern outward from Siberian and Sino-Korean-Tarim massifs.

– eugeosynclinal sediments, metamorphics of several grades, granitic bodies and migmatites – appear throughout South China and East China, from Hainan Island in the south to Hangzhou Bay in the northeast (Figure III-2; Pan Guoqiang, 1984). Guo Lingzhi et al. (1984) reported radiometric dates from this region as follows: 600 Ma for the metamorphics; 500 to 460 Ma, 480 Ma, and 477 Ma for three different bodies of migmatite. Episodes of deformation (or of cycles) in this huge region differed from area to area.

A Xingkaian geosyncline rimmed present-day China along its southern margin in the modern Himalaya, where Middle Cambrian granitic intrusives invade the geosynclinal section (radiometric dates of 550 to 500 Ma; Gansser, 1964; Mehta, 1977; Fuchs, 1982; Baud et al., 1982). Baud et al. (1982) postulated that the Middle Cambrian orogeny that tectonized the present Himalaya correlated with a Pan-African event, a postulate which they thought to be necessary because (1) they knew of no correlative northern European event (apparently they had not considered Norway or Asia) and (b) they noted that the Middle Cambrian orogeny (550 to 500 Ma) correlated with the terminal episode of the Pan-African orogeny (Baud et al., 1982, p. 184). Such postulates, however, are quite unnecessary in view of the known tectonic history of present conterminous Asia.

Xingkaian events, largely unreported along the southwestern sides of the Yangzi paraplatform, may yet be found to be widespread as investigations in the field continue. The Proterozoic rock exposures of Yunnan Province (Sun and Lu, 1985) may contain records of Xingkaian orogeny similar to (and on strike with) the Xingkaian events recorded far toward the northwest in the Himalaya (Huang Jiqing et al., 1987). Pre-Xingkaian and Xingkaian events occurred also in the Japanese islands (Matsumoto et al., 1968; Shibata et al., 1970; Sugisaki et al., 1972). No doubt similar events will be found on the Qinghai-Xizang Plateau, where Ordovician, Silurian, and Devonian rocks overlie the Pre-Cambrian unconformably (Chang and Pan, 1981, 1984).

Post-Xingkaian orogeny of the 'Caledonian' cycle is recorded in much of China, but is not as important as 'Variscan' orogeny. Within the Tian Shan-Da Hinggan geosynclinal system, two belts were intensely active, the Altay and Beishan geosynclines and their extensions (mainly in the U.S.S.R. and Mongolia) and the Kunlun Shan-Qilian Shan (Nan Shan)-Qin Ling Geosyncline, which was separated from the Xizang-Yunnan area by the developing (or already-developed) Jinsha Jiang-Hong He (Red River) fault zone (Huang Jiqing, 1978). The position of the Kunlun-Qilian-Qinling geosyncline was controlled in large part by the Kunlun Shan-Qin Ling deep fault zone. This geosyncline, once folded and intruded, shifted southward to the belt of the North Qaidam-South Qinghai Hu (Koko Nor)-North Qin Ling-North Huaiyang fault zone. The Altay geosyncline on the north extended beneath part of the Junggar platform. In eastern China, the South China geosyncline developed actively and repeatedly during the 'Caledonian' orogenic cycle (Huang Jiqing, 1978). Most of the rock associations that accumulated in the active geosynclines are eugeosynclinal in facies.

The best-developed part of the Kunlun Shan-Qilian Shan (Nan Shan)-Qin Ling geosyncline appears in the Nan Shan (Qilian Shan). Development of the Nan Shan (Qilian Shan) eugeosyncline is typical of the evolution of Chinese mobile belts during the 'Caledonian' orogenic cycle. The Nan Shan geosyncline thus is a polycyclic one, with (1) several phases of magmatism, deposition, and orogeny (Huang Jiqing, 1980, 1984) and (2) persistent southward migration from cycle to cycle (Wang and Liu, 1976). At least four ophiolitic belts are mapped, although because elements of these possibly were somewhat transported tectonically from one place to another as thrust nappes, they are not everywhere in place.

The Nan Shan (Qilian Shan) geosyncline developed in five cycles. Cycle I, Sinian (s.s.), includes a moderate development of ophiolites associated with large volumes of spilitic basalt. Many of the basalts are tholeiitic, and their effusions generally were followed by the deposition of thick deep-water sequences, very few of which are flysch deposits, unlike some subsequent cycles. At the end of the Sinian, folding and granitic intrusion took place. The early granitic plutons were syntectonic and therefore concordant. Later plutons were discordant. Cycle II developed in two stages. The first, subcycle II, belongs to the Middle Cambrian, and its geosynclinal sequence was deposited after the Xingkaian orogeny. Lower Cambrian is everywhere absent. The rock suites are much the same as those of Cycle I, except that the development of ophiolite is more advanced. In subcycle II_2, Late Cambrian through Early Ordovician, development of ophiolite reached a maximum. Development of flysch was ubiquitous. Granitic intrusions were larger, better developed, and folding was stronger. Cycle III rocks, Middle and Late Ordovician, indicate that the orogenic cycle had passed its peak. Development of ophiolite occurred mainly in the present South Nan Shan. Flysch, however, is better developed, as were granitic intrusive rocks and folding. Cycle IV, the last important tectogenic phase, includes the entire Silurian. Adesitic volcanism is important, and alternates with basaltic episodes. Ophiolites are minor, but the flysch is the best developed of all cycles. Folding and granitic intrusion reached their peaks. Cycle V, post-tectogenic, consists of a Devonian and Carboniferous molassic phase (Huang Jiqing, 1980) with only minor 'Variscan' geosynclinal development. (Huang Jiqing, 1980, gave the time of folding of the Kunlun-Qilian-Qin Ling system as end-Silurian; Chang and Pan, 1981, wrote that the Kunlun Shan geosyncline was folded between Early Devonian and Middle Devonian.)

Four prominent ophiolitic belts present in the Nan Shan become younger southward. These are Sinian, Middle Cambrian, and Early Ordovician. The ophiolitic belts range in length from 600 to 1,000 km, and each includes peridotite, harzburgite, tholeiite, spilitic basalt, radiolarian chert, siliceous limestone, and graywacke. All ophiolitic belts are reported by Wang and Liu (1976) to be in the hanging walls of thrust sheets and nappes, not in the footwalls. Belts of glaucophane schist are prominent.

Another 'Caledonian' sequence occurs as the Qin Ling series of troughs at the eastern end of the Kunlun Shan-Qilian Shan-Qin Ling geosyncline (Huang Jiqing, 1984a). The southern Qin Ling-northern Daba Shan, eugeosynclinal, is characterized by mafic and intermediate intrusive bodies with abundant andesitic extrusives in the Silurian. On the north, the main or central Qin Ling is underlain by thick miogeosynclinal carbonates, Paleozoic through Triassic. Many 'Variscan' granitic intrusions are present, especially farther north in the northern miogeosynclinal Qin Ling.

Late Paleozoic

Most geosynclinal development (almost all of it eugeosynclinal) in western and central China began after the orogeny, that is, in the Cambrian. Ordovician and Silurian volcanic and sedimentary accumulations, folded and intruded one or more times during the 'Caledonian' cycle, were succeeded by 'Variscan' sequences and orogeny. Within the Tian Shan-Da Hinggan geosynclinal system, the following eugeosynclines formed (the only exception being a miogeosynclinal sequence in the southern part of the Tian Shan geosyncline): Altay geosyncline, whose development continued well into the 'Variscan' cycle; Junggar belt, which persisted locally into the Carboniferous; Tian Shan geosyncline, eugeosynclinal in the north, miogeosynclinal in the south; Nei Monggol-Da Hinggan Ling geosyncline, including the Beishan trough; and Jilin-Heilongjiang geosyncline. Within several geosynclines, eugeosynclinal conditions changed to miogeosynclinal. In the Kunlun-Qilian-Qin Ling geosynclinal system, the geological history of the Qilian Shan (Nan Shan) has appeared in the preceding section. On trend toward the west and toward the east respectively, the Kunlun Shan and Qin Ling troughs, unlike the intervening Qilian Shan (Nan Shan) trough, developed only after the Xingkaian orogeny. The termination of cyclical geosynclinal development within this trough system also differed, with the Nan Shan cycles completed at the end of 'Caledonian' pulses and the other segments during 'Variscan' pulses. The southward shift of orogenic activities with the belts is typified best, perhaps, in the Qilian-Nan Shan system, which had shifted to the northern edge of the present Qaidam basin by the end of 'Caledonian' pulses and to the southern edge of the basin (in the eastern Kunlun Shan trough) during 'Variscan' pulses. The Kunlun Shan trough was deformed for the first time during the 'Caledonian' cycle, but also was deformed during the last part of the Early Devonian, and several times during the 'Variscan' cycle (Huang Jiqing, 1978, 1984a; Chang and Pan, 1981, 1984). The maximum time of development of several major geosynclines took place during the Devonian and Carboniferous as, for example, in the Tian Shan, Junggar, Altay, and Kunlun (Huang and Wang, 1984). In the Kunlun Shan, up to 10,000 m of Carboniferous alone reportedly has been measured. In contrast, the Qin Ling deformation lasted an even longer time, with the last - miogeosynclinal - stage of the Qin Ling trough extending into the Mesozoic, deformed by Indosinian orogeny (Huang Jiqing, 1978).

We should note that deformation of all these east-west ranges of Northwest China, North China, and Northeast China did not cease with the close of the geosynclinal histories of each. Indosinian, Yanshanian, and Himalayan compressive episodes continued through the post-orogenic history into what Chen Guoda (1984b) has called the 'diwa' stage, with cycles of compression and uplift that continue even today (Table III-1).

Other geosynclinal belts, both eugeosynclinal and miogeosynclinal, lay parallel with and south of the Northwest China orogenic belts. At the western end of the Lhasa foldbelt, or block, at Rutog (Figure III-6), deep-water turbidites recently were discovered - reportedly 3,500 m thick - with Late Carboniferous and Early Permian faunas (Hu Changming, 1984). Eugeosynclinal zones in Southwest China lie in southern Qinghai Province, southeastern Xizang (Tibet) Autonomous Region, and western Yunnan Province. From northeast to southwest they include the Burhan Budai Shan-A'nyemaqen Shan geosyncline (Burkhan Buddha Shan-Amne Machin Shan), the Jinsha Jiang (Jinsha River) eugeosynclinal belt in the western part of the Songpan-Garze foldbelt, and the Sanjiang foldbelt or eugeosyncline (Figures III-2, III-3 and III-6). The Burhan Budai Shan-A'nyemaqen Shan occur in the southern part of the Kunlun Shan-Qin Ling geosynclinal system, a large part of which, during the 'Variscan' orogenic cycle, was miogeosynclinal, with the Burhan Budai Shan-A'nyemaqen Shan as a notable exception (Huang Jiqing, 1978). The 'Variscan' eugeosynclinal facies of the Jinsha Jiang formed along the boundary between the Songpan-Garze and Sanjiang geosynclines, closely associated with the Jinsha Jiang-Hong He (Red River) deep fault zone. (Figures III-2, III-3, III-6, and III-7). This deep fault zone is attributed generally to Indosinian, or Mesozoic, events, but its movement began during the late Paleozoic or earlier. The Jinsha Jiang eugeosynclinal facies is associated with Permian and Triassic ophiolites (Chang and Pan, 1984). It is true, however, that much of the orogenic history of the Songpan-Garze and Sanjiang geosynclinal belts is Indosinian and younger. In other parts of China, ophiolitic belts become younger away from the shield or platform elements, and this area is no exception (Xiao and Wang, 1984; Figure III-8). The little that is known regarding Paleozoic and younger events in these mobile belts was summarized succinctly by Huang Jiqing (1978), Huang Jiqing et al. (1980, 1987), Chang and Pan (1981, 1984), and Zhang Qi et al. (1989).

The only other 'Variscan' belt that has been described from the mainland is the Southeast Maritime geosyncline which extends from Hainan Island along coastal East China (Figures III-2 and III-3). This geosyncline is offset southeastward and eastward from the older South China 'Variscan' geosyncline. More 'Variscan' events undoubtedly remain to be discovered on the mainland of China as geologic investigations proceed.

On Taiwan, Yen (1953) and Ho (1967a, 1967b) described Permian and possibly Carboniferous meta-limestones with fusulinids and other late Paleozoic fossils. These are associated with 'Variscan' ophiolites, eugeosynclinal sediments, cherts, and glaucophane schists. An ophiolite-bearing sequence of similar age is known from the southern Ryukyu island arc (Hanzawa, 1932, 1933; Foster, 1965; Ishibashi, 1969; Kizaki, 1986). Fusulinid-bearing meta-limestone is present in the ophiolitic-eugeosynclinal sequence. The same belt, or a closely similar belt, extends through the Philippine island arc and beyond (Gervasio, 1967; Holloway, 1982; Guo Lingzhi et al., 1984). Folding and mild metamorphism affected the entire region during a major 'Variscan' orogeny at the close of the Permian. The presence of more than one parallel or subparallel eugeosynclinal sequence along the later Paleozoic Pacific margin has not been discussed before in regional tectonic histories.

Huang Jiqing (1980) has described the complex history of a 'typical' geosyncline that developed during the middle and late parts of the 'Caledonian' orogenic pulses and the younger 'Variscan' orogenic cycle, and a few of the troughs that remained active until well into the Mesozoic Indosinian cycle (Table III-1). In China, the formation of these geosynclines began with the end of the Xingkaian orogeny: (in the U.S.S.R., the history of the Tian Shan geosyncline began in late Riphean time, unlike China; Kravchenko, 1979). During the Ordovician and the Silurian, deep and rapidly subsiding troughs developed, partly filled with thick flysch-like and other geosynclinal deposits and volcanics, many of them submarine. Ophiolitic suites with mafic to ultramafic intrusions are abundant, together with associated radiolarites. The western Junggar orogenic belt contains a particularly well-developed ophiolitic suite (Huang Jiqing, 1978). Huang Jiqing (1980) described development of the polyphase, multicycle Tian Shan geosyncline briefly (the northern or eugeosynclinal part), and we present here a summary of his description (Huang Jiqing, 1978, 1984a).

Cycle I, middle and late 'Caledonian' succeeded the Xingkaian orogeny, accompanied by extrusions of large masses of andesite and andesitic tuff, especially in the U.S.S.R. portion of the geosyncline. Late Silurian folding was accompanied by syntectonic and post-tectonic intrusion of plagiogranites, especially albite granites. Cycle II, early 'Variscan,' encompasses Devonian time. Andesite, geosynclinal and flysch-like sediments, and mafic/ultramafic rock associations are abundant. The sequence was folded and intruded by plagioclase (especially albite) granites. Cycle III, the middle 'Variscan,' took place during the Carboniferous. During this period, three stages of andesite and associated geosynclinal deposits

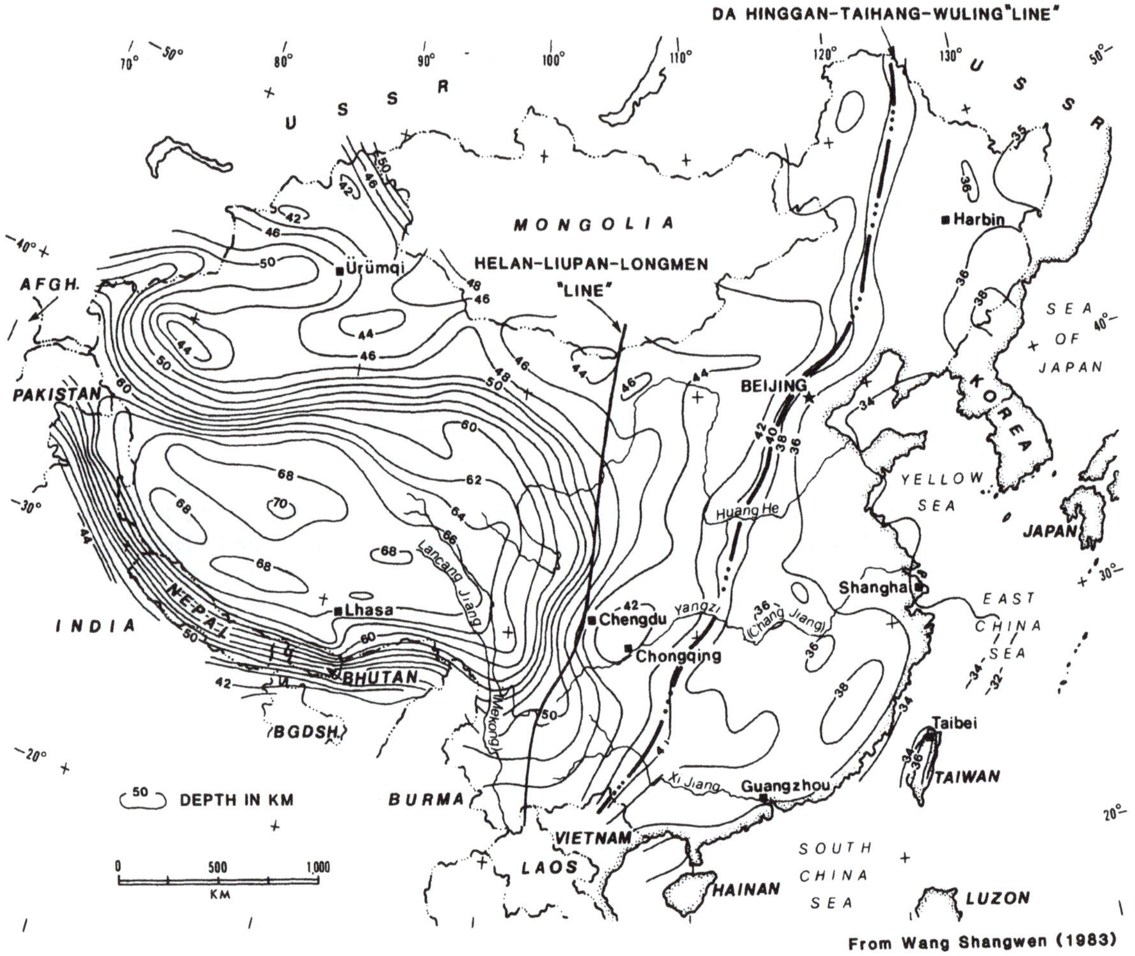

Fig. III–9. Map showing approximate depth to Mohorovicic discontinuity in China. Depth contours in kilometers.

accumulated, including one sequence of thick flysch-like sediments, two stages of mafic and ultramafic intrusion, and three stages of folding and intrusion, this time by more normal granitic rocks. Cycle IV, of the Permian, was accompanied by the accumulation of thick geosynclinal deposits, two stages of andesite and andesite-basalt outpouring, one flysch episode, two foldings, and the intrusion of potassic granite, alaskite, and some syenite. The andesite is perialkalic. After the final 'Variscan' paroxysm at the end of Permian time, Triassic post-geosynclinal molasse deposition set in, accompanied by uplift.

In this way the Pal-Asia continent or tectonic domain (Figure III–7) was thoroughly consolidated. Because consolidation took place in the north earlier than in the south, extensive swampy lowlands were able to spread across present Northeast China and North China. Thick coal measures developed on these lowlands during different intervals of the Carboniferous and the Permian. Marine waters almost vanished from northern China, and retreated to present Southwest China, Central-South China, and East China, with one small marine embayment along the northern flank of the present A'nyemaqen Shan in central Qinghai Province. By the end of the Permian, marine waters, with rare exceptions, had retreated from all of China except part of Southwest China (Wang and Liu, 1980; Cui Kexin, 1984; see Table III–1), and the first of the great inland intermontane basins (e.g. Junggar) had begun to form (Zhu Xia, 1984; Kamen-Kaye et al., 1988).

Mesozoic-Cenozoic

Introduction

The Mesozoic was a time of profound tectonic changes in China, changes that became more marked during the Cenozoic (Huang Jiqing, 1978, 1984a; Chen Guoda, 1984b; Liu Hefu, 1984). During the Mesozoic, this complex of geosynclines, deep fracture zones, and platforms was divided into two fundamentally different areas separated by a north-south transition zone 650 km to 720 km wide. The western zone occupied all of China west of a line of structure striking North-South to N 10° E. These include the present Helan Shan, Liupan Shan, and Longmen Shan (Figures III–5, III–7, III–9, and III–10). The eastern zone occupied all of China east of a N 10° E-striking zone that includes, from north to south, the Da Hinggan Ling (Great Khingan, Kin'gan), Taihang Shan, and Wuling Shan. This is the trend that Li Siguang

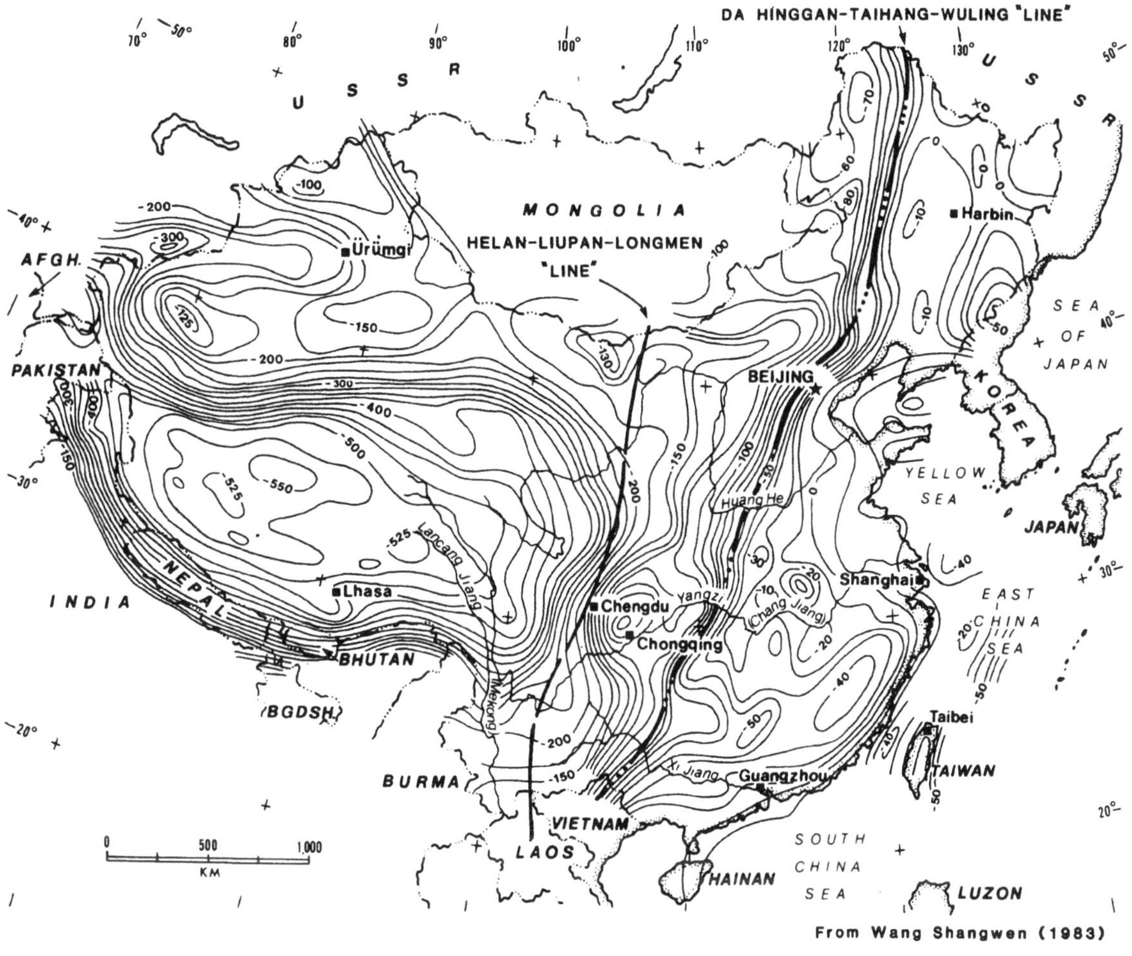

Fig. III-10. Bouguer gravity map of China, contour values in gal.

(J. S. Lee, 1939) called the Da Hinggan-Shanxi-Guizhou anticlinorium. Both the Helan-Liuping-Longmen trend and the Da Hinggan-Taihang-Wuling trend are marked today by a sharp change in the depth to the Mohorovicic discontinuity (Figure III-9) and by a steep gravity gradient (Figure III-10). These two N 10° E trends are considered to be quite old by Huang Jiqing (1978), but their true age(s) cannot be established except inferentially.

The *western* or *Xiyu zone* (Xi = west) was characterized by strong east-west structural trends, decreasing geosynclinal activity, north-south compression, uplift of the former geosynclines and mobile belts, creation of large continental, intermontane basins with triangular, rhomboid, and polygonal shapes, and marked strike-slip faulting. The *eastern* or *Cathaysian zone* was overprinted by northeast and north-northeast trending strike-slip and normal faults, compressional and tensile horst-and-graben structures (many geologists in China liken its structural development to that of the American Great Basin), continental volcanism and sedimentation associated with the fault trends, and the intrusion of major batholiths that trend northeast and north-northeast. The predominantly north-northeast trends of Meso-Cenozoic age are termed 'Neocathaysian.' Many large and important ore-bearing granitic bodies intruded existing rock where the Pre-Cambrian and Paleozoic east-west trends intersect the Neocathaysian fault trends. The *transition zone* between east and west includes the structurally dismembered interior continental basins of eastern Mongolia and a part of Nei Monggol Autonomous Region, the rectangularly-shaped Eerduos basin, and the rhomboidal-shaped Sichuan basin.

Mesozoic-Cenozoic geosynclinal and orogenic activities customarily are divided into three principal cycles, the Indosinian, the Yanshanian, and the Himalayan (Table III-1). Geosynclines of the Indosinian cycle persisted locally in North China and Northwest China, as in the Qin Ling. Most Indosinian geosynclinal activity took place in present Xizang (Tibet) Autonomous Region, Qinghai Province, and Yunnan Province. The former 'Variscan' Southeast Maritime geosyncline shifted farther southeast and east, with a new mobile zone very active from the Litian-Jinsha Jiang-Hong He (Red River) fault zone northeastward through the modern Taiwan Strait and Taiwan, thence beneath the present shelf area of the East China Sea to Japan. We term this mobile belt the 'China Sea geosynclinal belt,' mainly

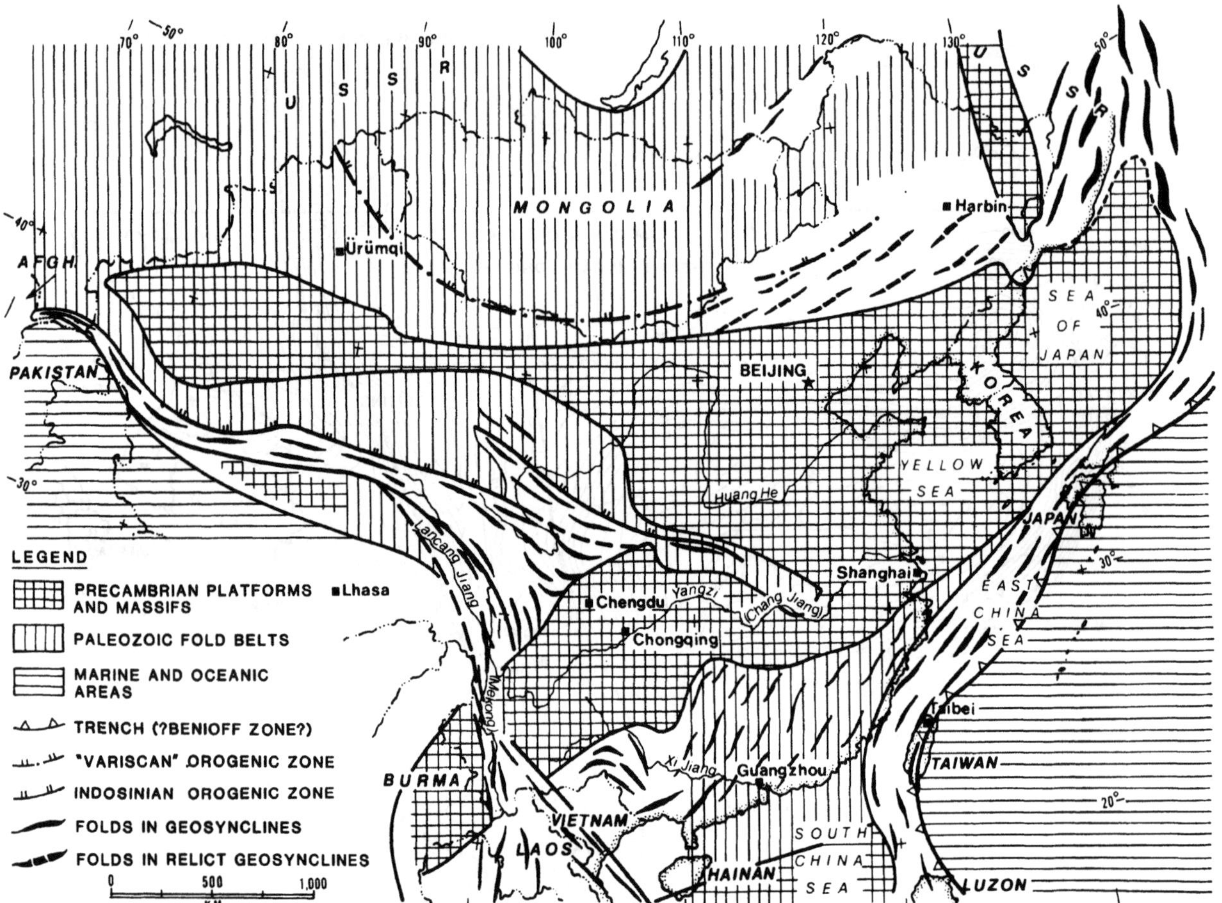

From Ren Jishun (1984)

Fig. III-11. Paleotectonic map of China and neighboring areas at the end of the Indosinian orogenic cycle (end of the Early Jurassic).

a Yanshanian feature. Figure III-11, from Ren Jishun (1984), presents a paleotectonic map of China after Indosinian orogeny. In addition to the active mobile belts, the map shows the marked and considerable extent of reactivization of old mobile belts in Northeast China, East China, and Central-South China.

During the Yanshanian orogenic cycle (Table III-1), the geosynclinal activity of the Xizang-Qinghai-Yunnan area shifted toward India, while the China Sea geosynclinal system shifted slightly toward the Pacific. Yanshanian orogenic activity affected the Nadanhada Ling, a southeast limb of the U.S.S.R.'s Sikhote-Alin geosyncline (Meyerhoff, 1981), and the Upper Heilongjiang miogeosynclinal foldbelt was deformed (this is a part of the Mongol-Okhotsk foldbelt; Meyerhoff, 1981). All geosynclinal activity ceased in the remainder of interior China, although processes of activization were intense in former mobile belts. Subsequent Himalayan activity was even more restricted in terms of geosynclinal development, with the only important geosyncline developed between Taiwan and Japan beneath the present Taiwan-Sinzi foldbelt.

The differentiation of China into highly distinctive structural/tectonic provinces increased during the Yanshanian and Himalayan orogenic cycles. Both western China and Cathaysian China continued to be distinct, but with some significant changes. Intense north-south-directed compression continued to activise western and central China as it had done since earliest geological history (Figures III-1 and III-4). The sites of former Xingkaian (and older), 'Caledonian,' 'Variscan,' and Indosinian mobile belts were folded and refolded, with enormous uplifts taking place in parts of every former geosyncline, possibly as a result of movements that Chen Guoda (1984b) called 'diwa' tectonism. Southwestern China became distinct from the remainder of China as the many parallel and subparallel mobile belts of that region were subjected to compression laterally and to repeated episodes of southward thrusting, particularly during the last stages of Himalayan orogeny. Of great significance is the circumstance that the Xizang-Qinghai-Yunnan geosynclinal system and subsequent foldbelts do not extend east of the Helan Shan-Liupan Shan-Longmen Shan 'line' (Figure III-7, III-9, and III-10), which also marks the southwestern margin of the Yangzi paraplatform. In addition, folding and large-scale overthrusts were directed from west to east against the Yangzi paraplatform (Misch, 1945a, 1945b, 1946; Chang Da, 1959). In contrast, the thrusts of the Qinghai-Xizang (Tibet) Plateau were directed southward toward the Indian shield.

The Xizang-Qinghai-Yunnan thrust zone thus stands

apart and distinct from the remainder of western China where the rocks of east-west trends were compressed and uplifted vertically. The clear distinction between these two parts of western China was emphasized even more when, during the last two million years of geologic time, the present plateau and mountain region took form by an uplift of about 3,500 to 4,000 m (Li Ji-jun et al., 1981).

In eastern or Cathaysian-Neocathaysian China, compressional grabens continued to form, together with tensional grabens (Huang Jiqing, 1978, 1984a). Many strike-slip faults became graben boundaries. Thrust faults, directed mostly toward the east but some toward the west, formed along and near these faults (Zhu Zhicheng et al., 1982). Many of these thrust features may be parts of 'flower structures' (Harding and Lowell, 1979) or of 'kobergen structure' (Taner and Meyerhoff, 1990; Meyerhoff et al., 1991). In general, graben development becomes younger eastward, with Jurassic and Cretaceous grabens on the west, Cretaceous and Tertiary grabens farther east, Tertiary grabens, mainly Eocene to Miocene, in the coastal zone and offshore, and mainly Neogene basins along and east of the continental shelf. There are, however, many exceptions (e.g. the grabens of Wei He, Feng He, and Hetao). These basins developed in a compressional stress field exemplified by the strongly folded China Sea geosynclinal belt, exposed on Taiwan and present beneath the Taiwan-Sinzi foldbelt (Figure III-2 and III-3). Another compressional system was present in the Ryukyus, a point disputed by some (e.g. Letouzey and Kimura, 1985), and in the Philippines (Gervasio, 1967).

Two tectonic provinces thus characterize the area of eastern China, and probably have done so since the beginning of the Mesozoic. These two provinces are: (1) a notably broad zone of horst-and-graben structure, part compressional, part tensile; and (2) compressional island-arc-related systems along and just east of existing continental margins. The Eerduous-Sichuan transition zone remained unchanged except for increasing uplift of the former foldbelts within them.

The Mesozoic-Cenozoic in China could, therefore, be regarded as the time when the dominance of the Pal-Asiatic tectonic domain gave way to the dominance of the Tethys-Himalayan and Marginal-Pacific tectonic domains (Huang Jiqing, 1978, 1984a). The extent of these domains after Paleozoic time, as interpreted by Huang Jiqing (1978), is shown on Figure III-7. Huang Jiqing regards the Pal-Asiatic domain as coming to an end with the close of the Paleozoic, only to be supplanted, during later time, by the new Marginal-Pacific and Tethys-Himalayan domains. We take quite a different view. In our opinion, China's tectonic history during the Phanerozoic is one of complex interaction between an approximately north-south-directed stress field and an approximately east-west-directed stress field. During the Paleozoic, the north-south stress field dominated, and resulting east-west trends were even more widespread (into eastern China) than is obvious today. A Marginal-Pacific domain in fact existed, however, during Paleozoic time, as is proved by the orientation of the fold and fault axes of foldbelts of the South China and Southeastern maritime areas, and by the north-south Paleozoic structures of Taiwan, the Philippines, the Ryukyus, and the Japanese arc (Wan Tianfeng, 1989). Then, during the Mesozoic and the Cenozoic, the strength of the north-south stress field diminished, and the area of the Pal-Asiatic tectonic domain decreased, while that of the Marginal-Pacific domain increased. From this it should be clear that we regard the Tethys-Himalayan domain as a continuation of the Pal-Asiatic domain, and that we believe that the term 'Tethys-Himalayan' should be dropped.

The reasons for reduction of area of the Pal-Asiatic domain are, in our opinion, directly related to the decreasing compressibility of western China between the Siberian, Tarim, and Indian platform massifs, and to the construction and rise of the lithosphere beneath the Qinghai-Xizang Plateau.

From the historical viewpoint, we should mention that the ideas proposed here by us were implied in some of the writings of Li Siguang (e.g. J. S. Lee, 1939). He proposed to call the area affected by northeast-southwest trends 'Cathaysia,' a name which he attributed to A. W. Grabau (J. S. Lee, 1939, p. 211). Li Siguang subdivided Cathaysian history into three stages of differing areal extents: (1) Paleocathaysia (Sinian-Paleozoic, p. 211-218), (2) Mesocathaysia (early Mesozoic, p. 218-220), and (3) Neocathaysia (Cretaeceous and younger, p. 220-227).

Indosinian Orogenic Cycle

This cycle (Table III-1) covered a period of transition from the dominant north-south-directed stress system of the Paleozoic time to the more balanced distribution between north-south and east-west stress fields that dominated the Mesozoic and the Cenozoic. This cycle also covered the time when the last of the great east-west-geosynclines of western and central China completed their development, to be replaced during the Mesozoic and the Cenozoic by geological histories of vertical uplift, high-angle thrusting, considerable amounts of folding, and regmagenesis (wrench faulting) (Figures III-2 and III-3). Huang Jiqing (1984b) expressed the opinion that Indosinian orogeny in China reflected Gondwanan events.

The major geosynclinal belt to be affected was the Kunlun Shan-Qin-Ling system, where miogeosynclinal conditions became much more general than during earlier times, although some eugeosynclinal conditions existed (Xiao and Wang, 1984), as shown in part on Figure III-8. The principal sectors of the Kunlun-Qin Ling system that underwent geosynclinal deposition, cyclical or episodic folding, and magmatic intrusion, were the central Kunlun Shan-Bayan Har Shan, southern Qilian Shan area, and the Qin Ling trough. The southern Qilian area includes the southern Qinghai Hu (south Koko Nor) and the A'nyemaqen troughs. Miogeosynclinal deposition predominated (Huang Jiqing, 1978, 1984a; Cui Kexin, 1984). The last deformations took place during the Middle and Late Triassic. On the west, in the Soviet Union, the Kunlun Shan geosynclinal trough (North Pamir) underwent the same history as it did in China, with a final late Indosinian orogeny (Kravchenko, 1979; Stocklin, 1980).

Elsewhere in western China, geosynclinal activity continued in Xizang (Tibet) Autonomous Region, southwestern Qinghai Province, and Yunnan Province west of the Yangzi paraplatform. One of the most powerful orogenies to affect the Qinghai-Xizang Plateau deformed the Qiangtang block late in the Triassic (Chang and Pan, 1981, 1984). We discuss the tectonic development of Southwest China in a separate section.

An important point to emphasize is a total absence of evidence for the existence of a Paleotethyan Ocean, as is required in some plate-tectonic reconstructions. This evidence now appears conclusive to many important geologists and, as Stocklin (1981, 1984) and Helmcke (1985) have warned geologists and geophysicists alike, is '...too persistent, too alarming to be disregarded, any longer, or to be explained away by subduction' (Stocklin, 1984, p. 76; see also Gansser, 1981).

The Helan Shan-Liupan Shan-Longmen Shan 'line' (Figures III-9 and III-10) was in existence before the onset of Indosinian orogeny. Liu Hefu (1984) wrote that this structural trend came into existence after the 'Variscan' cycle and before the Yanshanian cycle. In our opinion, it was associated almost certainly with a structural/topographic feature of some kind well before the Sinian, because many Proterozoic and Phanerozoic groups of lithofacies and biofacies are influenced by a north-south feature that coincides with the present trend of Helan Shan-Liupan Shan-Longmen Shan (e.g. Qian Li, 1984; Zhang and Wu, 1985). Field studies which have shown that the fractures are inherited from pre-existing basement structures (Zhang and Wu, 1985), suggest to us that this north-south feature probably is pre-Sinian. The distribution of facies within the geosynclinal realms of western Yunnan Province, southwestern Qinghai Province, and Xizang (Tibet) Autonomous Region further demonstrates the presence of a barrier along the Helan Shan-Longmen Shan 'line.' The large amount of Indosinian tectonism along this 'line' (e.g. Longmen Shan trough) is further evidence of the importance of this structural zone early in Phanerozoic history.

Several of China's large petroleum-bearing, post-orogenic basins began to take shape late in the 'Variscan' cycle and early in the Indosinian cycle. These include the Junggar, Turpan, Tarim, and Eerduos basins (Cui Kexin, 1984). Geosynclinal troughs that had completed their development during the 'Caledonian' and 'Variscan' cycles underwent gentle to strong activization during the Indosinian cycle.

In eastern China, activization also predominated, most particularly on the Yangzi paraplatform, where old foldbelts were rejuvenated to some extent, uplifted, refolded, and intruded locally by silicic to intermediate magmas. In the present offshore area, events on Taiwan are not clear, but a well-defined Indosinian geosynclinal/orogenic cycle affected the development of the Ryukyus (Ishibashi, 1969) and the Philippines (Gervasio, 1967; Holloway, 1982). Indosinian granitic intrusive rocks are ore-bearing in many areas of southeastern Asia, but seem to have been more important during this tectonic cycle in Malaysia and Indonesia than in China (Hutchinson, 1983). Ren Jishun (1984) notes some mineralization during Indosinian magmatism, and the occurrence of some folding of the China Sea geosynclinal belt.

Withdrawal of marine waters from present day China began during the late Palezoic. Withdrawal was all but complete by the end of the Triassic and by completion of the Indosinian cycle. Major marine incursions henceforth are found only under the present continental shelf and in Southwest China.

Yanshanian Orogenic Cycle

Continued compression, regmagenesis, and uplift during this cycle characterized both western China and the transition zone between itself and eastern China. Subsiding basins between rising mountain ranges developed foredeeps adjacent to those ranges, and a continuing compression produced repeated episodes of folding and faulting within the foredeeps (Huang Jiqing, 1978, 1984a; Liu Hefu, 1984). Compared with eastern China, Yanshanian orogenesis to the west was fairly mild (Liu Hefu, 1984).

In eastern China, regmagenesis, refolding, and uplift of 'Caledonian,' 'Variscan,' and Indosinian foldbelts also took place. Regmagenesis was much more intense, however, with a complex system of Neocathaysian faults developing along pre-existing lines of weakness themselves developed along Indosinian, Paleozoic, Xingkaian, and older orogenic trends. The northeast-trending and north-northeast-trending Neocathaysian faults include numerous wrench faults, parallel and associated with normal faults. Horst-and graben structure is pervasive. Old foldbelts, as elsewhere in China, were uplifted, refolded, and faulted. Unlike western and central China, however, important continental volcanism accompanied wrench faulting and graben formation. Andesitic to rhyolitic volcanic rocks, especially common, are associated with redbeds and lacustrine strata.

Horst-and-graben structures, and half-graben structures, Jurassic and younger, occur mainly east of the Da Hinggan Ling-Taihang Shan-Wuling Shan 'line' (Figures III-9 and III-10), but there are additional structures (e.g. Hailar, Erlian, Hetao-Yinchuan, Fen-Wei and Baise) west of this 'line.' These later tensional basins lie between the Da Hinggan Ling-Taihang Shan-Wuling Shan 'line' and the Helan Shan-Liupan Shan-Longmen Shan 'line' (Figures III-9 and III-10). Huang Jiqing (1978) wrote that the gravity gradient associated with the Da Hinggan Ling-Taihang Shan-Wuling Shan 'line' is post-late Mesozoic, while Wang and Quan (1984) favored a Late Triassic date. We suspect that the structure beneath the 'line' is related to a fault zone (or fault zones) much older than Late Triassic. Because the Wuling Shan segment is the site of a Proterozoic deep fault zone along the Jiangnan massif, itself reactivated during the Mesozoic in a tensional stress field.

Two geosynclinal systems penetrate slightly into eastern onshore China, one into northwest Heilongjiang Province from Mongolia and the other into southeastern Heilongjiang Province from the U.S.S.R. The former is related to the Mongol-Okhotsk geosynclinal complex; the latter forms part of the Sikhote-Alin geosyncline. Both date to the early Palezoic and/or the late Pre-Cambrian (Meyerhoff, 1981), and both were folded during the Yanshanian cycle.

In the present offshore region off eastern China, the geosynclinal belt of the China Sea developed. In Taiwan, Cretaceous eugeosynclinal rocks lie unconformably on tectonized Permian and older sequences (Yen, 1953; Ho, 1967a, 1967b; T. C. Huang, 1978). Known Cretaceous on the island includes Neocomian (?), Aptian, and younger rocks (T. C. Huang, 1978). Late Paleocene and younger strata are known below the Miocene, all in eugeosynclinal facies (T. C. Huang, 1977, 1979, 1980; Huang and Chi, 1979; Ting, 1979). Unconformities undoubtedly are present in Lower Cretaceous, Upper Cretaceous, and Paleogene sequences, and continuous exposures have not been observed on the surface or their equivalents penetrated in wells. The eugeosynclinal facies does underlie the entire western and central parts of the island, as far as is known. On the coastal plain, Miocene overlies Paleocene and Lower Cretaceous directly (T. C. Huang, 1978; Huang and Chi, 1979; Ting, 1979). In the coastal ranges, one granitic pluton, 86 Ma (Santonian-Campanian) was reported by Chai (1972). Thus the Yanshanian orogenic cycle is well represented on Taiwan, and at least two phases of orogeny are inferred. The Taiwan orogenic belt extends northeas tward to Japan beneath the present East China Sea shelf.

Cretaceous rocks are unknown from the Ryukyus (Kizaki, 1986), where late Eocene overlies older rocks with a pronounced angular unconformity (Letouzey and Kimura, 1985). The Cretaceous, and the Yanshanian cycle, are well represented in parts of the Philippines where at least three orogenic episodes have been recorded (Gervasio, 1967; Holloway, 1982).

From the very beginning of the Yanshanian cycle, China consisted of at least five distinct structural provinces: the Qinghai-Xizang geosynclinal province; the Northwest China area of reactivated foldbelts; the Eerduos-Sichuan 'transition' zone, the eastern horst-and-graben province; and the Taiwan-Japan eugeosynclinal province. A possible second northward extension to Japan via the submerged parts of the Ryukyus is likely. South of Taiwan, two eugeosynclinal branches have been established. The first passes through the Philippines (Gervasio, 1967). The second continues offshore to the Lilian-Jinsha Jiang-Hong He (Red River) fault zone southwest of Taiwan and Hainan. Wells drilled offshore from China in the South China Sea have, in a few localities, reached granites which in order of drilling yielded K-Ar dates ranging from 49 Ma to 92 Ma (early Miocene to Turonian; Zu Dowei, April 1984, oral commun; Huang and Chen, 1987, p. 81).

The Yanshanian orogenic cycle encompasses one of China's most important epochs of ore genesis. The great deposits of tin, tungsten, and related minerals of granites in eastern China (lower Yangzi, Nan Ling, etc.) formed during this cycle.

Of the five structural provinces named above, we discuss two in more detail after we present our summary of events in the main Himalayan orogenic cycle.

Himalayan Orogenic Cycle

The Himalayan orogeny, as the name indicates, is most prominent in Southwest China (Huang Jiqing, 1984a; Liu Hefu, 1984; Table III-1). In Northwest China and North China, folding, faulting, uplift, and regmagenesis continued. Most crustal shortening took place by folding and faulting in the foredeeps of the great interior basins as the towering ranges of Central Asia reached record elevations. Thrusts involving Pleistocene and even Recent sediments along the margins of the interior basins are common place. In Southwest China, tectonic events took place that terminated in the uplift of the Qinghai-Xizang Plateau.

In eastern China, new faults of Cathaysian trend appeared. Uplift, regmagenesis, and related deformation affected pre-existing Phanerozoic foldbelts. Old horsts-and-grabens, which had formed during the Jurassic and the Cretaceous, were rejuvenated; some underwent at least one new cycle of faulting and related deposition. The horst-and-graben development spread eastward, with progressively younger fault basins developing toward the South China Sea and the East China Sea. Very few exceptions to this statement are known, the most notable being the late Tertiary Shenxi (Wei-Fen) graben system (Liu Hefu, 1984; Zhang and Zhang, 1984). Tapponnier and Molnar (1977) related this graben system to tension resulting from sinistral movements on major east-west shears. In some of the older graben basins, such as the Songliao, orogenic activity had ceased before the onset of the Himalayan orogenic cycle. In almost all onshore basins and in many offshore basins, the last important episode of normal faulting took place by late Oligocene to early Miocene. Important normal faulting during and after the Miocene was restricted to the eastern most basins. Very few exceptions to this statement are known, the most notable being the Tertiary-Quaternary Wei-Fen and Helao-Yinchuan graben systems. Important fissure eruptions took place in eastern China during the Cenozoic, with formation of a few large volcanoes along the present border with Korea et al., 1938). Folding and thrusting created the Taiwan-Sinzi foldbelt at least from southeastern Taiwan to the Japanese islands. Compression continued on eastern and central Taiwan, where thrust faulting and great gravity-glide phenomena continue today. Attempts to explain the origins of the Philippine Sea and the Okinawa Trough in terms of regional tectonics have been mechanically unsuccessful. Investigators attribute the Okinawa Trough to 'backarc spreading' (Hilde et al., 1977; Herman et al., 1979; Letouzey and Kimura, 1985) behind a compression ('subduction') zone, two phenomena that physically are quite incompatible where they act upon the same body. An alternate mechanism is the topic of a subsequent section.

The Himalayan orogenic cycle continues to the present day. Deformation continues within the eastern China basins, although on a much reduced scale. Many deep fracture zones still are active (Figure III-5). The island arcs and the island of Taiwan are markedly mobile tectogenic belts. The Himalayan-Tibetan region stands out as one of the more spectacular orogenic systems in the world. In the next few sections, we consider several aspects of the active tectonics of China and of its tectonic regions.

Fracture Systems

Liu Defu et al. (1983) used Landsat images to prepare a map (Figure III-12) of the 'lineaments' of China. China is one of those unusual regions where most lineaments possess real structural (and not imaginary) significance, as defined by Wise (1982) and Wheeler and Wise (1983). Comparison of Figure III-12 with Figures III-1 thru III-7, and with Figure III-11, shows the profound influence of deep faults on the structural history of China.

Liu Defu et al. (1983) compiled the fracture patterns and the lineamental patterns of China from Landsat photographs, and they reached the following conclusions:

1. The predominant fracture directions of western China are NW, WNW, and ENE; those of eastern China are NE, NNE, and NNW. Thus eastern and western China are clearly separable on the basis of their fracture-lineament systems.
2. Lineaments and fractures in fold systems tend to be long and densely clustered. Those in platforms and stable massifs are not clustered, lie much farther apart, and occur in short segments (Figure III-12).

Additional possible conclusions:

3. Three east-west trends (actual strike: N 80° W) cross all of China. Li Siguang (1973) named them as (a) Tian Shan-Yin Shan, (b) Kunlun Shan-Qin Ling, and (c) Nan Ling. We discuss all three in detail in a subsequent section. Zhu Songnian (1984) recognized probable continuity of the Nan Ling fault zone

From Lu Defu et al. (1983)

Fig. III-12. Landsat lineament map of China.

through the Qinghai-Xizang Plateau, and named this fracture zone the 'Himalaya-Nan Ling' fault zone. We ourselves, however, believe it possible to be more specific, and to recommend that existing names be replaced by the term 'Yarlung Zangbo-Nan Ling' zone.

4. Neocathaysian overprinting begins east of the Helan Shan-Liupan Shan-Longmen Shan 'line,' and becomes dominant east of the Da Hinggan Ling-Taihang Shan-Wuling Shan 'line'. Despite overprinting, the older trends, largely Pre-Cambrian and Paleozoic, stand out clearly.
5. Between and parallel with the east-west fracture trends lie several stable blocks whose distribution suggests that they are parts of a much larger stable block covering most of China, formed before establishment of the east-west fracture zones. (Figure III-13; compare this figure with Figure III-1, on which Pre-Cambrian massifs are shown.) The stable blocks, like the east-west fracture zones, cross all of China, including one line of stable massifs from the Qiangtang-Nyainqentanglha block of northern Xizang (Tibet) Autonomous Region to the Jiangnan uplift in eastern China (locations are on Figures III-13 and III-1).
6. Fracture and lineament systems cross one another without great offsets.
7. These considerations together suggest that no part of China has undergone large-scale horizontal (lateral) translation since the earliest Pre-Cambrian. Corollaries of this conclusion are: (a) strike-slip offsets in China are minor; (b) the huge thrust displacements visible and, to some extent, measurable on the surface are superficial; and (c) the Qinghai-Xizang block was never far from its present position (Taner and Meyerhoff, 1990). We discuss this conclusion in more detail subsequently. Suffice it to say that, if the Yarlung Zangbo east-west shears are a continuation of the Nan Ling east-west zone, little movement of the Qinghai-Xizang area has taken place. The Yarlung Zangbo fractures most likely are a continuation of the Nan Ling zone because they are the same distance (ca. 850 km) from the Kunlun Shan-Qin Ling fault zone at both ends–in Southwest China and Central-South China–and they possess the same strike (Chinese Inst. Geol. Sci., 1975, 1976).

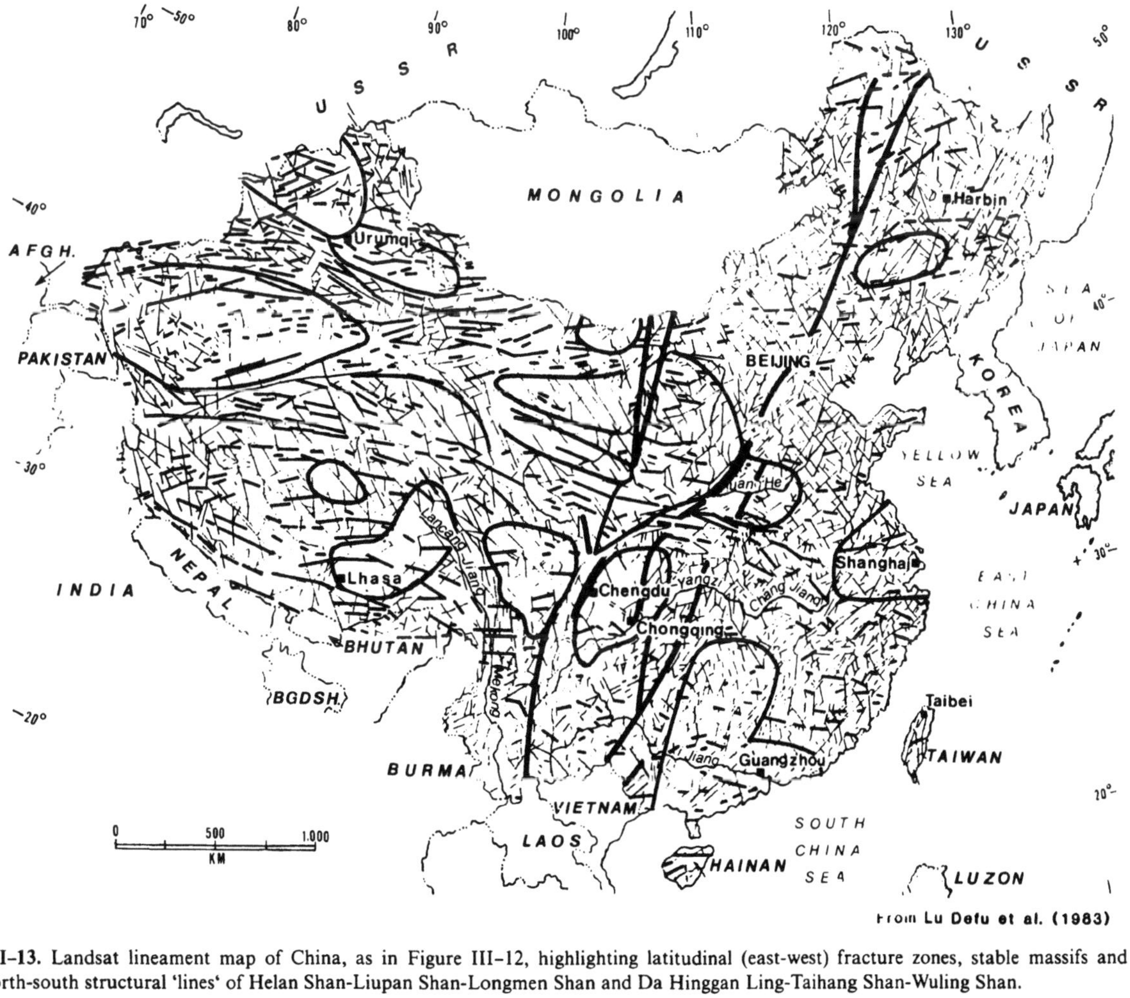

From Lu Defu et al. (1983)

Fig. III-13. Landsat lineament map of China, as in Figure III-12, highlighting latitudinal (east-west) fracture zones, stable massifs and the two north-south structural 'lines' of Helan Shan-Liupan Shan-Longmen Shan and Da Hinggan Ling-Taihang Shan-Wuling Shan.

8. The massifs of the present area of China once were much larger than has been suggested by most current workers in China with the exception of Zhang Wenyou and his colleagues. We do not imply by this that we are 'followers' of Zhang Wenyou's 'school of tectonics', but we are concerned to point out that Zhang Wenyou's view that horizontal (lateral) movements in China have been subordinate seems to be well founded in geological facts.

Close relations between the presently active faults of China and the fracture-lineament patterns are evident from comparison of Figures III-5, III-12, and III-13.

Those who disagree with the conclusions presented here may argue that the fracture patterns and the lineamental patterns of Figure III-12 are extremely young features and that, essentially, they tell us little about older geologic events. Such a conclusion ignores the evidence that the major geologic and structural trends brought out by Figure III-12 and III-13 are dated in the field as old. The east-west Pre-Cambrian trends that cross all of China serve most importantly as an example. The north-northeast trends of eastern China, or some of them at least, also are old, dating back to the Proterozoic in some cases (e.g. Tan-Lu fault zone; Sun Ronggui et al., 1983). More important, perhaps, is the long-established relationship that fracture traces mapped on satellite photographs and air photographs, properly (and, where possible) checked in the field, in most cases reflect diastrophic movements and/or compaction above old fracture zones and faults in the basement, an observation demonstrated by numerous authors (e.g. Wise, 1969; Plafker, 1976). The features mapped by Liu Defu et al. (1983) include very little that is not checked or verified independently, as outlined by Wise (1982).

Eastern China

The great extent of the eastern graben basins of China were revealed during the 1950s and 1960s in the search for oil. The first large grabens, other than small basins that contained oil shale and coal deposits in the Tertiary, were delineated by drilling in North China and Northeast China–in the North China basin complex and in the Songliao basin.

The formation of these graben basins generally is interpreted to decrease regularly in age from west to east, with Jurassic-Cretaceous basins closest to the Da Hinggan Ling-Taihang Shan-Wuling Shan 'line,' and with younger basins on the continental shelf, in the east. A careful study of the stratigraphic columns of these basins and of their depocenters, however, does not confirm the above statement. The Jurassic and Early Cretaceous graben basins are widespread east of the Da Hinggan Ling-Taihang Shan-Wuling Shan 'line' with additional occurrences west of this 'line' in Northeast China (e.g. Hailar, Erlian basins). The Late Cretaceous graben basins either were superimposed partly on earlier grabens or formed in the vicinity, but they do not show regular eastward shifting (see Geology Map of China, 1976). An important development occurred during Paleocene, Eocene and later times in a north-south direction, in contrast to the expected west-east direction. The Nanxiang, Jianghan, Dongting and Subei basins possess well-developed Paleocene and younger sections. To the south of these basins, the graben basins generally are filled by Paleocene and Eocene sections while post-Eocene sections are not deposited, except in the Maoming, Sanshui, Baise, and offshore basins. On the other hand, to the north of Nanxiang-Subei basins, the graben basins are filled by thick Eocene and younger sediments, the Paleocene sections, and in some grabens the Eocene sections, not having been deposited in them. During the late Tertiary and the Quaternary, two important grabens systems–Hetao-Yinchuan and Fen-Wei (Shenxi)–were formed west of the Da Hinggan Ling-Taihang Shan-Wuling Shan 'line,' on the Sino-Korean platform between the Tian Shan-Yin Shan and the Kunlun Shan-Qin Ling east-west trends.

Some bounding faults, all of which had been assumed to be normal faults, eventually were found to be thrust faults, and this evidence gave rise to the concepts of 'compressional grabens' and 'tensile grabens' (Huang Jiqing, 1978). Most authors continue to state, however, that the basins are wholly tensile in origin (Li Desheng, 1981, 1982, 1984; Ma Li et al., 1982; Li and Xue, 1983; and many others). Zhao Zhongyuan (1984), at the opposite extreme, stated that all are the products of compressional shear (compression along strike-slip faults). Zhu Zhicheng et al. (1982) reported the results of drilling along faults in southeastern China. Subsurface samples indicate that many of the boundary faults on the western sides of the basins are not normal faults, but high-angle reverse and thrust faults (Figure III-14). Similarly, but less commonly, samples show that some eastern boundary faults also are thrusts (Figure III-15). Abundant drill data, therefore, led to the conclusion that there are compressional grabens (ramp structures) in addition to tensile grabens.

Study of Figures III-14 and III-15 suggests that Zhao Zhongyuan (1984) is correct in one respect; at least *some* of the grabens are the products of compression along strike-slip faults. Most likely they have developed what are called 'flower structures' by R. F. Gregory (*in* Harding and Lowell, 1979) or 'kobergen structures' by Meyerhoff et al., 1991. The possibility that they are thrust faults, however, is not eliminated without more data; the structures on Figure III-15, for example, could easily be small backthrusts.

Jurassic and Cretaceous basins of eastern China (in particular, Northeast China) possess a basal section of coarse terrigenous redbed debris and intermediate to silicic volcanic rocks. During the later interval of Yanshanian orogeny, many of these basins were uplifted and eroded, only to begin a new cycle of graben development during the Tertiary (Ye et al., 1985). The Songliao basin is a notable exception, insofar as it underwent only one graben cycle, mainly during the Late

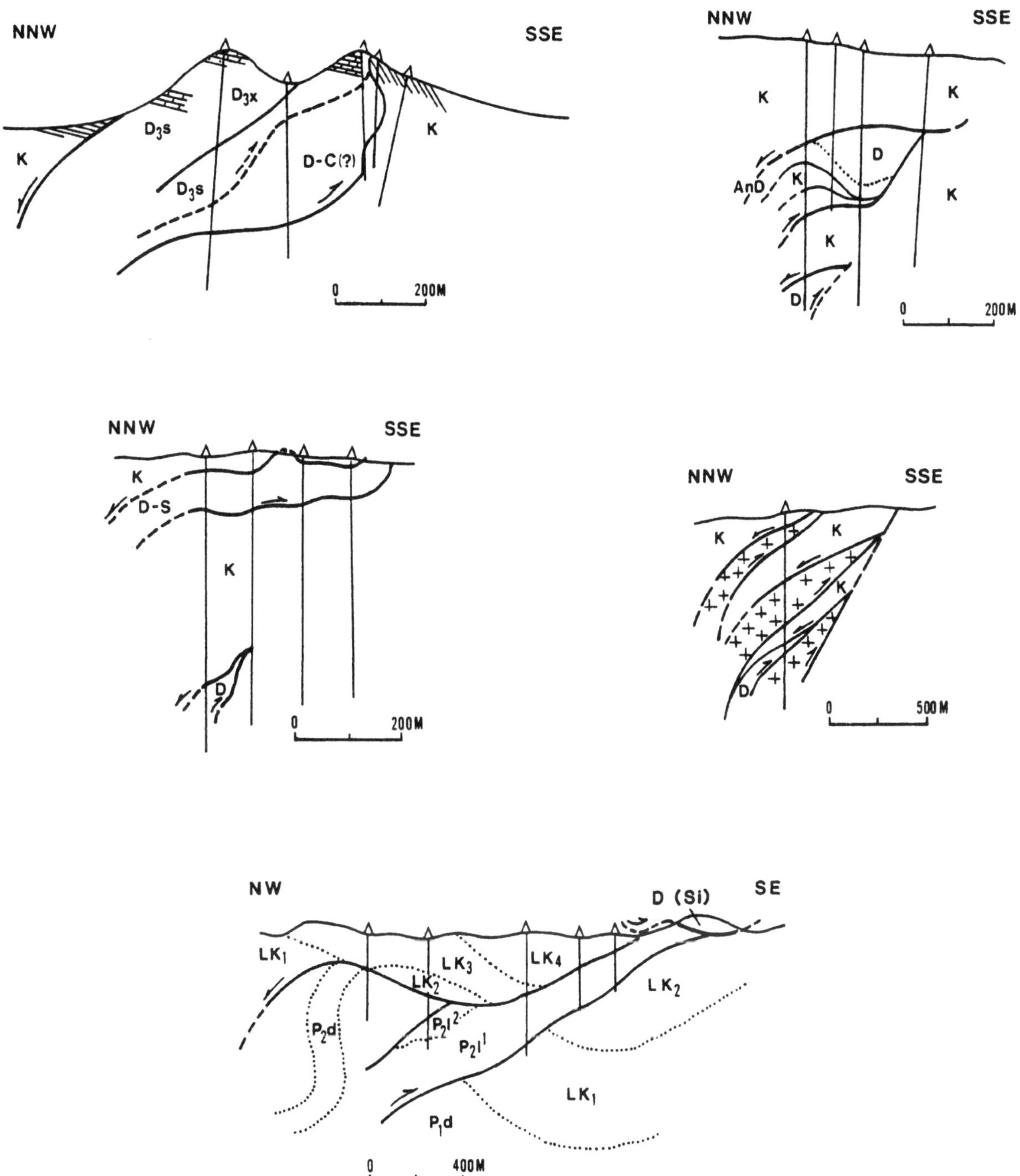

From Zhu Zhicheng (1982)

Fig. III-14. Thrust faults (probable half-flower structures) along western flanks of selected Late Jurassic-Cretaceous grabens, southeastern China.

Jurassic and the Cretaceous. Many basins passed through two cycles, one late Jurassic-Cretaceous, and the second, Eocene to Miocene or Pliocene. Tertiary basins, unlike Cretaceous basins, usually contain more mafic volcanic rocks, generally basalt. Boundary faults and internal faults that reach as deep as basement are controlled by older, largely Paleozoic to Triassic, lines of structural weakness (Liu Hefu, 1984). Many of these ancient lines of weakness that became active fault zones during the Mesozoic also were the sites of granitic intrusion. Granitic intrusives of the Yanshanian orogeny are widespread in all of Central-South China, East China, and Northeast China. They are particularly numerous where they intersect east-west fault zones, at which intersections large mineral deposits are associated with them (Wu and Qi, 1985; Zheng Xuezheng, 1985).

Liu Hefu (1984) recognized four overall stages of development during each basin-forming cycle. The first

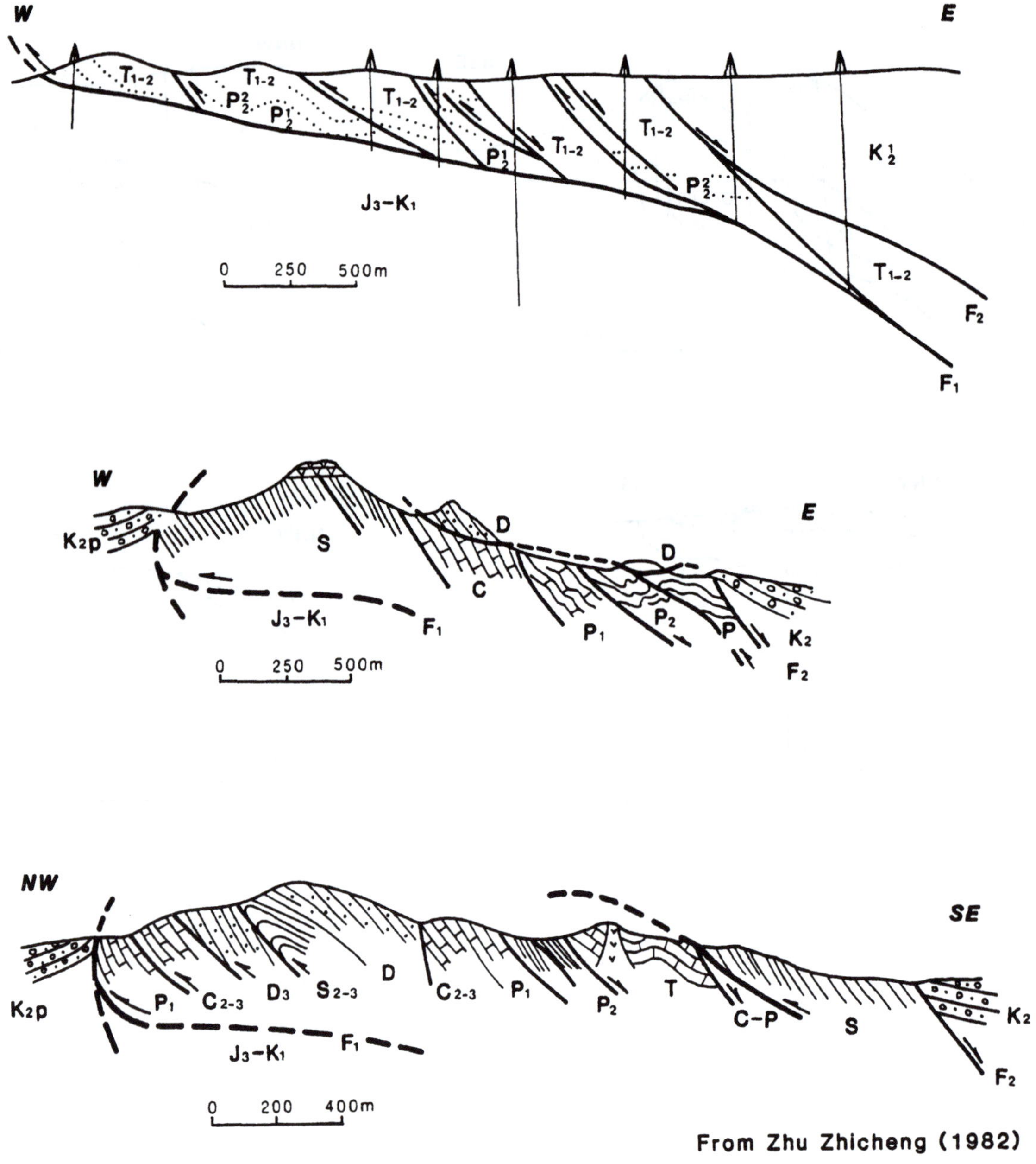

Fig. III-15. Thrust faults (probable half-flower structures) along eastern flanks of selected Late Jurassic-Cretaceous grabens, southeastern China.

is the formation of grabens and half-grabens, the latter forming the 'dustpan' structures of geology in China, a term derived from the appearance of the half-graben basins seen in plan and in cross-section. Several separate grabens and half-grabens are formed in each basin complex. These evolve into the second stage, the basin-and-range stage of Chinese geologists, when predominantly vertical motions take place and internal horsts and grabens develop. At the end of this second stage, the accumulated sedimentary sequences still are scattered in the several separate deeps or depressed zones created during the graben/half-graben stage. The third stage is the 'unifying stage,' when all faulting ceases and basin-fill begins to cover the internal horsts, thus creating within the youngest part of the section a single, unified graben complex in which the original deeps of the basin-and-range stage are hidden and the several original basins merge into one. The final stage is the masking of the graben complex with alluvium and other surficial deposits, and the effective hiding of basins.

The graben-boundary faults of Central-South China and East China have undergone some wrench movement, which, according to Liu Defu et al. (1983), was (and *is*, where an active fault is present) sinistral. The faults of southeastern China are numerous, complex, and strike mainly northeast, more so than the faults of Northeast China which strike more north-northeastward. The faults form broad zones that range in width from a few kilometers up to 40 km. Some of the faults are long, up to 1,000 km. They are associated with complimentary northwest-striking faults (Figure III–12), which theoretically were dextral shears. Liu Hefu (1984) believes that this fault system of southeastern China formed in a north-south stress field.

The faults of Northeast China are judged to be dextral, and their complimentary shears, with northwest or north-northwest strike are sinistral, as indicated by numerous fault-plane solutions (Gu Hao-ding, 1976; Qiu Qun, 1976; Liu Defu et al., 1983; Zhao Zhongyuan, 1984). Authors Huang Jiqing (1978), Liu Defu et al. (1983), and Zhang Buchun et al. (1985) wrote that the original stress field, like that in southeastern China, was oriented north-south, and that the northeast-trending faults of Northeast China originally were sinistral. Huang Jiqing (1978) and Liu Defu et al. (1983) claim that this change in direction of stress took place in relatively recent geologic time. Zhao Zhongyuan (1984), for example, stated that the change took place at the end of the Oligocene. His evidence is not clear, and the material that he published indicates that present dextral movement may be at least as old as late to middle Eocene. Zhang Buchun et al. (1985) wrote that the change took place at the beginning of the Cenozoic. This problem clearly requires more study, and regardless of what may emerge, Figure III–12 indicates that the *amounts* of wrench offsets are small, a few kilometers to 10 or 20 km at most. As in southeastern China, faults in Northeast China also can be long. The Tan-Lu fault system, for example, is 2,000 km long.

If one uses the fault-plane solutions and fault directions provided by Gu Hao-ding (1976) and Qiu Qun (1976), a simple computation shows that the present stress field is oriented N 75° E (using Moody and Hill, 1956, criteria; see also Moody, 1973). Li Qin-zu (1980) and Li Ziqiang et al. (1985) came to an identical conclusion as a result of their studies of the earth's active stress field, which they showed, quite independently, to be oriented N 75° E. Liu Hefu (1984), on theoretical grounds, showed that an east-west stress field is required on the basis of regional geology from the Jurassic onward. The presence of a NE-SW to ENE-WSE principal stress axis also accords well with the stress analysis published by Zhang Buchun et al. (1985).

A factor with an important bearing on the origin (1) of the Jurassic and younger graben systems of eastern China, onshore and offshore, as well as (2) of the wrench faults associated with them, involves great differences in depth to the Mohorovicic discontinuity and to the low-velocity zone (asthenosphere) of the upper mantle beneath Mesozoic and Cenozoic grabens, on the one hand, and the horsts between them, on the other. Tang (1982) showed that, beneath the graben areas, the 'Moho' rises to within 29 to 33 km of the surface; beneath the intervening horsts, the 'Moho' can lie as deep as 44 km, although its depth ranges general from 40 to 43 km. Similarly, depth to the upper-mantle low-velocity zone ranges from 50 to 60 km beneath the grabens to about 80 km beneath the horsts (Liu Yuanlong et al., 1984). High heat flow characterizes the graben areas, whereas heat flow beneath the horst blocks is more 'normal' (Wang Shangwen, 1983).

Several attempts have been made to explain the geologic development of eastern China and variable depths to the 'Moho' and the lithosphere. Tang (1982) thought that the tectonic history of eastern China during Meso-Cenozoic time reflects a period of continental accretion resulting from underthrusting of eastern China by the Pacific plate. One popular version of this hypothesis postulates that the progression of basin ages, older in the west and younger in the east, is explained by continental accretion. We believe, on the contrary, that because basement beneath the whole region to the present oceanic margin consists mainly of Proterozoic continental crust (Ren and Chen, 1989), the hypothesis in question is totally flawed. Similarly, the Nur and Ben-Avraham (1982) concept of 'Pacifica' in the form of fragments of accretionary terrane is contrary to all that is known of China's geology. Additionally, Ye et al. (1985), Hellinger et al. (1985), and others have attempted to explain formation of the basins wholly in terms of extensional tectonics. Hellinger et al. (p. 344) themselves admitted that their attempt was unsuccessful.

Another version involving plate tectonics was written by Li Desheng (1981, 1984). Li (1984, p. 993) wrote that, 'The severe convergence of the Indian plate with the Eurasia plate produced east-northeast-spreading of the South China Sea basin, which resulted in two triple junctions on its northern margins. The Pacific plate was subducted by downthrust beneath the Eurasia continental crust. The extension mechanism could be the rising of an upper mantle plume to produce two weak north-northeast-trending fracture zones. A series of intraplate and epicontinental rift-depression basins was formed.'

Liu Hefu (1984), in contrast, wrote that, 'Owing to northward convergence of the Indian massif and resistance of the Siberian massif, the crust of western China was subjected continuously to compression, and the crust of eastern China was forced toward the Pacific Ocean. Because this combination of compression and extension could not be uniform everywhere, diagonal strike-slip faults played an accommodating role in the crustal deformation' (p. 174). Liu's brief analysis states, in effect, that the north-south compression of western China between the Indian and Siberian cratons reached a stage beyond which the possibility of additional foreshortening was severely limited. Consequently, relief of new compressive stresses was accomplished by movements on diagonal wrench faults and by the movement of eastern China toward the Pacific. This Pacificward movement produced a tensile stress field. Liu did not provide a mechanism for this good mechanical analysis.

Meyerhoff and Meyerhoff (1977, 1982) provide a mechanism different from those listed above and term it the 'surge hypothesis.' 'Surge hypothesis' has been further developed (Taner and Meyerhoff, 1990; Meyerhoff et al., 1991) which will change the presented interpretation here to certain extent. According to this new hypothesis, softer materials in the upper mantle low-velocity zone (asthenosphere) move relatively eastward with respect to overlying lithosphere, a direct result of lag of the lithosphere as the earth rotates. Relative eastward movement of the upper mantle low-velocity materials is impeded and the flow is deflected where a low-velocity layer is thin or absent.

In theory, flow would seek and reach a steady-state condition, one which would be maintained if there were no changes in the shape and dimensions of the asthenosphere. Changes in the steady-flow state, however, would take place from time to time in any earth model and would require either a cooling or warming history, as most plate-tectonic models do (e.g. Kaula, 1979). Such changes in the flow rate would serve as triggering mechanisms for a change in the volume of asthenospheric flow, thereby producing asthenospheric 'surges' (or the reverse). The triggering events may take place cyclically or, more probably, episodically. Meyerhoff and Meyerhoff (1982) assume that the north-south compression of western China indicated abrupt and episodic changes in eastward asthenospheric flow. Growth through orogeny and accretion of the area of the Qinghai-Xizang Plateau increasingly restricted the possibility for unimpeded eastward flow of the asthenospheric mantle from western China. As the zone between the Kunlun Shan-Qin Ling fault zone and the Derbugan fault zone (north of the Altay foldbelt) more and more became the site of eastward asthenospheric flow (because of growing constrictions within the upper mantle south of the Kunlun-Qin Ling zone), the crust of present eastern China and its continental shelf were subjected to greater tension, because of eastward pull by asthenospheric material. The appearance of late 'Variscan' island arcs on the present sites of the Japan arc, the Ryukyus arc, Taiwan, and the Philippine arc further hindered the free eastward flow of mantle material; the result was the creation of a minor 'traffic jam.' The mantle material flowing eastward from beneath eastern China was slowed in its progress, began to 'back up,' and produced an arching of eastern China's lithosphere. The arching increased tension already present. Such arching, produced by less-than-solid material, is unstable, and leads to horst-and-graben formation, with magma ascending to the surface via Neocathaysian wrench and normal (graben-bounding) faults.

Flow from west to east, in the Meyerhoffs' opinion, explains eastward spread of the grabens, with the oldest in the west and the youngest in the east. Approximate synchroneity of episodes of faulting with episodes of orogeny of the Himalayan cycle is explained readily as responses to compression in the southwest. The growth of backarc basins such as, for example, the Okinawa Trough, is logically explained. Still more, the concentration of most of the world's true island arcs with backarc basins in the western Pacific also is explained. The flow directions required by the Meyerhoffs' hypothesis are the same as the directions found for the present stress field by Li Ziqiang et al. (1985), and by Zhang Buchun et al. (1985).

The already-existing east-west fracture systems of the Tian Shan-Yin Shan, Kunlun Shan-Qin Ling, etc. would serve as directional guides for the flow of asthenospheric mantle. Because of this directional flow (which had taken place well back in the Pre-Cambrian), the island arcs of the Pacific terminate where intersected by these east-west fracture zones (Meyerhoff and Meyerhoff, 1972, 1974, 1977; Li Siguang, 1973). We discuss these intersections further in a subsequent section.

The Meyerhoff and Meyerhoff hypothesis (1977, 1982), therefore, was presented to explain (1) eastward growth of the graben systems of onshore and offshore China; (2) the sudden appearance of extensive basin-and-range structure in the Mesozoic (and not earlier); (3) the appearance of the Ryukyu and Philippine arcs (apparently for the first time) south of the much older Japan arc; (4) the terminations of several western Pacific island arcs precisely where they intersect east-west fracture zones; (5) the approximate coincidence of episodes of graben faulting in eastern China with episodes of compression in the rest of China, especially in Southwest China; (6) the mechanism of arc formation as a series of straight-line segments (Ranneft, 1979); and (7) the occurrence of wrench faulting along numerous faults with a Neocathaysian strike, faults that ultimately became graben-bounding faults, many of them still undergoing wrench movements today. The Meyerhoffs also claimed that their hypothesis explains the chemical progression in volcanic rock generation within the grabens–more silicic to intermediate volcanic rocks in the earlier grabens and mafic volcanic rocks in younger grabens. They postulated that, during earlier graben formation, crustal materials were assimilated into the ascending lavas as the first faults formed; and later, after 'clean' conduits to the surface had been established, mafic lavas could pass through without plucking and assimilating large pieces of crustal material that impeded earlier ascents of magmas to the surface.

Okinawa Trough

This young trough is the easternmost of the Chinese post-Paleozoic graben systems (Figure III-16). Herman et al. (1979) interpreted the Okinawa Trough to be post-late Miocene to post-early Pliocene, but Letouzey and Kimura (1985) determined this trough to be pre-late Miocene, probably late Oligocene to middle Miocene. The crust beneath the Ryukyus, like the crust in the northern and central parts of the trough and west of it, evidently is continental. The crust beneath the southern part, the South Okinawa Trough, is 17 km thick, and is intermediate in type. In terms of popular Soviet tectonic theories and hypotheses, the Okinawa Trough is an area undergoing oceanization or basaltification (Beloussov, 1984). In terms of plate tectonics, the Okinawa Trough is a backarc basin produced by a small spreading center. In the parlance of Liu Hefu (1984), the Okinawa Trough is yet another example of the crust of eastern China being 'forced toward the Pacific Ocean' (p. 174). An interesting aspect of the associated Ryukyu arc is that the northern two thirds are underlain by Paleozoic to Cenozoic non-volcanic rocks; only the

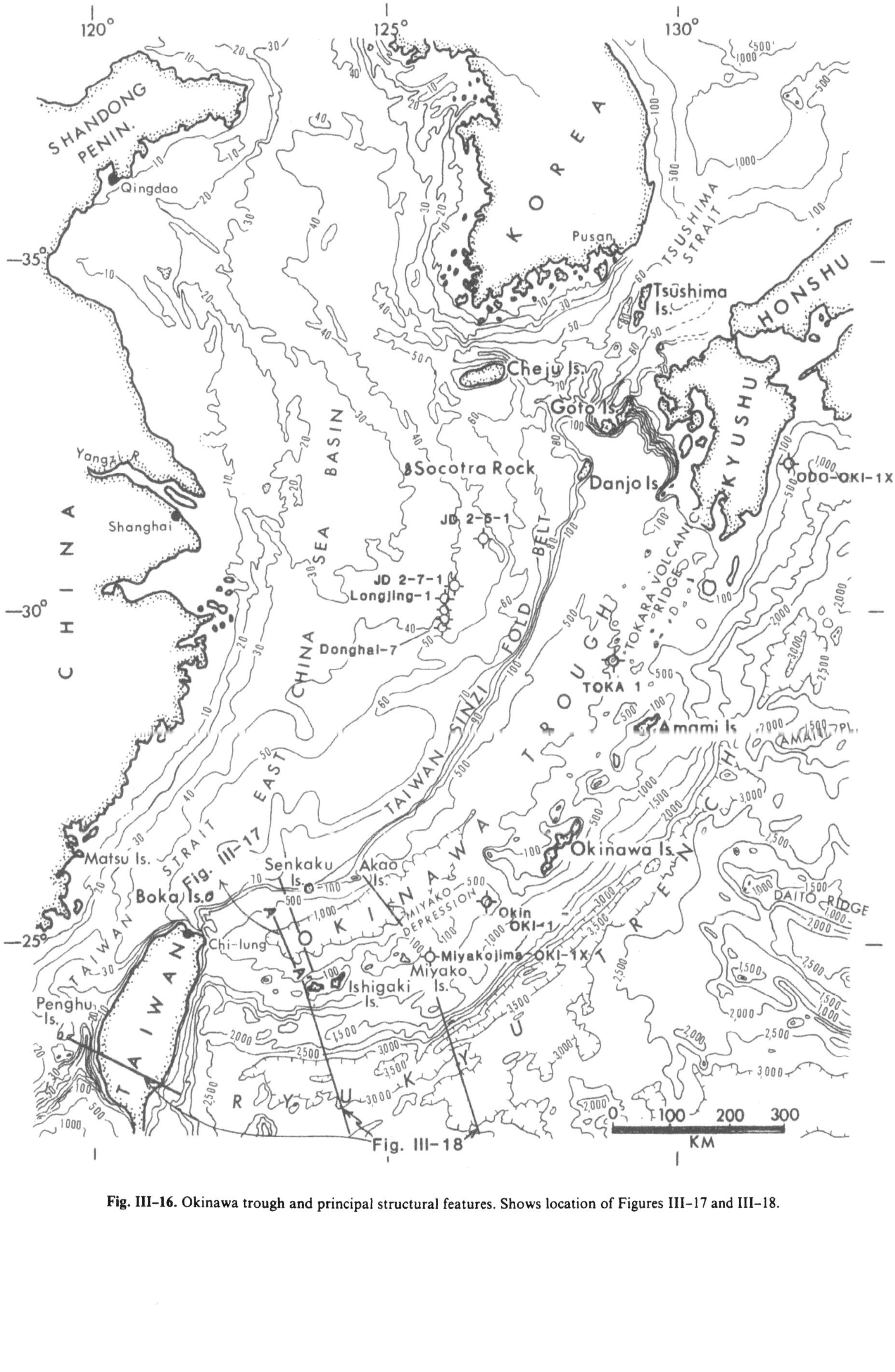

Fig. III-16. Okinawa trough and principal structural features. Shows location of Figures III-17 and III-18.

southern third is underlain by eugeosynclinal or volcanic-arc sections. Eugeosynclinal rocks include the mildly metamorphosed Permian and older section described by Hanzawa (1932, 1933) and Foster (1965). Letouzey and Kimura (1985) described the arc as a continental segment displaced by backarc spreading within, but close to the margin of, a continental mass.

In the Meyerhoff and Meyerhoff hypothesis (1977, 1982), the Ryukyu arc is the latest 'victim' of the most recent eastward surge of asthenospheric material that produced the entire Chinese horst-and-graben system farther west. The arcuate shape of the Ryukyus is a direct consequence of the circumstance that this tectonic unit was closest to the continental margin at the time that eastward-surging mantle material produced the arc. There was no competent massif east of the body that comprises the Ryukyus to oppose its eastward motion, unlike the graben systems farther west which lie much deeper in the continental interior.

Figure III–17 depicts a north-south seismic line from Herman et al. (1979). We have reinterpreted the geology, because in our view the original interpretation contains numerous errors. Figure III–18 shows three interpretative cross sections from Letouzey and Kimura (1985), the upper two being north-south or northwest-southeast sections across the Taiwan-Sinzi foldbelt, the Okinawa Trough, and the Ryukyu arc; the lower being a west-east structural cross section across Taiwan. The Taiwan-Sinzi foldbelt, according to Letouzey and Kimura (1985), is late Miocene and is transgressed unconformably by late Miocene (?) and/or early Pliocene strata. The foldbelt itself, folded in the Miocene, is underlain mainly by marine sedimentary rocks. Volcanic rocks of Miocene and younger ages are associated mainly with the Okinawa Trough, parts of Kyushu, and the Toka-1 dry hole drilled about 100 km northwest of Amami Oshima and 25 km southwest of Kyushu. Miocene rhyolitic volcanic rocks are abundant in the middle and upper Miocene sections of that well (Y. Uchiyama, written commun., May 6, 1981). Other wells drilled along the Ryukyu arc and in the East China Sea basin have penetrated marine and shallow-neritic to coastal sedimentary rocks without volcanics. These rocks produce hydrocarbons in several wells drilled by Chinese technicians (Figure III–16).

Figure III–17 across the deep South Okinawa Trough shows clearly the tensile origin of the trough. Magnetic anomalies attributed by Letouzey and Kimura (1985) to intrusions during backarc spreading are indeed associated with intrusive mafic rocks. These rocks, however, according to the seismic data, intruded along linear fault zones at different times in an apparently random, non-systematic manner and therefore are unrelated to backarc spreading, only to backarc tensile stress. The basin is, however, a zone of high heat flow (Herman et al., 1979).

The thickness of section in the trough ranges from perhaps 1,500 m in the south to 8,000 m (in the far northern part of the trough). In the Taiwan-Sinzi arc, the section is thrust slightly westward, according to Letouzey and Kimura (1985). Granitic intrusives are found in the foldbelt. One body is exposed where intruded into a Miocene coal-bearing sequence on the Senkaku Islands (Figure III–16). Granitic to rhyolitic early to middle Eocene intrusives of an underlying eugeosynclinal sequence provided a part of the detritus for Miocene sediments. These intrusives, dated at 47.5 Ma (= middle Eocene, Lutetian), (Letouzey and Kimura, 1985), are the same age as some of the intrusives found in wells in the Zhujiangkou basin (Pearl River Mouth basin) on the continental shelf of the South China Sea.

Qinghai-Xizang (Tibet) Plateau and Xizang-Yunnan Foldbelt

In preceding sections, we traced the geologic history, insofar as it is known, of southwestern China into the Mesozoic. During the Paleozoic, after the Xingkaian orogeny (Table III–1), the area of western and southwestern China existed as a platform on which shallow

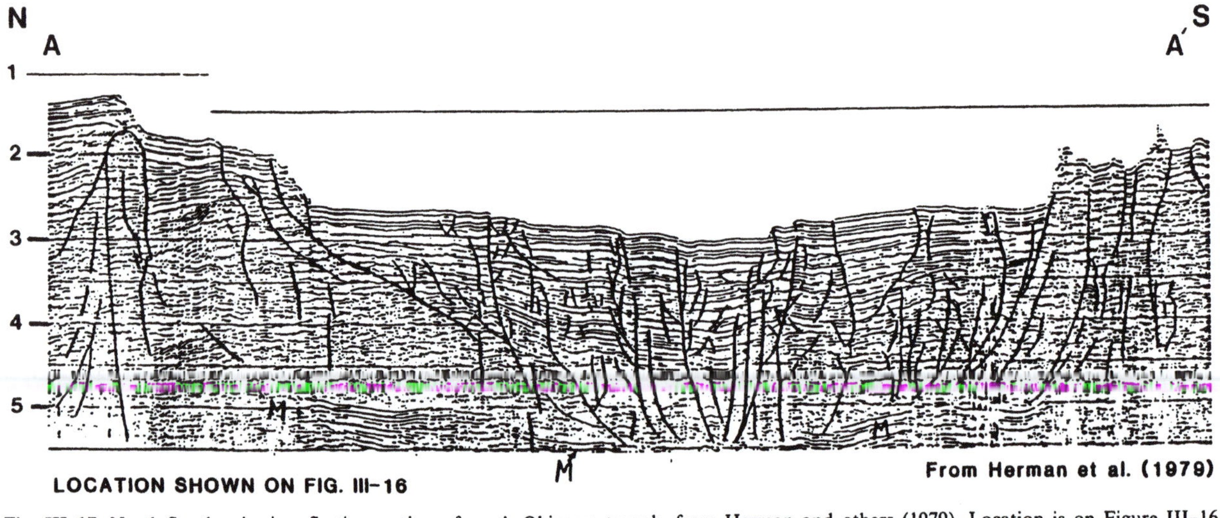

Fig. III–17. North-South seismic reflection section of south Okinawa trough, from Herman and others (1979). Location is on Figure III-16.

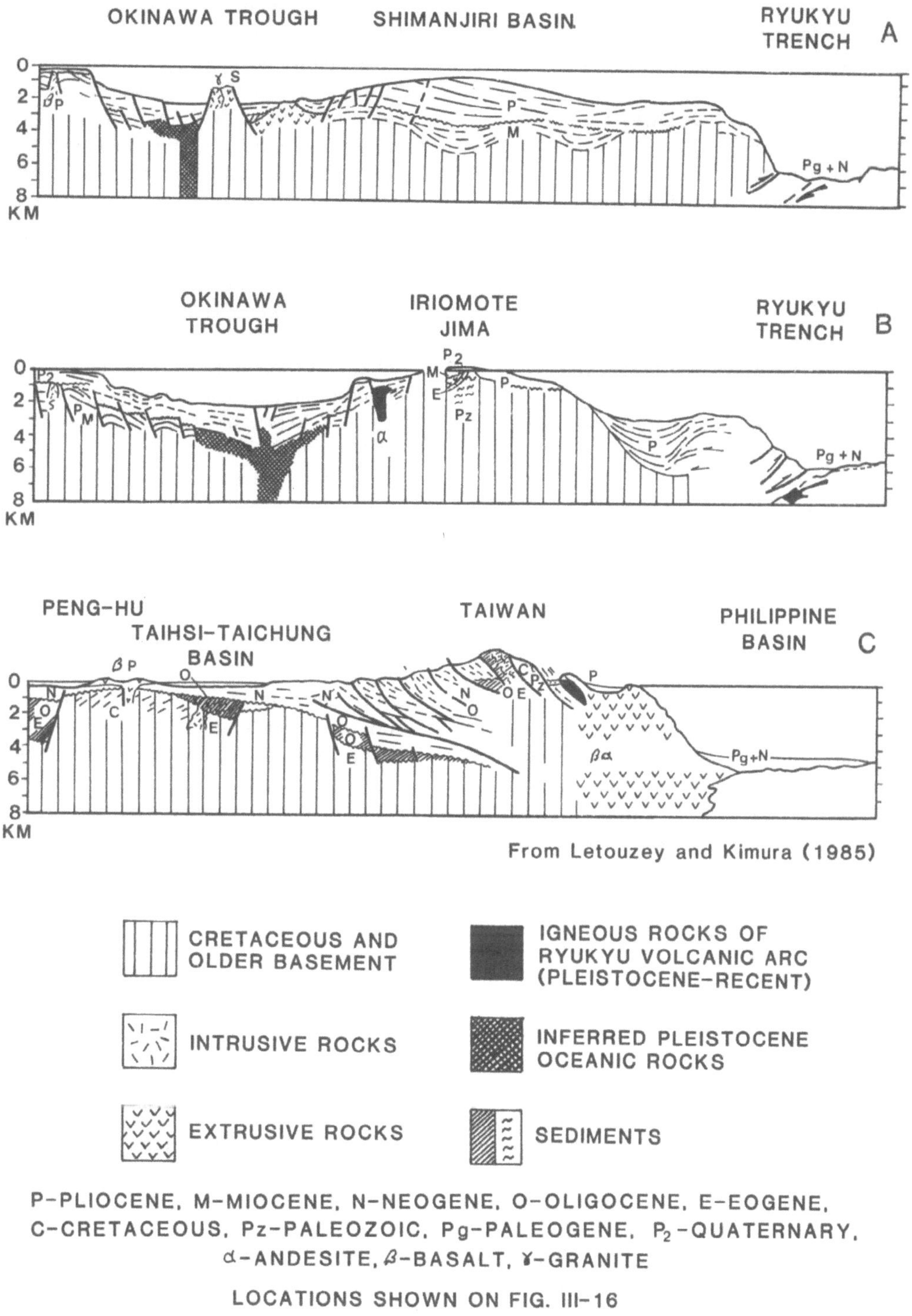

Fig. III-18. Above: two geologic cross-sections across south Okinawa trough. Below: WNE-ESE geologic cross-section of Taiwan. Locations are shown on Figure III-16.

seas were widespread from time to time. Eugeosynclinal conditions prevailed in the northern and eastern parts of the Xizang-Yunnan region during the early Permian on the site of the present South Kunlun fault zone (South Kunlun-A'nyemaqen fault zone = Xiugou-Maqen fault zone) and in regions along strike both on the west (India, Pakistan) and southeast (Yunnan) (Huang Jiqing, 1978; Chang and Pan, 1981, 1984; Fuchs, 1982a, 1982b; Cui Kexin, 1984; Taner and Meyerhoff, 1990). The South Kunlun fault zone is the northernmost of the several major fracture zones that cross the Qinghai-Xizang Plateau, and they bend southward through western Yunnan Province and into easternmost Burma. From north to south, the major units are (Figure III-6):

KUNLUN FOLDBELT
South Kunlun fault zone
(South Kunlun-A'nyemaqen fault zone).

HOH XIL-BAYAN HAR FOLDBELT
(= Songpan-Farze Foldbelt)
Litian-Jinsha Jiang fault zone
(Litian-Hoh Xil-Jinsha Jiang fault zone)
(= Hong He, or Red River, fault zone).

QIANGTANG BLOCK (QIANGTANG-SOUTHERN QINGHAI STRUCTURAL ZONE, TANGULA FOLDBELT
(in part, the SAN JIANG FOLDBELT)
Bangong-Nu Jiang fault zone
(North Xizang fault zone, Dongqiao fault zone, Salween fault zone).

LHASA BLOCK (LHASA FOLDBELT, GANGDISE BLOCK, LHASA-NYAINQENTANGLHA BLOCK)
Indus River-Yarlung Zangbo fault zone.

HIMALAYAN FOLDBELT and INDIAN PLATFORM/CRATON

Mid-Permian orogeny closed off marine waters next to the South Kunlun fault zone, and a new seaway appeared farther south in association with the Litian-Jinsha Jiang fault zone (Figure III-6), and its waters received late Permian through Triassic flysch-like turbidites and related deep-water facies. The trough extended into present southern Yunnan Province where, on the site of the present Ailao Shan, equivalent deep-sea deposits accumulated. On the west, in Ladakh, a similar belt of rocks is known from the Shyok Valley area (Huang Jiqing, 1978; Chang and Pan, 1981, 1984; Fuchs, 1982a). Also, in Ladakh, similar deep-water facies are found in the Upper Jurassic to Eocene section, strata evidently associated with both this fracture zone and the Indus River-Yarlung Zangbo fault zone farther south (Fuchs, 1982b). At Rutog close to the Bangong-Nu Jiang fault zone and close to the Kashmir border, Hu Changming (1984) found 3,500 m of deep-water Late Carboniferous and Early Permian turbidites.

Gondwanan facies is interbedded with Tethyan facies on the Qiangtang (Qantang) block, south of the Litian-Jinsha Jiang fault zone. In the larger geological picture, the zone across which Gondwana and Tethyan facies intertongue is nearly 1,000 km wide, and it extends from Madhya Pradesh on the Indian craton to the Jinsha Jiang at the northern edge of the Qiangtang block. We discussed changes in biofacies in the chapter on 'Stratigraphy,' and we made the point that the change is gradual, indicative of physical continuity (Wang Hongzhen, 1983; Wang and Mu, 1983). Moreover, faunal change also is not a drastic one, with many of the same elements in the Gondwanan and Tethyan areas, so that Waterhouse (1983), an early advocate of a broad Tethyan sea, now sees faunal evidence for only a narrow seaway. With the discovery of Gondwanan fauna and flora in the stratigraphic sequences of northeastern China (Gu Zhiwei et al., 1984) and those of the Soviet Maritime region (Zimina, 1967), any tectonic significance in the Gondwana biota has diminished greatly.

In a related study, Saxena et al. (1986) reviewed all geologic and geophysical data to arrive at a paleoposition for the Indian subcontinent. Their important conclusions are: (1) the Indian subcontinent has been part of Asia since at least the middle Proterozoic; (2) foreshortening within the region of central China (Kunlun Shan, Southwest China) to the Indian subcontinent has not exceeded 700 km and was not less than 300 km; and (3) different models of plate tectonics theory are not compatible with the field data (see also Haller, 1979; Auden, 1981; Tanner and Meyerhoff, 1990).

The Triassic deep-water basin was intensely deformed by Indosinian orogeny (Chang and Pan, 1981, 1984), and deep marine waters occupied the present zone of the Indus River-Yarlung Zangbo fault zone (Wang and Wang, 1981; Wu Hao-Ruo, 1981). A minor Jurassic marine basin formed along the intervening Bangong-Nu Jiang fault zone. The intense tectonism of the Indosinian cycle that deformed the Qiangtang block was not repeated again in the Xizang-Yunnan area until well into the Himalayan orogeny of the Tertiary (Huang Jiqing, 1978; Chang and Pan, 1981, 1984). As the center of activity shifted southward in the Qinghai-Xizang Plateau area, it shifted westward in present western Yunnan Province, from the Ailao Shan area to the Lancang Jiang, and then to the Gaoligong Shan of northwestern Yunnan Province along the China-Burma frontier. All three zones in Yunnan Province and adjoining Burma possess ophiolitic melange, mafic-ultramafic-radiolarite associations, omphacite, and eclogite with exotic blocks of shallow-water Devonian, Carboniferous, and Permian carbonates, together with wildflysch and Silurian through Permian shallow-water marine blocks. The ophiolitic-melange and associated rocks extend down the Hong He (Song Koi = Red River) (Misch, 1945a, 1945b; Huang Jiqing, 1978). The ophiolitic belts become younger toward the west and southwest (Xiao and Wang, 1984). The westernmost belt underlying the present Gaoligong Shan lasted into the Tertiary. Yanshanian and Himalayan folding affected the entire region of the Qinghai-Xizang Plateau and Yunnan Province. Late Jurassic thrust faults are associated with the Bangong-Nu Jiang fault zone (Girardeau et al., 1984a, 1985). Volcanic rock associations found in sections of the Triassic, the Jurassic, the Early Cretaceous, and the Late Cretaceous-Cenozoic were reviewed by Jin Cheng-wei (1981).

The Yarlung Zangbo fault zone deep-water facies extends for some 2,000 km from the Ladakh of India to Yunnan Province, and includes flysch, turbidites, and ophiolites which now are thrust southward, as are most of the rocks affected by thrusting in the Qinghai-Xizang Plateau area. Baud et al. (1982) called these southward-directed vergences backthrusts, in the belief that the original thrusting was northward, for which there is no observational evidence. The ophiolites, moreover, instead of occurring at the base of the sequence, are claimed by Li Yin-huai et al. (1981) to be one of the youngest units, near the top of the geosynclinal sequence. It is true, however, that the ophiolitic section is in fault contact with all other units. Gopel et al (1984), in addition, dated the ophiolites at 120 Ma, Early Cretaceous (Barremian). Srikantia and Bhargava (1981) wrote that the geochemistry of the sequence is not that of true ophiolite.

Fossils date the oldest strata in the Indus River-Yarlung Zangbo geosynclinal sequence as Late Triassic in China (Huang and Chen, 1987), and Middle or early Late Jurassic (Callovian) in the Ladakh of neighboring northeastern India (Fuchs, 1982b). The Bangong-Nu Jiang seaway was deformed by mild orogeny in the Late Jurassic, whereas the Yarlung Zangbo trough was not tectonized until the Eocene (Gansser, 1964, 1981a, 1981b).

The Qinghai-Xizang (Tibet) Plateau Today

This plateau today consists of a large, eye-shaped, symmetrical massif (Figures III-6, III-9, III-10, and III-19). Thrust faults around the margins verge outward, southward in the south, and northward in the north, and eastward in the east (Huan Wenlin et al., 1980, 1981; Yan Jia-quan et al., 1981; Zhou Jiu et al., 1983). Depth to the 'Moho' everywhere is greater than normal (Figure III-9). Symmetrical gravity (Liu Defu et al., 1983) is difficult to reconcile with northward-directed or southward-directed subduction, obduction, or other mechanisms producing uniaxial stress (Zhou Jiu et al., 1983). The existing stress field also is symmetrical (Figure III-19), with a radial stress pattern (Huan Wenlin et al., 1980, 1981; Yan Jia-quan et al., 1981; Zhou Jiu et al., 1983; Li Ziqiang et al., 1985). Present symmetry closely mirrors geological symmetry, with a Xingkaian orogenic belt enclosing the plateau on both the north (Kunlun trend) and south (Himalayan trend).

The depth to the Mohorovicic discontinuity is extremely variable. Figures III-10 and III-19, which show

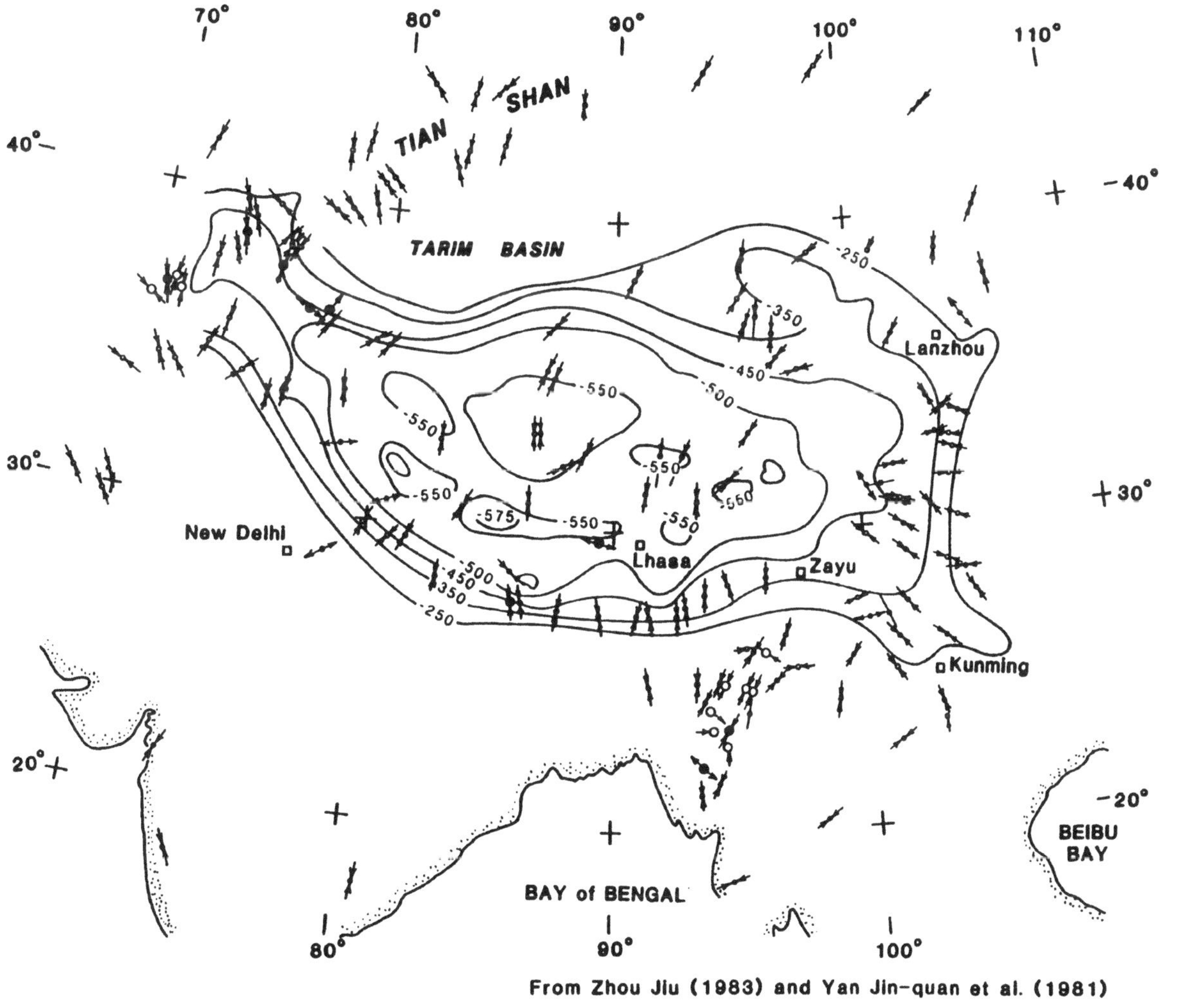

Fig. III-19. Qinghai-Xizang (Tibet) Plateau: Bouguer gravity map (from Zhou Jiu et al., 1983). Stress-field vectors (from Yan Jin-quan et al., 1981) are superimposed.

Bouguer gravity for the plateau, give some indication of the possible variety, because they show close parallelism with the 'Moho' surface calculated from seismic studies (Figure III–9). Careful studies by gravity, magnetics, refraction and reflection seismology, and magnetotelluric currents indicate that the 'Moho' on the northern side of the Indus-Yarlung Zangbo fault zone is 70 to 73 km deep. South of the fracture zone, the 'Moho' is 68 km deep, and generally it rises southward to about 45 km under the Ganga plain of northern India (Inst. Geophysics, 1981; Tang Ji-wen et al., 1983). Nowhere in the general region discussed here has any evidence been found for a depth to the 'Moho' of less than 60 km except possibly in a 100,000 km^2 area of the northern Qiangtang block (Brandon and Romanowicz, 1986). Li Yin-huai et al. (1981) and Hirn et al. (1984) found no suggestion from either gravity or from seismic data that one layer of crust underlies another, or that any major body dips northward beneath the southern margin of the Qinghai-Xizang Plateau. The Institute of Geophysics (1981) team and Yang Bing-ping et al. (1981) discovered a 10 km thick low-velocity zone between about 20 and 30 km under the plateau, while Pham et al. (1986) interpreted the existence of multiple low-velocity zones (low-resistivity zones) from magnetotelluric data.

Allegre et al. (1984), in their description of a refraction-seismic traverse of 500 km across the plateau, correlated some refractor, or refractors, with the 'Moho' (Figure III–20), to indicate offsets of up to 25 km. They also claim that superimposed 'Mohos' exist on their refraction profiles, and they suggest that this is a result of southward thrusting or northward dipping subduction. We, on the contrary correlated the refraction profile of Allegre et al. with the geo-electric profile of Pham et al. (1986) (Figure III–20), and we suggest that the vertical offsets and some of the refractors are related to low-velocity layers in the plateau's crust.

Southward-directed thrusts at the surface have been known for decades (Gansser, 1964). With the advent of the hypothesis of plate tectonics, it became necessary to interpret the southward directed thrust nappes as backthrusts, as Allegre et al. (1984), Baud et al., (1982), and many others have done (or assumed). Extensive

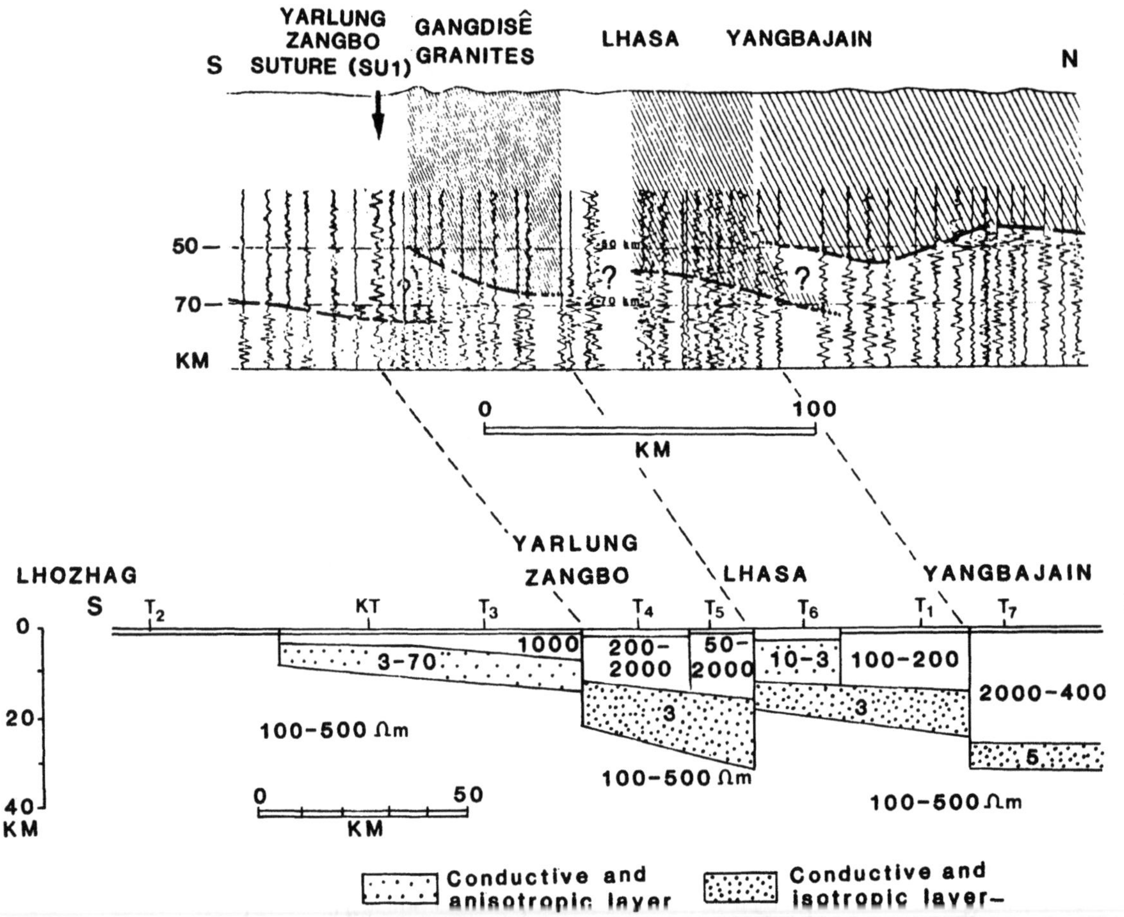

Fig. III–20. South-to-north refraction (above) and magnetotelluric (geo-electrical) (below) profiles from the High Himalaya to the Nyainqentanglha Range north of the Yanbajain geothermal field of Xizang (Tibet). Refraction data from Allegre et al. (1984); magnetotelluric data from Pham et al. (1986).

surface mapping of the Himalaya, the Lhasa block, and the Qiangtang block has demonstrated that southward-directed thrusting was a pervasive phenomenon of Indosinian, Yanshanian, and Himalayan orogenic movements (e.g. Gansser, 1964; Fei Ding et al., 1981a, 1981b; Girardeau et al., 1984a, 1984b; and many others). Northward-directed thrusts are present but are rare (Girardeau et al., 1985). Fei Ding et al. (1981a, 1981b) pointed out that surface geology requires that the Indian block be obducted northward over the Eurasian landmass to satisfy a solution in terms of plate tectonics. This concept, and that of backthrusting, might be acceptable were it not for a mass of contrary evidence.

Studies of earthquake epicenters/hypocenters have been made in recent years by several groups from the Institute of Geophysics (Beijing) and by other organizations. Figure III–21 is a map showing the locations of epicenters from Southwest China and adjacent regions for the period 1969 to 1976 (Huan Wenlin et al., 1980, 1981). The great seismic zones of the Hindu Kush and of Burma dip away from the Indian shield (Figure III–22). In contrast, all zones of earthquake hypocenters that lie directly north of India dip southward (Figure III–22; Huan Wenlin et al., 1980, 1981; Fei Ding et al., 1981a, 1981b). Huan Wenlin et al. (1980, 1981) showed that the Indus-Yarlung Zangbo fault zone dips southward at an angle ranging from 40° to 70°, with hypocenters extending downward 200 to 280 km (below the surface). The south-dipping South Kunlun fault zone dips 45° to 60° to depths ranging from 220 to 300 km. Farther north, the great fracture zone beneath the southern Tian Shan dips southward at an angle of 60° with hypocenters to 110 km. Even smaller faults than these, as in the Damxung area 140 km north of the Yarlung Zangbo, give hypocenters dipping southward to a depth of 250 km (Yang Bing-ping et al., 1981). Despite this clear evidence, geophysicists such as Teng Ji-wen (1981) doubt the reality of the seismic, gravity, magnetic, and geologic data, insisting instead that the data must fit predictions of the hypothesis of plate tectonics. Teng Ji-wen (1981), it is true, presented a weak and badly documented case to show that the south-dipping fracture zone of the Indus River-Yarlung Zangbo is matched on the southern side of the Himalaya with a north-dipping fracture zone, creating in this manner a wedge-shaped zone of seismicity beneath the Hima-

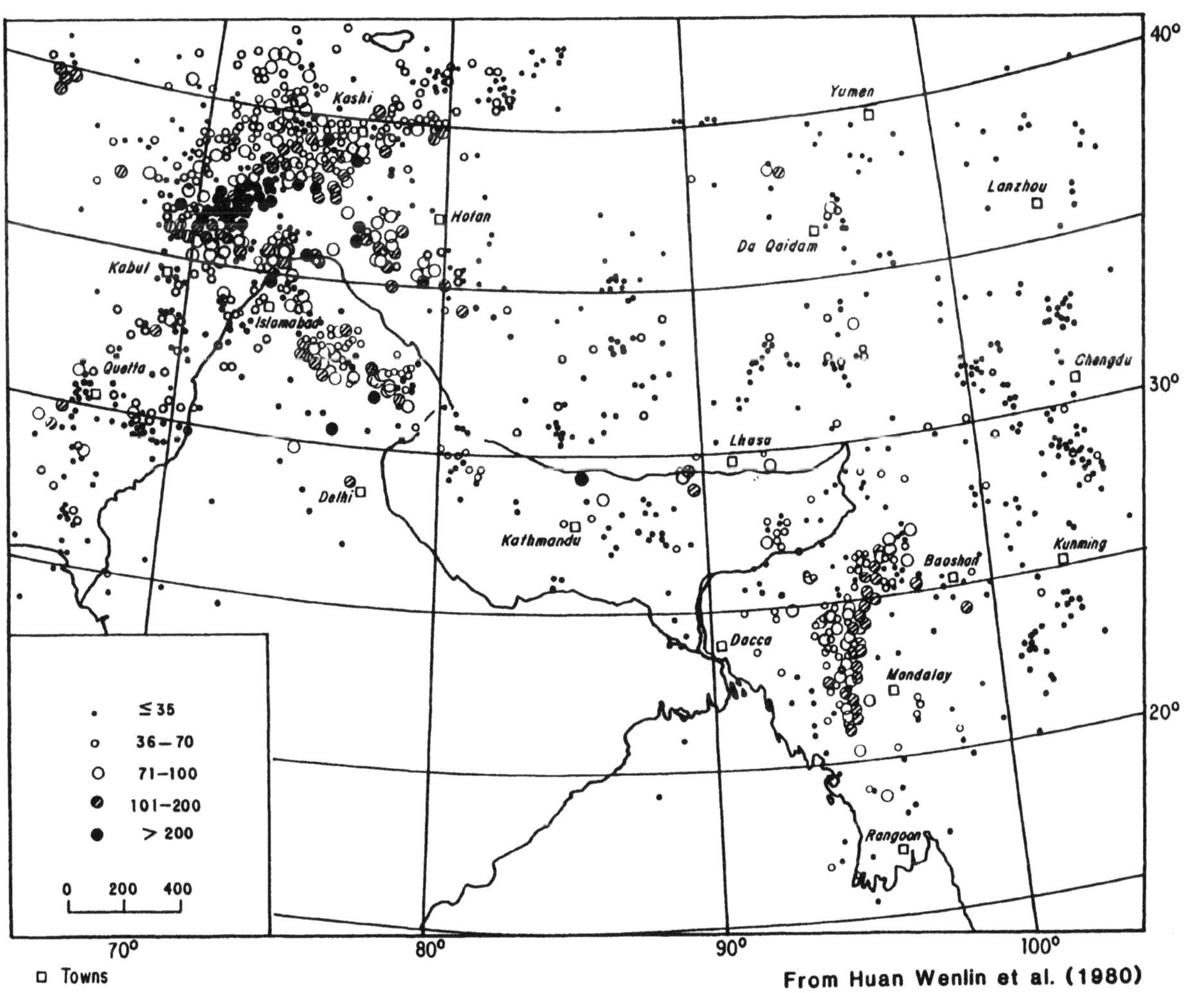

Fig. III-21. Qinghai-Xizang (Tibet) Plateau: earthquake epicenters for 1969-1976.

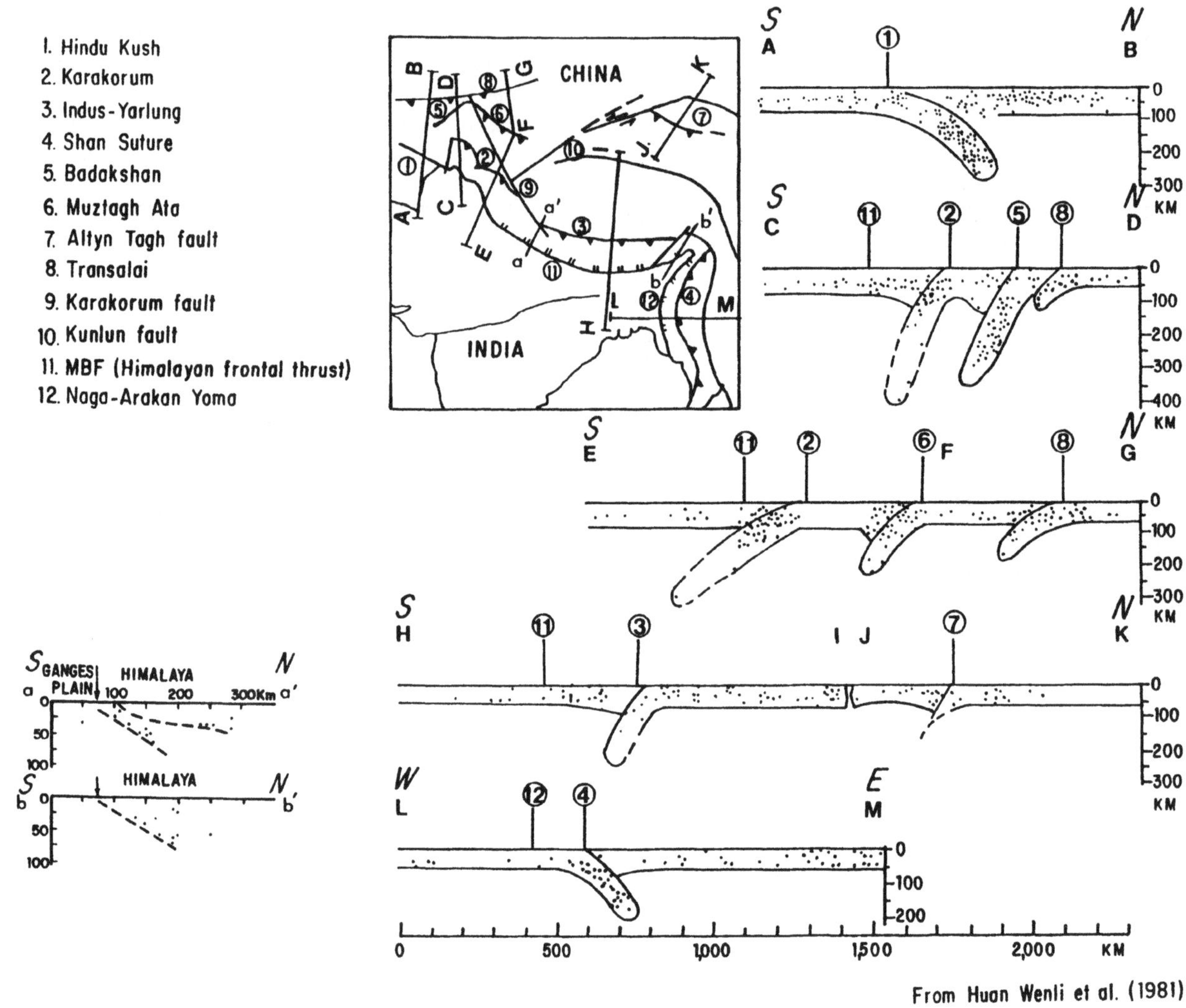

Fig. III–22. Qinghai-Xizang (Tibet) Plateau and surrounding regions: earthquake hypocenters and directions of dip of fault planes.

layas. He postulated the existence of a similar wedge beneath the Kunlun block around the northern rim of the Qinghai-Xizang Plateau. Teng Ji-wen (1981) ignored all hypocenters below about 70 km, and cited in his support the paleomagnetic evidence of abrupt changes in declination from one side of the Indus River-Yarlung Zangbo fault zone to the other (Zhu Zhi-wen et al., 1981).

Aeromagnetic data from the Himalaya block, from the Lhasa block, and from the Qiangtang block show that the Indus River-Yarlung Zangbo fault zone comprises an unusually deep double fault, about 2,000 km long. Fei Ding and his colleagues (1981a, 1981b) demonstrated that this fracture zone is associated with two east-west, parallel zones, 15 to 20 km apart. The Yarlung Zangbo fault zone is discontinuous, being offset by north-south, left-lateral faults. Modeling of the anomalies showed that they are associated with southward-dipping bodies, intrusive in nature, and essentially nonmagnetic. The anomalies are not, according to Fei Ding et al., related to the presence of ultramafic rocks. Fei Ding et al. (1981a) therefore postulated that the intrusive bodies are intermediate to silicic. Far toward the west, Sharma (1983) studied the Gaik Granite, part of the composite Ladakh-Deosai batholith which intrudes the Indus River-Yarlung Zangbo fault zone. He found that the radiometric age of the oldest part of the batholith is 235 ± 13 Ma (Triassic, Scythian to Norian). Most of the remaining parts of the batholith which Sharma studied gave dates from 48 to 21 Ma (middle Eocene, Lutetian, to lower Miocene, Burdigalian). Bhat (1984) pointed out the tectonic implication of the presence of this 2,400 km² body that the batholith formation coincides with the formation (rifting stage) of the Indus River-Yarlung Zangbo fracture zone.

Aeromagnetic data show that the Bangong-Nu Jiang fault zone has no magnetic expression. Fei Ding et al. (1981a, 1981b) therefore concluded that the fault is not a profound one. The associated ultramafic rocks may or may not be true ophiolites and, in any case, they are severely dismembered along the trace of the fracture zone.

A widespread tenet of plate tectonics is that the Steinmann 'trinity,' or ophiolitic association of ultramafic-mafic-radiolarite, is one of the principal criteria (if not *the* principal criterion) for the existence of an active (or formerly active) subduction zone. We believe, on the contrary, that this tenet has been disproved (1) as a consequence of the study of numerous ophiolitic associations in deep ocean basins far from any active or formerly active orogenic belt, and (2) in some continental areas, where essential components such as the radiolarites may be missing. Yet many geologists and geophysicists have accepted the tenet, and they attempt to place a subduction zone along any fracture with mafic and ultramafic rocks in association. The problem that results from such attempts is a very real one, because it has caused confusion among many earth scientists who apparently are unaware that criteria for various types of ultramafic-mafic associations have yet to be established. The presently used criteria, which pay little attention in many cases to geochemistry, and which may or may not include radiolarites, are wholly inadequate.

Opinions of the various workers in the plateau region show an extreme range of interpretation. Wu Hao-ruo (1981), Allegre et al. (1984), Baud et al. (1982), Girardeau et al. (1984a, 1984b, 1985), and Hirn et al. (1984a, 1984b), for example, accept the concept that the ultramafic belts of the Indus River-Yarlung Zangbo and Bangong-Nu are bona fide ophiolitic belts. The same view is adopted by most geologists and geophysicists, who tend to be somewhat uncritical, largely because the literature treats almost all such belts as subduction zones - just as uncritically. Examples are found in the works of Gansser (1964, 1981a, 1981b, and many other papers), Huang Jiqing (1978, 1984a), Huang Jiqing et al. (1980), and Teng Ji-wen (1981 and many other papers). Li Yin-huai et al. (1981) and Srikantia and Bhargava (1981) state that the Indus River-Yarlung Zangbo ultramafic-mafic-radiolarite association does not constitute an ophiolitic zone. Deng Wan-ming (1981) accepts the consensus, but observes that the rock association is not typical. Wu Hao-ru (1981) also accepts the consensus, but is dubious about the amount of movement that has taken place. Fei Ding et al. (1981a, 1981b) observed the discontinuous nature of the Bangong-Nu Jiang fault zone ultramafic and mafic bodies and the southward dip of the Indus River-Yarlung Zangbo zone. They concluded that one is a subduction zone, if India is being obducted, and that the other is not a subduction zone, because it is not a deep fault (on the basis of an aeromagnetic survey). The situation regarding this group of deep fractures in the Qinghai-Xizang Plateau area thus is confused and unsettled.

These same fracture zones show evidence of strike-slip movement, one of the best indications being the remarkably straight traces of long segments of the east-west fracture zones. Molnar and Tapponier (1975) and Molnar et al. (1981) stated that strike-slip movement is left-lateral. Valdiya (1981), Yan Jia-quan et al. (1981), and many others concur in this opinion. Baud et al. (1982), however, wrote that movement on the Indus River-Yarlung Zangbo is dextral, not sinistral. Seemingly sound evidence supports both interpretations.

We have mentioned the possibility that the Nan Ling fault zone (which cuts east-west across Central-South China and East China) and the Indus River-Yarlung Zangbo fault zone originally were one and the same fracture zone. Both are ultradeep fault zones that extend well into the mantle; both are associated with and intersected by granitic intrusions. The Nan Ling fault zone probably was once the site of dextral movement. The possibility that the two faults are segments of an originally single fracture zone requires further investigation, if for no other reason than the fact that both are deep fault zones almost precisely on strike and parallel with other old, deep, fracture zones. Important differences include the presence of an ophiolite-like suite of rocks in the Indus River-Yarlung Zangbo fault zone, together with moderate seismicity and high heat flow (Francheteau et al., 1984).

The great elevation of the Qinghai-Xizang Plateau is a characteristic that this mass of rock acquired only in extremely late geological time (beginning with the Himalayan orogenic cycle), mainly during the last 2 to 3 Ma (Figure III-23). Many geologists have reached this conclusion from different sets of data (e.g. Li Ji-Jun et al., 1979, 1983; Guo Shang-Xing, 1981; Xu Ren, 1981; Liu Dongsheng et al., 1986). In a recent study, we compiled all available data (paleontologic, paleoecologic, paleoclimatologic, depositional rates, volcanics, geothermal, radiometric dates and magmatism, structure and geophysical) and concluded that the uplift of the Qinghai-Xizang Plateau started by the late Miocene and accelerated by the Pleistocene, similar to the history of uplift shown on Figure III-23. All plate-tectonic models, in contrast project an uninterrupted and uniform history of uplift since the Eocene for this plateau. The inconsistency between the history of elevation and the postulated history of plate-tectonic models is not explained by their authors. Zhao and Morgan (1985) thus flatly assume that a plate-tectonic interpretation of the Pliocene fossils is wrong, '...perhaps due to special events at the sampling locations or changes in climatic factors...' (p. 359). Because their assumption is wrong that paleoclimatic data are bad, the rest of their paper is wrong, based as it is, wholly on theoretical speculation. They even wrote (p. 359) that, if the paleoclimatic data are correct, '...the continental convergence of India and Eurasia cannot be the main cause of the plateau formation.'

Li Tingdong et al. (1986) proposed a different plate-tectonic model in which they interpret the plateau to be shortened and thickened, due to northward subduction of the Indian subcontinent from the late Eocene to the Pliocene, with negligible, uplifting but with remarkable deepening of the upper mantle. Then they mention that the plateau has been uplifted rapidly and greatly since the early Pleistocene, due to relaxation of stress and to isostatic adjustment. Though their timing of uplift is correct, the mechanism that they suggest requires uplift of the plateau since the late Eocene,

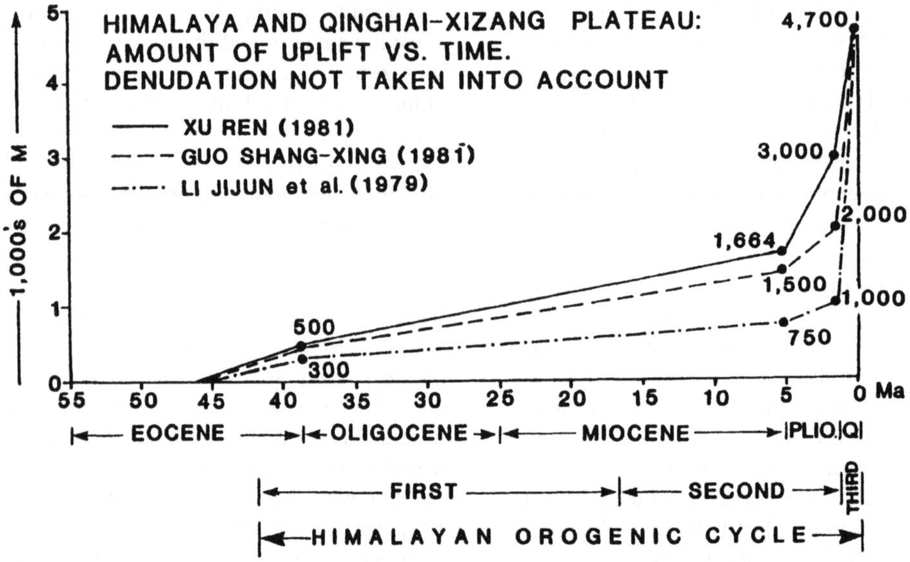

Fig. III-23. Three curves showing the timing of the uplift of the Himalaya and of the Qinghai-Xizang (Tibet) Plateau. The curves are based on data from Li Ji-jun et al. (1979, 1981), Guo Shang-xing (1981), Song and Liu (1981), and Xu Ren (1981).

therefore as unacceptable as the mechanism in other plate-tectonic models.

Another important characteristic of the Himalaya-Qinghai-Xizang Plateau region is the presence of Plio-Pleistocene grabens that cut at right angles to the east-west tectonic trends. These contain up to 3,000 m of coarse rubble, and their evidence confirms the time of uplift given by Pliocene fauna and flora. The cross-cutting grabens have been illustrated by several workers, including Hirn et al. (1984a) and Girardeau et al. (1984b). One graben sequence in Nepal was described by Fort et al. (1981). Valdiya (1976) postulated that the approximately north-south striking grabens in the overthrusts of the Himalaya are related genetically to faults and folds of the same trend on the upper surface of the Indian craton, beneath the frontal thrust sheets and nappes of the Himalayas, and beneath the Tertiary-Quaternary sedimentary fill of the east-west-striking Ganga basin. Some of the north-south faults are tear faults with a reported dextral component of wrench movement (Valdiya, 1976). Most have undergone a history of normal faulting. Several appear still to be active, including the graben system in which the Yangbajain geothermal field of Xizang (Tibet) Autonomous Region is located, and some normal fault earthquake solutions have been obtained (Tapponnier et al., 1981). Similar, on-strike, and parallel north-south tear faults with a presumed sinistral rather than dextral sense of motion cut the double-intrusive band of the Indus-Yarlung Zangbo fault zone (Fei Ding et al., 1981a, 1981b). Sinistral tears appear to be older than dextral tears.

The presence of young (Pliocene and younger) grabens striking at right angles to the east-west structural trends indicates that east-west extension both has taken place and continues to take place. Tapponnier et al. (1981) relate the east-west extension to uplift of the Qinghai-Xizang Plateau, which began mainly in the Pliocene. Their concept may well be correct. The presence of these north-south active trends, however, definitely raises a group of related geometric problems, which never have been addressed to the best of our knowledge.

In the most popular plate-tectonic 'models,' India drifts northward from a location in the present Southern Hemisphere, 6,000 to 7,500 km south of its present position (Dewey and Burke, 1973; Molnar and Tapponnier, 1975; Stoneley, 1975; Johnson et al., 1976; Mattauer et al., 1981; Molnar et al., 1981; Huang Jiqing, 1984a, 1984b; and many others).

From the time of the alleged 'collision' of India with Asia, the position and shape of the Asian continental margin are unknown, but it is popularly believed that the alleged 'collision' produced (1) the present indented shape of pre-'collision' Asia and (2) the great uplift of the Qinghai-Xizang Plateau. The ophiolitic zones that 'wrap around' the northwestern, northern, and eastern flanks of the Indian craton are assumed to owe their present spectacular geometric form to that supposed 'collision.'

Neither the original position nor the shape of the Asian continental margin can be determined, although some reasonable limitations can be placed on the latter. In the simplest case, the present Himalaya parallels the original margin but has undergone straight northward displacement between the present Baluchistan ranges together with Afghanistan on the west, and the Yunnan-Burma ranges on the east. The original length is equivalent to the distance between the Indus River and central Arunachal Pradesh (the distance between the Owen and Ninety-East fracture zones), or about 3,200 km. At the opposite extreme, the original length was equivalent to the combined lengths of the Baluchistan ranges, the Himalaya, and the Arakan Yoma (700 + 3,200 + 1,000) or about 4,900 km. If the latter 'model' is correct, the present shape and position may be closely similar to the originals. The truth may lie somewhere between the two 'models.' Consequently a large continent presumably was present before India drifted northward. That India 'docked,' so to speak, in this very indentation of approximately correct shape and size (and which happened to be at the right place) would amount to remarkable coincidence. The ophiolites of the Baluchistan ranges, Indus-Yarlung Zangbo fault zone, and the Arakan Yoma then would have formed as the result of an as-yet-unknown formative process, because of the unsolved problem of how a subduction zone can form simultaneously on three sides of a geometric body. Somewhat similar geometric relations have been described from other parts of the world, as in parts of the Indonesian arc, for example, but no one has seen fit to explain the workings of this problem in fundamental mechanics and geometry.

The problem is, in fact, even more complex. In Burma, and in China's neighboring Yunnan Province, three, and more likely four, ophiolitic zones and associated melanges are present, younger from east to west. East of the youngest mobile belt, the Arakan Yoma, a Yanshanian zone is present on the Yunnan-Burma border, and was described by Huang Jiqing (1978, 1984a) from the Gaoligong Shan. Farther east on the Lancang Jiang lies a still older Yanshanian ophiolitic complex. Still farther east lies the Ailao Shan ophiolitic complex, emplaced during late 'Variscan' orogenesis and reactivated during the succeeding Indosinian orogenesis.

Possible analogues (in part) of the Burma-Yunnan belts are the ophiolitic and melange complexes in western Pakistan and eastern and central Afghanistan. These appear to become younger in a direction opposite from that in Burma and Yunnan Province (i.e. from west to east, instead of from east to west). The easternmost occurrence consists of the Bela-Muslim Bagh-Zhob belt; farther west lies the Chaman-Moqur belt. West again lies the Vashir-Helmand belt in central Afghanistan (Weippert et al., 1970; Shareq, 1981). Many reconnaissance studies of the Pakistan belt (or belts) have been published (late Yanshanian orogeny, in Chinese parlance). The major papers are by Ahmad and Abbas (1979), Allemann (1979), Asrarullah and Abbas (1979), DeJong and Subhani (1979), and Gansser (1979). Few first-hand accounts of the Afghanistan ophiolitic complexes are available (cf. Weippert et al., 1970, and Shareq, 1981), but the presence of various subparallel ophiolitic belts of different ages on both the east and the west demonstrates that the generally accepted model of a simple, one-time subduction definitely is incorrect. Stoneley (1975), Haller (1979), Auden (1981), Bhat (1984), and Stocklin (1984, 1989), on quite different grounds in fact, presented a most convincing cases to show that subduction models and their variants are not applicable in this region.

In the light of the preceding, the north-south grabens of the Himalayas and of the southern Qinghai-Xizang Plateau could be tensional zones caused by east-west extension that resulted from stretching (during 'collision') of the former pre-'collision' model adopted, and could involve extension of 600 to 800 km. Less than a dozen of these garbens are present, and the maximum measurable east-west stretching may not exceed 100 to 200 km. The Pliocene and younger age of these features, moreover, like the Pliocene age of uplift and magmatism of the Qinghai-Xizang Plateau, is not compatible with any 'collision' model yet proposed. Compression has admittedly taken place on a large scale, with perhaps as much as 500 to 700 km of foreshortening, but anything in excess of these figures violates known surface geology (Meyerhoff and Meyerhoff, 1972, 1974a, 1978; Crawford, 1979; Auden, 1981; Gansser, 1981a; Stocklin, 1981, 1984, 1989; Waterhouse, 1983, Saxena et al., 1986).

We are well aware that the above statements contradict widely-accepted interpretations of magnetic anomalies in the ocean basins. Alternate hypotheses have been discussed elsewhere (Meyerhoff and Meyerhoff, 1974b; Agocs et al., in press). The popular belief, however, that the 'negative' and 'positive' linear anomalies of the ocean basins were caused by magnetic field reversals within the earth appears to us to be without a factual basis. Nowhere has the source of the anomalies been identified, sampled, and studied. The only concrete data that pertain directly to the problem of India's relationship with Asia through geological time are the data of field geology, paleontology, and continental earthquake seismology. These disciplines provide the fundamental multiple bases for the doubts expressed here.

As for India's position with respect to Asia, the opinions of the geologists who have worked in the region can be summarized as follows. For pre-Late Permian times, most of China's geologists maintain that India was a part of Asia. Chang and Pan (1984a, p. 455) stated the prevailing viewpoint when they wrote, 'In consideration of biogeography there are good indications that until the Early Permian there was no paleogeographic discontinuity between the Indian subcontinent and northern Xizang.' Separation, if any, began at the end of the Early Permian and lasted, according to which author one reads, until the end of the Triassic, the beginning of the Cretaceous, or, more commonly, the earliest Tertiary. Most geologists in China, and other earth scientists who have specialized in the area, do not believe that the separation of India from Eurasia ever exceeded 700 km; the majority give figures that range from 300 to 700 km. Only a few follow the Molnar and Tapponnier (1975) hypothesis of 6,000 to 7,500 km of separation. With notably few exceptions, the distance of separation between Eurasia and India appears to us to be related directly to the amount of work, especially field work and paleontology, that the individual authors have done in the region. Many of China's geologists and geophysicists–probably the majority–favor plate-tectonic models, but very few accept pre-Permian separation (i.e. Huang and Chen, 1987). Even those who do accept such separation generally favor the small distances mentioned above.

The geologists, oceanographers, and geophysicists involved seem to fall into three categories: (1) rigid fixists (of which there are very few), (2) a group that favors lateral movements of the order of 300 to 700 km (with and without plate tectonics), and (3) a group that bases its judgments on the oceanic magnetic anomalies, the results of paleomagnetism, and some paleontological/paleoclimatic data. This last-named group favors the widest separations, and almost universally ignores the rigorous results of detailed field mapping.

Of the first group, the most notable are Petrushevsky (1971) and Beloussov (1984). The second group, including supporters and non-supporters of some form of plate tectonics, includes Gansser (1964, 1981a), Saxena (1971, 1978), Meyerhoff and Meyerhoff (1972, 1974a), Chang and Zheng (1973a, 1973b), Li Siguang (J. S. Lee, 1973), Chen Guoda et al. (1975), Chang Chengfa et al. (1977), Crawford (1979), Huan Wen-lin et al. (1980, 1981), Auden (1981), Chang and Pan (1981, 1984), Fei Ding et al. (1981a, 1981b), Li Yin-huai et al. (1981), Srikantia and Bhargava (1981), Stocklin (1981, 1984, 1989), Fuchs (1982a), Kozur and Gupta (1983), Liu Defu et al. (1983), Waterhouse (1983), Zhang and Zhang (1983), Zhou Jiu et al. (1983), Bhat (1984), Chen Guoda (1984b), Liu Hefu (1984), Yang Weiran et al. (1984), Wenyou et al. (1984), Dickins (1985), Kapoor and Takuoke (1985), Khudoley and Prasorovskaya (1985), and Huang and Chen (1987). The third and last group includes Dewey and Burke (1973), Molnar and Tapponnier (1975), Stoneley (1974), Valdiya (1976, 1981), Johnson et al. (1976, 1978), Tapponnier and Molnar (1977), DeJong and Subhani (1979), Deng Wan-ming (1981), Jin Cheng-wei (1981), Mattauer et al. (1981), Molnar et al. (1981), Teng Ji-wen (1981), Zhu Zhi-wen et al. (1981), Nur and Ben-Avraham (1982), Teng Ji-wen et al. (1983), Baud et al. (1982), Girardeau et al. (1984a, 1984b, 1985), Huang Jiqing (1984a, 1984b), Ren Jishun (1984), and Zhu Xia (1984). Ren Jishun is one of several who have promoted the 'accordion' hypothesis of drift, the Chinese version of a so-called 'Wilson cycle.' What Yang Weiran et al. (1984) called the 'closing'-and-'opening' hypothesis is a greatly scaled-down version.

The above lists, particularly that of the third group, do not contain nearly all of the names associated with each viewpoint, but the lists do give a fair idea of the cross section of individuals and specialties involved (see Bibliography).

The weight of the evidence today is well summarized by Gansser (1981a, p. 119): 'All these facts demand a reappraisal for the drift history of greater India. *The observations require that peninsular India was never very far distant from the southern front of the Eurasian continent* and that the Tethyan "ocean" now consumed along the northern suture zone was actually a zone of complicated island arcs, internal basins and irregular slices of continental rocks (Crawford, 1979).' This last statement represents a great step for Gansser, who began conservatively, swung strongly to large-scale movements, of the order of thousands of kilometers, between India and Eurasia (Gansser, 1966), and then gradually changed back to his present conservative course based on decades of experience in the field.

We ourselves welcome this most important statement of Asian geotectonics by Gansser. On the whole, however, our position on the relationship between peninsular India and Asia remains largely as presented more than once in this chapter. We believe firmly that India and Asia have remained united (unseparated) from the Pre-Cambrian to the present.

Bibliography

The bibliography contains references that are published in the earth science periodicals of China. These periodicals are mainly published in the Chinese language, with an English abstract or summary. A list of the most cited periodicals in the book are presented below. Other references published in Chinese are indicated after each reference.

1. Acta Geologica Sinica, in Chinese with English abst.
2. Acta Geophysica Sinica, in Chinese with English abst.
3. Acta Paleontologica Sinica, in Chinese with English abst.
4. Acta petrolei Sinica, in Chinese with English abst.
5. Acta Sedimentologica Sinica, in Chinese with English abst.
6. Acta Stratigraphica Sinica, in Chinese with English abst.
7. Chinese Acad. Geol. Sciences Bull., in Chinese with English abst.
8. Earth Science - Jour. Wuhan College of Geology, in Chinese with English abst.
9. Geological Review, in Chinese with English abst.
10. Mineral Deposits, in Chinese with English abst.
11. Oil and Gas Geology,in Chinese with English abst.
12. Oil Geophysical Prospecting, in Chinese with English abst.
13. PalAsiatica, in Chinese with English abst.
14. Petroleum Exploration and Development, in Chinese with English abst.
15. Professional Papers of Stratigraphy and Paleontology, in Chinese with English abst.
16. Scientia Geologica Sinica, in Chinese with English abst.
17. Bull. Geol. Soc. China, in English.
18. Geochemistry (Beijing), in English.
19. Scientia Sinica, in English.

Abbas, S. G. and Ahmed, Z., 1979, The Muslimbagh ophiolites. *In*: Farah, A. and De Jong, K. A. (eds.) Geodynamics of Pakistan: Quetta, Geological Survey of Pakistan, p. 243–249.

Academia Sinica (China Academy of Sciences), 1958, A regional stratigraphic table of China: Washington, U. S. Dept. Commerce, Joint Pub. Research Service, JPRS 18, 539 OTS 63-21499, April 4, 1963, p. 511.

Ahmad, F., 1987, Indus-Tsangpo suture - fact or fiction? *In*: McKenzie, K. G. (ed.) Shallow Tethys 2: Boston, A. A. Balkema, p. 113-125.

Allegre, C. J., Courtillot, V., Tapponier, P., Hirn, A., Mattauer, M., Coulon, C., Jaeger, J. J., Achache, J., Scharer, U., Marcoux, J., Burg, J. P., Chengfa, Chang, Guangqin, Li, Baoyu, Lin, Jiwen, Teng, Naiwen, Wang, Guoming, Chen, Tonglin, Han, Xibin, Wang, Wanming, Den, Huaibin, Sheng, Yougong, Cao, Ji, Zhou, Hongrong, Qiu, Peisheng, Bao, Songchan, Wang, Bixiang, Wange, Zaoxiu, Zhou and Ronghua, Xu, 1984, Structure and evolution of the Himalaya-Tibet orogenic belt: Nature, v. 307, p. 17–22.

Allemann, F., 1979, Time of emplacement of the Zhob Valley ophiolites and Bela ophiolites. *In*: Farah, A. and DeJong, K. A. (eds.) Geodynamics of Pakistan: Quetta, Geol. Survey of Pakistan, p. 215–142.

Allen, C. R., Gillespie, A. R., Yuan, Han, Sieh, K. E., Buchun, Zhang and Chengnan, Zhu, 1984, Red River and associated faults, Yunnan Province, China; Quarternary geology, slip rates and seismic hazard: Geol. Soc. America Bull., v. 95, no. 6, p. 686–700.

An Sanyuan and Lu Xinxiang, 1984, The geologic features of the porphyry association and its relationship with mineralization in eastern Qinling Mountains. *In*: Xu Qexin and Tu Guangchi (eds.) Geology of granites and their metallogenic relations: Beijing, Science Press, p. 595–609.

An Taixiang, 1981, Recent progress in Cambrian and Ordovician conodont biostratigraphy of China. *In*: Teichert, C., et al. (eds.) Paleontology in China, 1979: Geol. Soc. America, Spec. Paper, 187, p. 209–225.

Anderson, E. M., 1951, *The dynamics of faulting and dyke formation with applications to Britain*, 2nd ed.: Edinburgh, Oliver and Boyd, 206 p.

Andrews, R. C., 1932, The new conquest of Central Asia; a narrative of the explorations of the Central Asiatic expeditions in Mongolia and China, 1921–1930. *In*: Natural history of Central Asia: Amer. Mus. Natural History, v. 1, p. 678.

Anonymous, 1982, International field workshop and seminar on phosphorite: China, Kun-ming shih, 2 vols., p. 389 (in Chinese, English abstracts).

Antropov, P. Ya., 1958, On some achievements of geological surveying in the Chinese People's Republic: Washington, U. S. Dept. Commerce, Joint Pub. Research Service Document L-10008-N (reprinted by Am. Geol. Inst., 1960, Internatl. Geol. Review, v. 2, no. 12, p. 1071–1077).

Argand, E., 1924, La tectonique de l'Asie. International geological Congress (13th), Brussels 1924, Comptes Rendus 5, p. 171–372.

Arkell, W. J., 1956, *Jurassic geology of the world*: New York, Nafner Publ. Co., 806 p.

Armijo, R., Tapponier, P. and Han Tonglin, 1989, Late Cenozoic right-lateral strike-slip faulting in southern Tibet: Jour. Geophysical Research, v. 94, no. 83, p. 2787–2838.

Armijo, R., Tapponier, P., Mercier, J. L. and Tonglin, Han, 1982, A field study of Pleistocene rifts in Tibet: *Eos* 63, p. 1093.

Arthurton, R. S., Alam, G. S., Anisuddin-Ahmad, S. and Iqbal, S., 1979, Geological history of the Alamreg-Mashki Chah area, Chagai District, Baluchistan. *In*: Farah, A. and DeJong, K. A. (eds.) Geodynamics of Pakistan: Quetta, Geological Survey of Pakistan, p. 325–331.

Arthurton, R. S., Farah, A. and Ahmed, W., 1982, The Late cretaceous-Cenozoic history of western Baluchistan Pakistan - the northern margin of the Makran subduction complex. *In*: Leggett, J. K. (ed.) Trench-forearc geology: sedimentation and tectonics on modern and ancient active plate margins: Geological Society of London, Special Publication 10, p. 373-385.

Asrarullah, Ahmad, Z. and Abbas, S. G., 1979, Ophiolites in Pakistan: an introduction. *in* Farah, A. and DeJong, K. A. (eds.) Geodynamics of Pakistan: Quetta, Geol. Survey of Pakistan, p. 181-192.

Atlas Publications, 1977, Zhonghua renmin gongheguo fen sheng dituji (Atlas of Provinces of the People's Republic of China): Beijing, Xinhua Book Company and Atlas Publication, 51 maps, 169 p. (geographic names in Pinyin romanization).

Auden, J. B., 1974, Afghanistan-West Pakistan. *In*: Spencer, A. M. (ed.) *Mesozoic-Cenozoic orogenic belts*, Geological Society of London, Special Publication 4, p. 235-253.

Auden, J. B., 1981, India's former crustal neighbours: Proceedings of Indian National Sci. Acad., v. 47, Part A, no. 6, p. 588-630.

Bailey, G. B. and Anderson, P. D., 1982, Applications of landsat imagery to problems of petroleum exploration in Qaidam basin, China: Am. Assoc. Petroleum Geol. Bull., v. 66, no. 9, p. 1348-1354.

Bain, H. F., 1933, Ores and industry in the Far East; the influence of key mineral resouces on the development of oriental civilization: New York, Council on Foreign Relations, 288 p. (chapter on petroleum by W. B. Heroy).

Bakhadur, S. R. and Nemkov, G. I., 1984, Tectonic zoning and geologic history of the Nepal Himalaya: Internat. Geol. Rev., v. 26, no. 3, p. 339-347.

Bakirov, A. A., Barentsov, M. I. and Bakirov, E. A., 1971, Neftegazonosnyye provintsii i oblasti zarubezhnykn stran (Petroleum-bearing provinces and regions of foreign countries): Moscow, Izd. Nedra, 542 p.

Bally, A. W., Allen, C. R., Geyer, R. B., Hamilton, W. B., Hopson, C. A., Molnar, P. H., Oliver, J. E., Opdyke, N. D., Plafker, G. and Wu, F. T., 1980, Notes on the geology of Tibet and adjacent areas - report of the American plate tectonics delegation to the People's Republic of China: U. S. Geol. Survey, Open-File Rept., 80-501, 101 p.

Bannert, D. and Helmcke, D., 1981, The evolution of the Asian plate in Burma: Geologische Rundschau, v. 70, p. 446-458.

Barazangi, M. and Ni, J., 1982, Velocities and propagation characteristics of Pn and Sn beneath the Himalayan arc and Tibetan Plateau: possible evidence for underthrusting of Indian continental lithosphere beneath Tibet: Geology v. 10, p. 179-185.

Bassoullet, J. P., Colchen, M., Jutfau, Th., Marcoux, J., Mascle, G. and Reibel, G., 1983, Geological studies in the indus suture zone of Ladakh (Himalayas). *In*: Gupta, V. J. (ed.) 1983, Stratigraphy and structure of Kashmir and Ladakh Himalaya: Delhi-India, Hindustan Publishing Corp., p. 96-124.

Bassoullet, J. P., Colchen, M., Marcoux, J. and Mascle, G., 1981, Field evidences for continental rifting in Triassic time in the Ladakh part of the Indus suture zone. *In*: Liu Dong-sheng (ed.) Geological and ecological studies of Qinghai-Xizang Plateau, v. 1, Geology, geological history and origin of Qinghai-Xizang Plateau: Beijing, Science Press; and New York, Gordon and Breach, Science Publishers, Inc., p. 579-585.

Basu, A. R., Wang-ming, Deng, Wen-you, Zhang, Mehnert, H. and Hannah, J., 1984, Isotope study of Tertiary alkalic volcanism in Changtang, northern Tibet: Geological Society of America Abstracts with Programs 16, p. 439.

Bates, R. L. and Jackson, J. A. (eds.) 1980, Glossary of geology: Falls Church, Virginia, Amer. Geol. Institute, 749 p.

Baud, A., Gaetani, M., Garzanti, E., Fois, E., Nicora, A. and Tintori, A., 1984, Geological observations in southeastern Zanskar and adjacent Lahul area (northwestern Himalaya): Eclogae Geologicae Helvetiae, v. 77, no. 1, p. 171-197.

Bazhenov, I. I., Leonenko, I. A. and Kharchenko, A. K., 1959, Ugolnaya promyshlennost' Kitayskoy Narodnoy Respubliki (Commercial resources of the PRC): Moscow, Izd. Gosgortekh., 479 p.

Bei Geng, 1982, A study on organic matter and the evaluation of petroleum generation potential of Upper Triassic Maantang Formatin and Xiaotangzi Formation in northwestern Sichuan: Oil and Gas Geology, v. 3, no. 1, p. 67-74.

Beka, K. and Vysotskiy, I. V., 1976, Geologiya nefti i gaza (Geology of oil and gas): Mowcow, Izd. Nedra, p. 592.

Beloussov, V. V., 1962, *Basic problems in geotectonics*: New York, McGraw-Hill, 816 p.

Beloussov, V. V., 1984, Certain problems of the structure and evolution of transition zones between continents and ocean basins: Tectonophysics, v. 105, no. 1-4, p. 79-102.

Beloussov, V. V., Volvovsky, B. S., Volvovsky, T. S., Khamrabaev, I. H., Kaila, K. and Marussi, A., 1984, deep structure of Central Asia along the Tian-Shan-Pamir-Himalayas geotraverse. *In*: Tectonics of Asia, Moscow, International geological Congress, 27th, colloquium 05, Reports, v. 5, p. 29-38.

Ben-Avraham, Z. and Uyeda, S., 1973, The evolution of the China basin and the Mesozoic paleogeography of Bornea: Earth and Planetary Science Letters, v. 18, p. 365-376.

Bender, F., 1983, *Geology of Burma*: Berlin, Gebruder Borntraeger, 293 p.

Berkey, C. P. and Morris, F. D., 1927, Geology of Mongolia. *In*: Natural History of Central Asia: Amer. Museum of Natural History, v. II, p. 475.

Bhat, M. I., 1984, Evolution of the southern passive margin of the Himalayan Tethys: model implications and constraints-1: Journal of Petroleum Geology, v. 7, p. 341-350; 11: Journal of Petroleum Geology v. 7, p. 429-436.

Billington, S., Isacks, B. L. and Barazangi, M., 1977, Spatial distribution and focal mechanisms of mantle earthquakes in the Hindu Kush-Pamir region: a contorted Benioff zone: Geology, v. 5, p. 699-704.

Bor-ming, Jahn, 1974, Mesozoic thermal events in southeast China: Nature, v. 248, p. 480-483.

Bosum, W., Burton, G. D. and Hsieh, S. H., 1970, Aeromagnetic survey of offshore Taiwan: U. N. Econ. Comm. Asia Far East (ECAFE), Comm. Coord. Joint. Prospect. Miner. Resour. Asian Offshore Areas, Tech. Bull., v. 3, p. 1-34.

Boucot, A. J. and Gray, J., 1987, The Tethyan concept during the Paleozoic. *In*: McKenzie, K. G. (eds.) Shallow Tethys 2: Boston, A. A. Balkema, p. 31-47.

Bowman, J. D., 1974, Petroleum developments in Far East in 1973: Am. Assoc. Petroleum Geologists Bull., v. 58, no. 10, p. 2124-2156.

Brandon, C. and Romanowicz, B., 1986, A 'no-lid' zone in the central Chang-Thang platform of Tibet: evidence from pure path phase velocity of long period Rayleigh waves: Journal of Geophysical Research, v. 91, p. 6547-6564.

Brasier, M. D., 1989, China and the Palaeotethyan belt (India, Pakistan, Iran, Kazakhstan and Mongolia). *In*: Cowie, J. W. and Brazier, M. D. (eds.) The Precambrian-Cambrian boundary: New York, Oxford University Press, p. 40-74.

Brod, I. O. and Vysotskiy, I. V., 1965, Neftegazonosnyye basseyny zemnogo shara: Moscow, Nedra, 598 p.

Brunel, M., 1986, Ductilee thrusting in the Himalayas: shear sense criteria and stretching lineations: Tectonics, v. 5, p. 247-265.

Brunel, M. and Kienast, J. R., 1986, Etude petro-structurale des chevauchements ductiles himalayens sur la transversale de l'Everest-Makalu (Nepal oriental): Canadian Journal of Earth Sciences, v. 23, p. 1117-1137.

Buffetaut, E., 1989, The contribution of vertebrate palaeontology to the geodynamic history of South East Asia. *In*: Sengor, A. M. C. (ed.) Tectonic Evolution of the Tethyan Region: London, Kluwer Academic Publishers, p. 645-653.

Buffetaut, E., Jaeger, J. -J. and Rage, J. -C., (orgs.), 1985, Paleogeographie de l'Inde, du Tibet et du sud-est Asiatique: confrontation des donnees paleontologiques avec les modeles geodynamiques: Soc. Geol. France, Memoir 147, 194 p.

Burbank, D. W. and Li Jijun, 1985, Age and palaeoclimatic significance of the loess of Lanzhou, North China: Nature, v. 316, p. 429-431.

Burtman, V. S., 1975, Structural geology of Variscan Tienshan, U. S. S. R.: Am. Jour. Science, v. 275A, p. 157-186.

Cai Qianzhong, 1983, Geological conditions of Permo-Carboniferous coal-formed gas in North China platform and its prospects: Oil and Gas Geology, v. 4, no. 1, p. 34-44.

Cai Qianzhong, 1985, The earlier petroleum exploration in Huabei basin: Oil and Gas Geology, v. 6, no. 4, p. 395-401.

Cai Shuntian and Guo Yiqiu, 1984, An outline of petroleum geology of Baise basin: Oil and Gas Geology, v. 5, no. 4, p. 362-371.

Cai, Z. Y. and Li, X. X., 1980, Classification and correlation of Devonian continental strata in China: Nanjing Inst. Geol. Paleont., Academia Sinica, p. 128-141 (in Chinese).

Cao Congzhou, 1989, The Ophiolite belts of northeastern China: Jour. Southeast Asian Earth Sci., v. 3, nos. 1-4, p. 233-236.

Cao Rong-long, 1981, Lithological features and geological significance of Yarlung Zangbo River ophiolite belt and trench sediments in Xizang Plateau. *In*: Liu Dong-sheng (ed.) *Geological and econogical studies of Qinghai-Xizang Plateau*, v. 1, *Geology, geological history and origin of Qinghai-Xizang Plateau*: Beijing, Science Press; and New York, Gordon and Breach, Science Publishers, Inc., p. 611-620.

Cao Ruiji, 1983, Microflora of the Precambrian, influence on the morphology of stromatolites: Nanjing Institute of Geology and Palaeontology, Academia Sinica, Bull., v. 6, p. 1-7 (in Chinese, English abstract).

Cao Zhengyan, 1984, Oil shale in China; resources and utilization: Oil and Enterprise, p. 44-45.

Carey, S. W., 1955, The orocline concept in geotectonics: Royal Society of Tasmania Proceedings, 89, 255-88.

Carey, S. W., 1958, The tectonic approach to continental drift (conv., S. W. Carey. *In*: Continental drift, a symposium: Hobart, University of Tasmania, Geology Department, p. 177-355.

Central Intelligence Agency, 1971, People's Republic of China - Atlas: Washington, D. C., U. S. Government Printing Office, 82 p.

Central Intelligence Agency, 1975, China: energy balance projections: Washington, U. S. Library of Congress, Document Expediting Project, Exchange and Gift Division, 33 p.

Central Intelligence Agency, 1977, China; oil production prospects: Washington, D.C., DOCEX, Exchange and Gift Division, Library of Congress, ER 77-10030U (June), 28 p.

Chai, B. H. T., 1972, Structure and tectonic evolution of Taiwan: Amer. Jour. Science, v. 272, p. 389-422.

Chai, H. -S. and Pan, Y. S., 1975, The magnetic model study of the northwest offshore of Taiwan: Taipei, Petroleum Geol. Taiwan, no. 12, p. 131-140.

Chandra, U., 1981, Focal mechanism solutions and their tectonic implications for the eastern Alpine-Himalayan region. *In*: Gupta, H. K. and Delany, F. M. (eds.) Zagros-Hindu Kush-Himalaya geodynamic evolution, American Geophysical Union, Geodynamics Series 3, p. 243-271.

Chang Cheng-fa and Cheng Xi-lan (Cheng Hsi-lan), 1973, Some tectonic features of the Mt. Jolmo Lungma area, southern Tibet, China: Scientia Sinica, v. 16, p. 257-265.

Chang Cheng-fa and Pan Yu-sheng, 1981, A brief discussion on the tectonic evolution of Qinghai-Xizang Plateau. *In*: Liu Dong-sheng (ed.) Geological and ecological studies of Qinghai-Xizang Plateau, v. 1, Geology, geological history and origin of Qinghai-Xizang Plateau: Beijing, Science Press; and New York, Gordon and Breach, Science Publishers, Inc., p. 1-18.

Chang Chengfa and Pan Yusheng, 1984, The structure and evolution of the Qinghai-Xizang (Tibet Plateau). *In*: Su Zongwei (ed.) Developments in Geoscience, Contribution to 27th International Geological Congress, 1984, Moscow: Beijing, Science Press, p. 453-462.

Chang Cheng-fa, Zheng Xi-lan and Pan Yu-sheng, 1977, The geological history, tectonic zonation and origin of uplifting of the Himalayas: Beijing, Academic Sinica, p. 17.

Chang Chengfa, Chen Nansheng, M. P. Coward, Deng Wangming, J. F. Dewey, A. Gansser, N. B. W. Harris, Jin Chengwei, W. S. F. Kidd, M. R. Leeder, Li Huan, Lin Jinlu, Liu Chengjie, Mei Houjun, P. Molnar, Pan Yun, Pan Yusheng, J. A. Pearce, R. M. Shackleton, A. B. Smith, Sun Yiyin, M. Ward, D. R. Watts, Xu Juntao, Xu Ronghua, Yin Jixiang and Zhang Yuquan, 1986, Preliminary conclusions of the Royal Society and Academia Sinica 1985 Geotraverse of Tibet: Nature, v. 323, no. 6088, p. 501-507.

Chang Chengfa, Pan Yusheng and Sun Yiying, 1989, The tectonic evolution of Qinghai-Tibet Plateau: a review. *In*: Sengor, A. M. C. (ed.) Tectonic evolution of the Tethyan Region: London, Kluwer Academic Publishers, p. 415-476.

Chang Chengfa, Shackleton, R. M., Dewey, J. F. and Yin Jixiang (leaders: geotraverse of the Qinghai-Xizang plateau), 1988, The Geological evolution of Tibet: London, The Royal Society, 413 p.

Chang Chiyi, 1981, Alluvial-fan coarse clastic reservoirs in Karamay. *In*: Mason J. F. (ed.) Petroleum geology in China: Tulsa, PennWell Publ., p. 154-170.

Chang Da (Ch'ang Ta), 1959, The geology of China: Beijing, 944 p. (in Chinese).

Chang Da (Ch'ang Ta), 1963, The geology of China: Washington, U. S. Dept. Commerce, Joint Publ. Research Service, JPRS, 19, 209, OTS 63-21820, May 16, 1963, p. 623.

Chang, K. (Zhang, G.), 1978, The antiarchs from the Early Devonian of Yunnan: Vertebrata Palasiatica, v. 16, no. 3, p. 147-186.

Ch'ang K'eng (Chang Keng), Ch'eng Ch'ing-ta (Cheng Qingda) and Zabarinskiy, P. O., 1958, Neftanyye i gazovyye mestorozhdeniya Kitayskoy Narodnoy Respubliki: Moscow, Izd. Gos. Nauchno-Technicheskoyo, Neftyanoye i Gorno-Toplivnoy Literaturi, 112 p.

Chang, L. S., 1975, Biostratigraphy of Taiwan. *In*: Geology and paleontology of Southeast Asia: Tokyo, Univ. Tokyo Press, v. 15, p. 337-361.

Chang Lu-Chin (Zhang Lujin), 1983, On the age of Badaowan Formation in northern Xinjiang: Scientia Sinica, ser. B, v. 26, no. 7, p. 774-781.

Chang, S. S. L., 1971, Regional stratigraphy and petroleum possibilities of Miocene formations in northwestern Taiwan, China: Am. Assoc. Petroleum Geologists Bull., v. 55, no. 10, p. 1838-1865.

Chang, S. S. L., 1987, Petroleum geology and petroleum exploration on the continetal shelf of China: Geological Society of China, Memoir 8, p. 259-278.

Chang, S. S. L. and Liu, H. S., 1984, Geochemistry of the geothermal fields in eastern Taiwan, China. *In*: Watson, S. T. (ed.) Transactions of the third Circum-Pacific Energy and Mineral Resources Conference, August 22-28, 1982, Honolulu, Hawaii: American Association of Petroleum Geologists, p. 383-391.

Chang Wenyou, Zhong Dalai, Ma Fuchen, Sun Huanchang and Zhong Jiayou, 1981, Fault block tectonics and oil resources in China. *In*: J. F. Mason (ed.) Petroleum geology in China: Tulsa, PennWell Books, p. 116-131.

Chang, W. T., 1962, *Ordovician of China*: Beijing, Science Press, 161 p. (in Chinese).

Chang, W. T., 1980, A review of the Cambrian of China: Jour. Geol. Soc. Australia, v. 27, nos. 1, 2, p. 137-150.

Chang, W. T., 1987, Tethys in the early Palaeozoic. *In*: McKenzie, K. G. (eds.) Shallow Tethys 2: Boston, A. A. Balkema, p. 61-64

Chang Xiaozhe, Zhen Yichuen, Xie Chenkan and Yang Fuxin, 1981, The prospect of Permo-Carboniferous coal-formed gas in North China: Oil and Gas Geology, v. 2, no. 4, p. 341-350.

Chao, C. K., Chen, C. and Liang, H. L., 1962, Triassic of China: Beijing, Science Press, p. 132 (in Chinese).

Chao, E. C. T., 1961, Progress and outlook of geology science in communist China: Am. Assoc. Advancement Sci., Publ. 68, p. 497-522.

Chao, K., 1965, The Permian ammonoid-bearing formations of south China: Scientia Sinica, v. XIV, no. 12, p. 1813-1825.

Chao Tsungtu, 1984, Ophiolite and continental suture: Scientia Geologica Sinica, no. 4, p. 359-372.

Chao (Zhao), Y. T. and Huang, T. K., 1931, The geology of the Tsinlingshan (Qin Ling) and Szechuan (sichuan): Geol. Soc. China, Mem., ser. A, no. 9, 230 p.

Chappell, B. W. and White, A. J. R., 1974, Two contrasting granite types: Tokyo, Tokai University Press, Pacific Geology 8, p. 173-174.

Chase, C. G. and Wallace, T. C., 1986, Uplift of the Sierra Nevada of California: Geology, v. 14, p. 730-733.

Chatelain, J. L., Roecker, S. W., Hatzfeld, D. and Molnar, P., 1990, Micro-earthquake seismicity and fault plane solutions in the Hindu Kush region and their tectonic implications: Journal of Geophysicall Research, v. 85, p. 1365-1387.

Chatterjee, S., 1984, The drift of India: a conflict in plate tectonics.

In: Paleogeographie de l'Inde, du Tibet et du sud-est Asiatique: confrontation des donnees paleontologiques avec les modeles geodynamiques (organisateurs, E. Buffetaut, J. -J. Jaeger and J.-C. Rage), Societe geologique de France, Memoire no. 147, p. 43–48 (nouvelle serie).

Chatterjee, S. and Hotton III, M. 1986, The paleoposition of India: Jour. Southeast Asian Earth Sci., v. 1, no. 3, p. 145-189.

Chaudhri, R. S., 1983, Provenance of the Siwalik sediments of Nepa Himalaya. *In*: Sinha, A. K. (ed.) Contemporary geoscientific researches in Himalaya, v. 2, Geochemistry-petrology-mineralogy-sedimentology-geomorphology-metallogeny and geo-technical study: Dehra Dun, Bishen Singh Mahendra Pal Singh, p. 85–90.

Che Zhicheng, 1986, The formation and evolution of Qaidam basin in relation to the uplift of Qinghai-Tibet Plateau: Oil and Gas Geology, v. 7, no. 1, p. 87–94.

Chen Binghua, 1980, On the relation between the distribution of the buried-hill oil fields and the evolution of paleostructures and paleogeomorphological features in the middle Hebei: Acta Petrolei Sinica, v. 1, no. 4, p. 49–56.

Chen Binghua, 1985a, Division of structural zones in the Saihantala depression of Erlian basin and its relation to oil and gas accumulation: Acta Petrolei Sinica, v. 6, no. 4, p. 41–45.

Chen Binghua, 1985b, Tectonic characteristics of Manite depression in Eren basin and prediction of hydrocarbon enrichment zones: Oil and Gas Geology, v. 6, no. 3, p. 335–340.

Chen Changming, Huang Jiakuan, Chen Jingshan, Tian Xingyou, Chen Ruijun and Li Li, 1981, Evolution of sedimentary tectonics of Bohai rift system and its bearing on hydrocarbon acumulation: Scientia Sinica, Series B, v. 24, no. 4, p. 521–529.

Chen, C., 1980, Non-marine setting of petroleum in the Songliao basin of Northeastern China: Jour. Petroleum Geology, v. 2, no. 3, p. 233–264.

Chen, C. -H., 1984, A simple geological model for geothermal systems in the Central Range of Taiwan. *In*: Watson, S. T. (ed.) Transactions of the Third circum-Pacific Energy and Mineral Resources Conference, August 22-28, 1982, Honolulu, Hawaii: American Association of Petroleum Geologists, p. 393–397, 694.

Chen Fajing, 1986, Types, tectonic evolution and hydrocarbon occurrences of petroliferous basins of China: Earth Science, Journal of the Wuhan College of Geology, v. 11, no. 3, p. 221–230.

Chen Fajing, Li Mingcheng and Sun Jiazhen, 1983, The regional structural setting of Creteaceous-Early Tertiary evaporite and source rock sediments in Eastern China: Oil and Gas Geology, v. 4, no. 2, p. 133–140.

Chen Fajing, Li Sitian and Li Chen, 1982, The distribution of source rock, coal and evaporites in Mesozoic and Cenozoic basins in eastern China and the factors governing this distribution: Earth Sciences, no. 1, p. 167–179 (in Chinese) (in Engl. as Rept. No. 798672, Geol. Survey, Canada).

Chen Fu-min, 1981, On the evolution of Yarlung Zangbo River abyssal fault as viewed from the space-time distribution pattern of the Gangdise magmatic complex belt. *In*: Liu Dong-sheng (ed.) Geological and ecological studies of Qinghai-Xizang Plateau, v. 1, Geology, geological history and origin of Qinghai-Xizang Plateau: Beijing, Science Press; and New York, Gordon and Breach, Science Publishers, Inc., p. 387–393.

Chen Gong, Deng Jingui, Tian Ru and Chen Daisheng, 1986, Discussion on the mineralization conditions and the genetic model for uranium deposits in Meso-Cenozoic basins of China. *In*: Huang Jiqing (ed.) Proceedings of the symposium of Mesozoic and Cenozoic geology in connection of the 60th anniversary of the Geological Society of China: Beijing, Geological Publishing House, p. 725–741.

Chen Guoda, 1984a, The diwa-type metallogenesis in East China. *In*: Su Zongwei (ed.) Developments in geoscience; contribution to 27th International Geological congress, 1984, Moscow: Beijing, Science Press, p. 433–441.

Chen Guoda, 1984b, The southern Himalayan Diwa region. *In*: Su Zongwei (ed.) Developments in geoscience; contribution to 27th International Geological Congress, 1984, Moscow: Beijing, Science Press, p. 443–451.

Chen Guoda, 1988, The diwa-type metallogenesis in East China: Journal of Central-South Institute of Mining and Metallurgy, v. 19, no. 4, p. 349–356.

Chen Guoda, 1989, *Tectonics of China*: Beijing, International Academic Publishers; Oxford, Pergamon Press, 258 p.

Chen Guoda and Xue Jiamon, 1978, Some characteristics of geotectonics in China: Scientific Papers on Geology for International Exchange, 1 (in Chinese).

Chen Guoda, Chen Jiachao, Wei Bailin, Xue Jiamou, Liu Yixuan, Wen Shanji, Wei Zhoulng and Hu Huoyen, 1975, A brief review on the geotectonics of China: Scientia Geologica Sinica, no. 3, p. 205–219.

Chen Heli and Tang Xiyuan, 1983, A discussion on the compaction of argillaceous rocks and primary migration of oil and gas in Dongying sag of Shandong Province: Acta Petrolei Sinica, v. 4, no. 2, p. 9–16.

Chen Huanjiang, Sun Zaucai and Zhang Yuchang, 1986, The framework of Chinese petroliferous basins: Journal of Petroleum Geology, v. 9, no. 4, p. 451–462.

Chen Jinbiao, 1985, Preface – an explanatory note on Proterozoic stratigraphic nomenclature used in the People's Republic of China: Precambrian Research, v. 29, nos. 1–3, p. 3–4.

Chen Jinbiao, Zhang Huimin, Xing Yusheng and Ma Guogan, 1981, On the upper Precambrian (Sinian suberathem) in China: Precambrian Research, v. 15, nos. 3–4, p. 207–228.

Chen Jinhua, 1982, Mesozoic transgressions, regressions and bivalve provinces in China: Acta Geologica Sinica, v. 56, no. 4, p. 334–346.

Chen Jinshi, 1984, Carbonisotopic variation in carbonates at the boundary between the Permian and Triassic in China. *In*: Su Zongwei (ed.) Developments in geoscience; contribution to the 27th International Geological Congress: Beijing, Science Press, p. 247–254.

Chen Jun-yuan and Qi Dun-lun, 1981, Upper cambrian cephalopods from western Zhejiang. *In*: Teichert, C., Lu Liu and Chen Pei-ji (eds.) Paleontology in China, 1979: Geological Society of America, Special Paper 187, p. 137–141.

Chen Junyuan and Teichert, C., 1983, Cambrian cephalopods: Geology, v. 11, no. 11, p. 647–650.

Chen Junyuan and Teichert, C., 1983, Cambrian Cephalopoda of China: Plaeontographica, Abt. A, Bd. 181, Lf. 1–3, p. 102.

Chen Jun-yuan, Qian Yi-yuan, Zhang Jun-ming, Lin Yao-kun, Yin Yei-ming, Wang Zhi-hao, Wang Zong-he, Yang Jie-dong, and Wang Ying-Xi, 1988, The recommended Cambrian-Ordovician global boundary stratotype of the Xiaoyanggiao section (Dayangcha, Jilin Province), China: Geological Magazine, v. 125, no. 4, p. 415–444.

Chen Jun-yuan, Liu Geng-wu and Chen Ting-en, 1981, Silurian nautiloid faunas of central and southwestern China: Nanjing Inst. Geol. and Paleontology Memoirs, no. 13, p. 94–262.

Chen Jun-yuan, Tsou Siping, Chen Tingen and Li Dunluan, 1979a, Late Cambrian cephalopods of North China – *Plectronocerida, Protactinocerida* (Order nov.) and *Yanhecerida* (Ord. nov.): Acta Palaeont. Sinica, v. 18, no. 1.

Chen Jun-yuan, Tsou Siping, Chen Tingen and Qi Dunluan, 1979b, Late Cambrian *Ellesmerocerida* (cephalopods) of North China: Acta Palaeont. Sinica, v. 18, no. 2.

Chen Kaihui, 1984, Genetic characteristics of kaolin deposits in China: Scientia Sinica, Ser. B, v. 27, no. 7, p. 744–755.

Chen Keqiao, Yu Tinggao, Zhang Yongee and Peng Zhizhong, 1982, Tetraauricupride, Cu Au, discovered in China: Scientia Geologica Sinica, no. 1, p. 111–116.

Chen Ke-zao, Yang Shao-xiu and Zhen Xi-yu, 1981, The saline lakes on the Qinghai-Xizang Plateau. *In*: Liu Dong-sheng (ed.) Geological and ecological studies of Qinghai-Xizang Plateau, v. 2, Environment and ecology of Qinghai-Xizang Plateau: Beijing, Science Press; and New York, Gordon and Breach, Science Publishers, Inc., p. 1719–24.

Chen Mu-qiu (Muqiu), 1985, Microscopic characteristics of Late Carboniferous coal in China. *In*: Crore, A. T. (ed.) Neuvieme congres international de stratigraphie et de geologie du Carbonifere: Southern Illinois University Press, p. 555–559.

Chen Moxiang, Huang Gehsan, Wang Jian, Deng Xiao and Wang Jiyang, 1984, A preliminary research on the geothermal characteristics in the Bohai area: Scientia Geologica Sinica, no. 4, p. 392–401.

Chen Pei-ji (Piji), 1983, A survey of the non-marine Cretaceous in

China: Cretaceous Research, v. 4, no. 2, p. 123–143.

Chen Pei-ji (Piji), 1987, cretaceous paleogeography in China: Paleogeography, Palaeoclimatology, Palaeoecology, v. 59, no. 1–3, p. 49–56.

Chen Peiji and Shen Yanbin, 1981, Paleogene conchostracan faunas of China. *In*: Teichert, C. et al., eds. Paleontology in China, 1979: Geol. Soc. America, Spec. Paper, 187, p. 193–201.

Chen Peiji and Wang Zhen, 1984, On the non-marine Cretaceous-Tertiary boundary of China. *In*: Su Zongwei (ed.) developments in geoscience; contribution to the 27th International Geological Congress: Beijing, Science Press, p. 129–134.

Chen Piji, Li Wenben, Chen Jinhua, Ye Chunhi, Wang Zhen, Shen Yanbin and Sun Dongli, 1982, Stratigraphical classification of Jurassic and Creteaceous in China: Scientia Sinica, Series B, V. 25, no. 11, p. 1227–1248.

Chen Quanmao and Dickinson, W. R., 1986, Contrasting nature of petroliferous Mesozoic-Cenozoic basins in eastern and western China: Amer. Assoc. Petrol. Geol. Bull., v. 70, no. 3, p. 263–275.

Chen Ruchen, 1981, The developments of biogas utilization in China: United Nations Natural Resources Forum, v. 5, no. 3, July, Notes, 2.0., p. 277–282.

Chen Shaozhou, Gao Xinchen and Qiu Dongzhou, 1982, A preliminary study on China's Eogene transitional facies: Oil and Gas Geology, v. 3, no. 4, p. 343–350.

Chen Shi, 1984, The exploration and exploitation of natural gas in Sichuan before the Qing Dynasty: Oil and Gas Geology, v. 5, no. 2, p. 183–192.

Chen Shuzhu, 1988, A discussion on carbonate sedimentary facies of Chuenhuazhen Formation, Jiyang depression: Petroleum Exploration and Development, v. 15, no. 2, p. 41–47.

Chen Sizhong and Wang Ping, 1980, Geology of Gudao oil field and surrounding areas: Am. Assoc. Petrol. Geol. Me. 30, p. 471–486.

Chen Sizhong, Qian Kai and Li Zesong, 1982, Character and rules of distribution stratigraphic oil pools in Jiyang depression: Acta Petrolei Sinica, v. 3, no. 3, p. 23–30.

Chen, W., 1988, Mesozoic and Cenozoic sandstone-hosted copper deposits in South China: Mineralium Deposita, v. 23, no. 4, p. 262–267.

Chen Wang-yong, 1981, Natural environment of the Pliocene basin in Gyirong, Xizang. *In*: Liu Dong-sheng (ed.) Geological and ecological studies of Qinghai-Xizang Plateau, v. 1, Geology, geological history and origin of Qinghai-Xizang Plateau: Beijing, Science Press; and New York, Gordon and Breach, Science Publishers, Inc., p. 343–352.

Chen Xiaodong, Zhou Jizhi, Qiu Yanghui and Guo You, 1984, Petroleum geology features of Dongming depression: Oil and Gas Geology, v. 5, no. 4, p. 373–384.

Chen Xu, 1984, The Siliurian graptolite zonation of China: Canadian Jour. Earth Sci., v. 21, no. 2, p. 241–257.

Chen Xu, 1984, Influence of the late Ordovician glaciation on basin configuration of the Yangtze (Yangzi) platform in China: Lethaia, v. 17, p. 51–59.

Chen Yuchuan, 1984, Metallogenic series of non-ferrous, rare-metallic and rare-earth ore deposits associated with Yenshanian granites, South China. *In*: Janelidze, T. V. and Tvalchrelidze, A. G. (eds.) 1982, Proceedings of the Sixth Quadrennial IAGOD Symposium: E. Stuttgart, Schweizerbart'sche Verlagsbuchhandlung, p. 263–268.

Chen Yunlin, 1984, A preliminary analysis on the types of river system depositional characteristics and development results of the Guantao reservoir in Gudao oil field: Petroleum Exploration and Development, v. 11, no. 5, p. 66–72.

Chen Zhaobo, Xie Youshin, Wan Guoliang, Ji Shufan, Wang Canlin and Fang Xiheng, 1982, Uranium deposits in Mesozoic volcanics in south-east China: Acta Geologica Sinica, v. 56, no. 3, p. 235–243.

Chen Zhi-ming, 1981, Structural origin on lakes on the Xizang Plateau. *In*: Liu Dong-sheng (ed.) Geological and ecological sutdies of Qinghai-Xizang Plateau, v. 2, Environment and ecology of Qinghai-Xizang Plateau: Beijing, Science Press; and New York, Gordon and Breach, Science Publishers, Inc., p. 1769–1775.

Chen Zongqing, 1982, The formation of Carboniferous gas pools in east Sichuan: Acta Petrolei Sinica, v. 3, no. 1, p. 23–28.

Chen Zongqing, 1983, Enriching factors of natural gas in Carboniferous System of East Sichuan: Oil and Gas Geology, v. 4, no. 3, p. 310–317.

Cheng Jiming, 1988, Types and metallization patterns of gold placers in China: Geophysical Prospecting for Petroleum, v. 24, no. 3, p. 3–4 (in Chinese, English abstract).

Cheng Keming and Zhang Yirong, 1981, Thermal evolution of organic matter in Meso-Cenozoic continental basins in China: Acta Petrolei Sinica, v. 2, no. 4, p. 17–24.

Cheng Senqiang, Liu Zuhui, Liu Zhaoshu, He Shanmou, Huang Siliu, Yuan Hengyong and Zheng Yixiang, 1981, Features of gravity and magnetic anomalies in central and northern parts of the South China Sea and their geological interpretation: Scientia Sinica, v. 24, no. 9, p. 1271–1284.

Cheng Xirong, 1986, Textual research of oil wells in ancient China: Oil and Gas Gaology, v. 7, no. 1, p. 104–106.

Cheng Xueru, 1982, Positions of plate tectonics in the Songliao basin and characteristics of occurrence of hydrocarbons: Petroleum Exploration and Development, v. 9, no. 6, p. 39–48.

Cheng Xueru, 1987, Features of the early stage rift valley and its oil-bearing characteristics of Songliao basin: Petroleum Exploration and Development, v. 14, no. 1, p. 1–9.

Cheng Yuqi, Bai Jin and Sun Dazhong, 1982, The Lower and Middle Precambrian of China; *in* Chinese Academy of Geological Sciences (ed.) An outline of the stratigraphy of China; Beijing, Geological Publishing House, 46 p.

Cheng Yuqi, Chung Fudao and Su Yongjun, 1973, The pre-Sinian of North and Northeast China: Acta Geologica Sinica, no. 1, p. 72–81 (in Chinese).

Cheng Yuqi, Dong Shenbao, Shen Qihan, Sun Dozhong and Lu Liangzhou (eds.) 1986, metamorphic map of China, 1:4,000,000 scale with 162 p. text: Beijing, Geol. Publ. House.

Cheng Zhengwu, 1981, Permo-Triassic continental deposits and vertebrate faunas of China. *In*: Cresswell, M. M. and Vella, P. (eds.) Gondwana Five: Holland, Rotterdam, A. A. Balkema, p. 65–77.

Cheng Zhengwu, 1982, Permo-Triassic continental deposits and vertebrate faunas of China. *In*: Cresswell, M. M. and Vella, P. (eds.) Gondwana Five: Rotterdam, A. A. Balkema Publ., p. 65–70.

China Academy of Sciences, 1958, A regional stratigraphic table of China: Washington, U. S. Dept. Commerce, Joint Pub. Research Service, JPRS 18,539 OTS 63–21499, April 4, 1963, 511 p.

China, State Bureau of Seismology, Institute of Seismology; and China, Institute of Geology, 1982, The album of typical satellite imagery of active structures of China: Beijing, Seismological Press, 157 p. (in Chinese).

Chinese Academy of Geological Sciences, 1973, Atlas of Provincial geologic map of the People's Republic of China: Beijing, Geological Map printing House of China, scales mainly 1:2,000,000 and 1:3,000,000 (twenty-seven Provincial geologic maps with separate texts), 149 p. (in Chinese).

Chinese Academy of Geological Sciences, 1975, Geologic map of Asia (1:5,000,000): Beijing, Cartographic Publishing House (in Chinese).

Chinese Academy of Geological Sciences, 1975, Map of tectonic systems of the People's Republic of China: Beijing, Cartographic Publishing House (in Chinese), 1:4,000,000.

Chinese Academy of Geological Sciences, 1976, Geological map of the People's Republic of China: Beijing, Cartographic Publishing House, 1:4,000,000 (in Chinese).

Chinese Academy of Geological Sciences, 1982, An outline of the stratigraphy in China. *In*: Stratigraphy of China, v. 1: Beijing, Geological Publishing House, 445 p.

Chinese Academy of Geological Sciences, Institute of Geology, 1987, *Guide to the metallogenic map of endegenic ore deposits of China*: Beijing, Cartographic Publishing House, p. 142.

Ching Yu-kan (Yukan) and Liao Zhuo-ting (Zhuoting), 1985, Carboniferous brachiopod faunas of China. *In*: Washington, D. C., and Urbana-Champaign, Illinois, International congress on Carboniferous stratigraphy and geology, 9th, 1979: Compte rendu, v. 5, plaeonto., paleoecol., paleogeog., p. 245–250.

Chou, J. T., 1969, A petrographic study of the Mesozoic and Cenozoic rock formations in the Tungliang well TL-1 of the Penghu Islands, Taiwan, China: United Nations ECAFE, CCOP, Tech. Bull., v. 2, p. 97–115.

Chow, J. T. and Lin, C. C., 1974, Geology of Taiwan: Taiwan Provincial Documentary Committee, 450 p. (in Chinese).

Chu, T. O., 1924, The oil fields of China: Am. Assoc. Petrol. Geol. Bull., v. 8, no. 2, p. 169–177.

Chun, K. -Y. and McEvilly, T. V., 1986, Crustal structure in Tibet: high seismic velocity in the lower crust: Journal of Geophysical Research, v. 91, p. 10405–10411.

Chung-Chien Young (Zhongjian), 1945, A review of the fossil fishes of China, their stratigraphical and geographical distribution: Am. Jour. Science, v. 243, no. 3, p. 127–137.

Chung-Hsiang P'an, 1982, Petroleum in basement rocks: Am. Assoc. Petroleum Geol. Bull., v. 66, no. 10, p. 1597–1643.

Chu-husiang P'ang (Zhuxiang Pang) and Ryabukhin, G. E., 1963, The geologic structure of the intermontane basins of Central Asia and the presence of gas and oil therein: International Geology Review, v. 5, no. 8, p. 985–998.

Clapp, F. L. and Fuller, M. L., 1926, Oil fields of China; acknowledgements and correlations: Am. Assoc. Petrol. Geol. Bull., v. 10, no. 4, p. 449.

Clark, A. L., Dorian, J. P. and Pow-Foong, Fan, 1987, An estimate of the mineral resources of China: Resources Policy, v. 13, no. 1, p. 68–84.

Colbert, E. H., 1969, Evolution of the vertebrates: A history of the backboned animals through time, 2nd ed.: New York, John Wiley & Sons, Inc., 535 p.

Coleman, R. G., 1989, continental growth of Northwest China: Tectonics, v. 8, no. 3, p. 621–635.

Compilation Group of the Geological Map of Asia, Chinese Academy of Geological Sciences, 1975, The geologic map of Asia: Beijing, Topographic Publishing House of China, scale 1:5,000,000, twenty sheets (in Chinese).

Compilation Group of the Geological Map of China, 19976, An outline of the geology of China: Peking, Chinese Academy of Geological Sciences, 22 p.

Coney, P. J., 1970, The geotectonic cycle and the new global tectonics: Geological Society of America Bulletin, v. 81, no. 3, p. 739–747.

Connell, H. R., 1974, China's petroleum industry – an engima: Am. Assoc. Petrol. Geol. Bull., v. 58, no. 10, p. 2157–2172.

Conway, M. H., 1985, Gold mining and production: The China Business Review, v. 12, no. 3, p. 8–12.

Cook, P. J. and Shergold, J. H., 1984, Phosphorus, phosphorites and skeletal evolution at the Precambrian-Cambrian boundary: Nature, v. 308, p. 231–236.

Cordier, H., 1920, Ser Marco Polo, notes and addenda to Sir Henry Yule's edition, containing the results of recent research and discovery: London, John Murray, 161 p. (see Sir H. Yule, v. 3).

Cowie, J. W. and Braiser, M. D. (eds.) 1989, The Precambrian-Cambrian boundary: New York, Oxford University Press, 213 p.

Crawford, A. R., 1974a, A greater Gondwanland: Science, v. 184, p. 419–426.

Crawford, A. R., 1974b, The Indus suture line, the Himalaya, Tibet and Gondwanaland: Geological Mazazine 111, p. 369–383,

Crawford, A. R., 1979, The myth of a vast ocean Tethys, the India-Asia problem and earth expansion: Journal of Petroleum Geology 2, p. 3–9.

Cressey, G. B., 1934, China's geographic foundations; a survey of the land and its people: New York, McGraw Hill.

Cui Kexin, 1984, On the origin and development of China's continent. *In*: Su Zongwei (ed.) Developments in geoscience, Contribution to 27th International Geological contress 1984, Moscow: Beijing, Science Press, p. 493–508.

Cui Kexin and Zhen Yongyi, 1984, The Paleoclimates from Sinian to Permian in China: Scientia Geologica Sinica, no. 1, p. 1–12.

Cui Shengqin and Li Jinreng, 1986, On the Indosinian movement of China's peri-Pacific tectonic belt. *In*: Huang Jiqing, chief ed., Proceedings of the Symposium on Mesozoic and Cenozoic Geology: Beijing, Geological Publishing House, p. 273–281.

Cui Wenyuan, Wang Shiqi, Wei Chunjing and Tao Quan, 1989, Characteristics and evolution of the Lower Precambrian basement of the North China platform: Jour. Southeast Asian Earth Sci., v. 3, nos. 1–4, p. 284–291.

Cummins, W. A., 1971, Oceanic expansion and continental contraction: Tectonophysics, v. 12, no. 4, p. 273–281.

Curray, J. R. and Morre, D. G., 1971, Growth of the Bengal deep-sea fan and denudation in the Himalayas: Geological Soceity of America Bulletin, v. 82, p. 563–573.

Curray, J. R., Emmel, F. J., Moore, J. G. and Raitt, R. W., 1982, Structure, tectonics and geological history of northern Indian Ocean. *In*: Nairn, A. E. M. and Stehli, F. G. (eds.) The Ocean basins and margins, v. 6, The indian Ocean: New York, Plenum Press, p. 399–450.

Dai Jingxing, 1980, Preliminary research on natural gas in coal series in China: Acta Petrolei sinica, v. 1, no. 4, p. 27–38.

Dai Jingxing, 1981, The geographical distribution of oil and gas discovered in ancient China: Oil and Gas Geology, v. 2, no. 3, p. 292–298.

Dai Jingxing, 1983, Gas in Yang-Xin (Yangxin) Formation in Sichuan Province is originated from coal containing formation – a discussion on this problem with Mr. Huang et al.: Petroleum Explor. Development, v. 10, no. 4, p. 69–75.

Dai Jingxing and Qi Houfa, 1982, Types of gas pools in China: Acta Petrolei Sinica, v. 3, no. 4, p. 13–19.

Dai Wentian, 1983 (1984), The Sinian System in China; an outline of stratigraphy, glaciation and mineral deposits: Zentralblatt Geol. und Palaont, Teil 1, Allg., ang., Region., Hist., Geologie, 1983, (May 1984). no. 9–10, p. 1156–1164.

Dana, J. D., 1873, On the origin of mountains: American Jour. Sci., v. 3, no. 5, p. 347–350.

Dana, J. D., 1873, On some results of the earth's contraction from cooling including a discussion of the origin of mountains and the nature of the earth's interior: American Jour. Sci., v. 3, no. 5, p. 423–443, 474–475; no. 6, p. 6–14, 104–115, 161–172, 304, 381–382.

Dang Xin and Khou Mingzhen, 1964, The vertebrate-bearing Early Tertiary of South China; a review: Peking (Beijing), Vertebrata Palasiatica, v. 8, no. 2, p. 119–133.

Das Gupta, S. K., 1975, A revision of the Mesozoic-Tertiary stratigraphy of the Jaisalmer basin, Rajasthan: Indian Journal of Earth Sciences, v. 2, no. 1, p. 77–94.

Debon, F., Zimmerman, J. -L., Liu Guohue, Jin Chengwei and Xu Ronghua, 1985, Time relationships between magmatism, tectonics and metamorphism in three plutonic belts in southern Tibet: new K-Ar data: Geologische Rundschau, v. 74, no. 2, p. 229–36.

Defense Mapping Agency, 1979, Gazetteer of the people's Republic of China; pinyin to Wade-Giles, Wade-Giles to Pinyin: Geographic Names Data Base Division, Washington, D. C., 919 p.

DeJong, K. A. and Subhani, A. M., 1979, Note on the Bela ophiolites, with special reference to the Kanar area. *In*: Farah, A. and DeJong, K. A. (eds.) Geodynamics of Pakistan: Quetta, Geol. Survey of Pakistan, p. 263–269.

Den Aisong, 1987, Impact of mine development on geological environment in tropical and subtropical areas of South China and proposal for future management: Bangkok, Association of Geoscientists for International Development, AGID News, no. 50 (January), p. 16–24.

Deng Qidong, Sung Fengmin, Zhu Shilong, Li Mengluan, Wang Tielin, Zhang Weiqi, Burchfiel, B. C., Molnar, P. and Zhang Peizhen, 1984, Active faulting and tectonics of the Ningxia-Hui Autonomous Region, China: Jour. Geophys. Research, v. 89, no. 86, p. 4427–4445.

Deng Wanming, 1981, A preliminary study on the petrology and genesis of the Yarlung Zangbo ophiolite belt. *In*: Liu Dong sheng (ed.) Geological and ecological studies of Qinghai-Xizang Plateau, v. 1, Geology, geological history and origin of Qinghai-Xizang Plateau: Beijing, Science Press; and New York, Gordon and Breach, Science Publishers, Inc., p. 529–538.

Deng Yonggao and Cai Keqin, 1983, The relationships between salinization from Cretaceous to Tertiary and fault structure in eastern China: Earth Science - Journal of Wuhan College of Geology, no. 2 (Tol. 20 (sic), p. 75–84.

Department of Coal Teaching and Researches, Wuhan Geologic College, 1979, *Coal Geology*: Beijing, Geology Press, v. 1, 280 p. (in Chinese).

Department of Coal Teaching and Researches, Wuhan Geologic College, 1981, *Coal Geology*: Beijing, Geology Press, v. 2, 296 p. (in Chinese).

Department VII, Institute of Geology, Academia Sinica, 1978, The deltaic systems of lower Oligocene of Liaohe western depression: Petroleum Exploration and Development, 6 (in Chinese).

Desio, A. M. Gaetani and Martinis, B., 1982, Aspetti geologici del Tibet; il primo symposium sull' Altopiano Tibetano: Boll. Soc. Geol. Italiana, v. 101, no. 1, p. 27–42.

Dewey, J. F. and Burke, K. C. A., 1973, Tibetan, Variscan and Precambrian basement reactivication: products of continental collision: Journal of Geology, v. 81, no. 6, p. 683–693.

Di Hengshu, 1984, Overthrust belts on the northern margin of Qaidam basin and their oil prospecting: Oil and Gas Geology, v. 5, no. 2, p. 79–88.

Dickins, J. M., 1985, Paleobiofacies and palaeobiography of Gondwanaland from Permian to Triassic. *In*: Nakazawa, K. and Dickins, J. M. (eds.) 1985, The Tethys: Tokyo, Tokai University Press, p. 83–92.

Dong, Z., 1980, On the Dinosaurian faunas and their stratigraphic distribution in China: Jour. Stratigraphy, v. 4, no. 4, p. 267–263 (in Chinese).

Dorian, J. P. and Clark, A. L., 1986, Tectono-stratigraphic terranes and mineral resources of China: Resource Systems Institute, Honolulu, Rep. No. RSI-86-8, 46 p.

Dorian, J. P. and Fridley, D. G. (eds.) 1988, *China's energy and mineral industries; current perspectives*, 6th edition: Boulder, Colorado, Westview Press, 162 p.

Dovzhikov, A. E., et al. (eds.), 1963, The geological map of North Vietnam, 1:500,000 (in Vietnamese and Russian).

Dronov, V. I. and Abdullah, Sh., 1984, Tectonics of the Afghan-South Tajdik sector in the Asian part of Tethys. *In*: Reports of the 27th International Geological Contress, 1984, Moscow, Colloqium 05, Tectonics of Asia: Moscow, Nauka, v. 5, p. 97–110.

Drummond, M. and Hirn, A., 1984, Crustal thickening in Himalayas and Caledonides: Nature, v. 308, p. 497–498.

Du Guoqing, 1983, On the colour alteration of conodonts and maturity of organic matter in Permian and Triassic Systems, Hubei: Acta Petrolei Sinica, v. 4, no. 4, p. 11–18.

Du Jinshuang, Yi Chichang, Zhou Zeming, Huang Zhongfan and Liu Yuexin, 1984, Types of the contemporaneous faults in the eastern China basins and their mechanism of formation: Petroleum Exploration and Development, v. 11, no. 3, p. 8–17.

Du Younglin, 1985, A brief discussion on the conditions for the formation of oil fields in Eastern China: Petroleum Exploration and Development, v. 12, no. 5, p. 6–13.

Du Yonglin and Sheng Zhiwei, 1984, The stratigraphic characteristics of the Erlian basin: Petroleum Exploration and Development, v. 11, no. 1, p. 1–5.

Duan Jinying, 1983, Discovery of Yangzi-type trilobites in an Early Ordovician formation from a deep borehole in Donggai, northern Jiangsu: Petroleum Exploration Development, v. 10, no. 2.

Duan Weiwu and Lei Zuoqi, 1984, Tertiary stratigraphy of Zhujiangkou basin: Oil and Gas Geology, no. 2, p. 102–112.

England, P. and Searle, M., 1986, The Cretaceous-Tertiary deformation of the Lhasa block and its implications for crustal thickening in Tibet: Tectonics, v. 5, p. 1–14.

Fan Pu, Luo Binjie, Huang Ruchang, Shen Ping, Hui rongyao, Shao Hongshun, Wang Youxiao and Rong Guanghua, 1980, Formation and migration of continental oil and gas in China (1) - a fundamental geological condition for the formation of continental oil and gas: Scientia Sinica, v. 23, no. 10, p. 1286–1295.

Fan Pu, Luo Binjie, Huang Ruchang, Shen Ping, Hui Rongyao, Shao Hongshun, Wang Youxiao and Rong Guanghua, 1980, Formation and migration of continental oil and gas in China: Part II, Scientia Sinica, v. 23, no. 11, p. 1417–1427.

Fan Pu, Wang Youxiao, Rong Guanghua, Shen Qixiang and Lu Dexuan, 1980, The formation and evolution of Paleozoic oil and gas in a large sedimentary basin in North China platform: Geochimica, 2, p. 148–159.

Fan Shaoyi, Zhang Xiaozhi, Chen Renjie and Hu Juwen, 1983, A preliminary study on oil prospects in Xiling depression: Oil and Gas Geology, v. 4, no. 2, p. 192–201.

Fang, T. C., Xia, T. L. and Liu, H. L., 1979, *Paleontological stratigraphy*: Beijing, Geological Press, 308 p. (in Chinese).

Fang Zukang, 1981, Discussion on the main periods and the threshold temperature (depth) of oil generation in the Songliao basin: Acta Petrolei Sinica, v. 2, no. 3, p. 103–106.

Farah, A., Abbas, G., Dejong, K. A. and Lawrence, R. D., 1984, Evolution of the lithosphere in Pakistan: Tectonophysics, v. 105, p. 207–227.

Fei Ding, Cheng Shi-xing and Hao Chun-rong, 1981a, On the structural (features of) the central part of Xizang (Tibet) and obduction of the Indian plate: Acta Geophysica Sinica, v. 24, p. 394–403.

Fei Ding, Cheng Shi-xing, Hao Chun-rong, Yin Shu-jie, Ren Rui, dong Yun, Ma Jun-ru, Chang Xue-nu and Wang Men-ying, 1981b, On the structural features in the central part of Xizang and obduction of the Indian plate. *In*: Liu Dong-sheng (ed.) Geological and ecological studies of Qinghai-Xizang Plateau, v. 1, Geology, geological history and origin of Qinghai-Xizang Plateau: Beijing, Science Press; and New York, Gordon and Breach, Science Publishers, Inc., p. 747–756.

Fei Qi and Wang Xiepei, 1984, Significant role of structural fractures in Renqiu buried-hill oil field in eastern China: Amer. Assoc. Petrol. Geol. Bull., v. 68, no. 8, p. 971–982.

Feng Zengzhao and Wu Shenghe, 1988, Potential of oil and gas of Qinglong Group of Lower-Middle Triassic in Lower Yangtze River region from the viewpoint of lithofacies paleogeography: Acta Petrolei Sinica, v. 9, no. 2, p. 1–11.

Ferrar, G., Lombardo, B. and Tonarini, S., 1983, Rb/Sr geochronology of granites and gneisses from the Mount Everest region, Nepal Himalaya: Geologische Rundschau, v. 72, p. 119–136.

Fort, M., Freytet, P. and Colchen, M., 1981, The structural and sedimentological evolution of the Thakkola-Mustang graben (Nepal Himalaya) in relation to the uplift of the Himalayan Range. *In*: Liu Dong-sheng (ed.) Geological and ecological studies of Qinghai-Xizang Plateau, v. 1, Geology, geological history and origin of Qinghai-Xizang Plateau: Beijing, Science Press; and New York, Gordon and Breach, Science Publishers, Inc., p. 307–313.

Foster, H. L., 1965, Geology of Isigaki-shima, Ryukyu-retto: United States Geological Survey Professional Paper 399-1, 119 p.

Francheteau, J., Jaupart, C., Shen Xian-jie, Kang Wen-hua, Lee De-lu, Bai Jia-chi, Wei Hung-pin and Deng Hsia-yen, 1984, High heat flow in southern Tibet: Nature, v. 307, no. 5946, p. 32–36.

Frazier, S. B., Choi, S. O., Kim, B. D. and Schwartz, D., 1976, Marine petroleum exploration of Huksan platform, Korea: Am. Assoc. Petrol. Geol. Mem. 26, p. 268–275.

Fu Jia-mo, 1980, Origin of hydrocarbons and early Paleozoic rocks in China (abs.): Geol. Soc. America Abs. with Programs, v. 12, no. 7, p. 430.

Fu Jiamo and Sheng Guoying, 1986, Organic geochemical characteristics of major-type sedimentary formations of oil and gas-bearing basins in China. *In*: Tu Guangzhi (ed.) Advances in science of China, v. 1, Earth Sciences: Beijing, Science Press; New York, John Wiley and Sons, p. 251–286.

Fuchs, G., 1982a, *Explanations of the geologic-tectonic map of the Himalaya*, Vienna: Geologische Bundesanstalt, 50 p.

Fuchs, G., 1982b, The geology of western Zanskar: Jahrbuch der geologischen Bundesanstalt, v. 125, p. 1–50.

Gallagher, J. J. Jr., 1981, Tectonics of China; continental scale cataclastic flow: Amer. Geophys. Union, Monograph, v. 234, p. 259–273.

Gansser, A., 1964 (published, December 1965), *The geology of the Himalayas*: London, Interscience Publishers, 189 p.

Gansser, A., 1966, The Indian Ocean and the Himalayas; a geological interpretation: Eclogae Geol. Helvetiae, v. 59, no. 2, p. 831–848.

Gansser, A., 1974, Himalaya. *In*: Spencer, A. M. (ed.) *Mesozoic-Cenozoic orogenic belts*, Geological Society of London, Special Publication 4., p. 267–278.

Gansser, A., 1977, The great suture zone between Himalaya and Tibet, a preliminary account. *In*: *Ecologie et geologie de l'Himalaya*, Paris: Centre National de la Recherche Scientifique, Colloques Internationaux 268, p. 181–191.

Gansser, A., 1979, Reconnaissance visit to the ophiolites in Baluchistan and the himalaya. *In*: Farah, A. and DeJong, K. A. (eds.) Geodynamics of Pakistan: Quetta, Geol. Survey of Pakistan, p. 193–214.

Gansser, A., 1980, The significance of the Himalayan suture zone: Tectonophysics, v. 62, p. 37–52.

Gansser, A., 1981a, The geodynamic history of the Himalaya. *In*: Gupta, H. K. and Delany, F. M. (eds.) Zagros-Hindu Kush-Jimalaya geodynamic evolution: American Geophysical Union and Geological Society of America, Geodynamics Series 3, p. 111–121.

Gansser, A., 1981b, The timing and significance of orogenic events in the Himalaya. *In*: Liu Dong-sheng (ed.) Geological and ecological studies of Qinghai-Xizang Plateau v. 1, Geology, geological history and origin of Qinghai-Xizang Plateau: Beijing, Science Press; and New York, Gordon and Breach, Science Publishers, Inc., p. 23–30.

Gansser, A., 1983a, The morphogenic phase of mountain building. *In*: Hsu, K. J. (ed.) Mountain building processes, London: Academic Press, p. 221–228.

Gansser, A., 1983b, Geology of the Bhutan Himalaya: Denkschriften der Schweizerischen Naturforschenden Gesellschaft, v. 96, p. 1–181.

Gao, J. S., Xiong, Y. S. and Gao, P., 1931, Preliminary notes on the Sinian stratigraphy of North China: Geol. Soc. China, Bull., v. 13, no. 2, p. 257–275.

Gao Lianda, Wang Zengji and Wu Xianghe, 1983, The Middle Carboniferous of China. *In*: Martinez-Diaz, C., gen. ed., 1983, The Carboniferous of the World, v. 1, China, Korea, Japan and S. E. Asia: IUGS Publication no. 16, p. 56–86.

Gao Ruiqi, 1980, Characteristics of the continental Cretaceous in the Songliao basin: Acta Geologica Sinica, v. 1, no. 1, p. 9–22.

Gao, R. Q. and Zhao, C. P., 1976, Pollen and spore assemblage of Late Cretaceous from the Songliao basin: Beijing, Science Press, 83 p. (in Chinese).

Gao Shi-yang and Li Bing-xiao, 1981, Borate minerals on the Qinghai-Xizang Plateau. *In*: Liu Dong-sheng (ed.) Geological and ecological studies of Qinghai-Xizang Plateau, v. 2, Environment and ecology of Qinghai-Xizang Plateau, Beijing: Science Press; and New York, Gordon and Breach, Science Publishers, Inc., p. 1725–1731.

Gao Xingchen, 1985, The Triassic-Jurassic sedimentary facies of Tarim basin in relation to tectonics: Oil and Gas Geology, v. 6, no. 4, p. 369–379.

Gao Zhengzhong and Duan Taizhong, 1986, Submarine fan of terrigenous debris and allochthonous carbonate sediments on the base of slope in the Devonian trough in west Yunnan: Oil and Gas Geology, v. 7, no. 4, p. 308–323.

Gao Zhengzhong and Liu Huaibo, 1983, Early Triassic carbonate gravity flow along the north margin of Shiwandashan (Shiwan Dashan) basin and its geological significance: Oil and Gas Geology, v. 4, no. 1, p. 55–66.

Garth, B. G. (ed.) 1975, *China's changing role in the world economy*: New York, Praeger Publ., 222 p.

Ge Zhizhou, Rong Jiayu et al., 1977, The data of ten measured sections from Southwest China: Professional Papers of Stratig. and Paleont., no. 8, p. 92–111.

Geng, L., 1980, A preliminary study of the continental Jurassic-Cretaceous boundary in Asia: Jour. Stratigraphy, v. 4, no. 1, p. 13–28 (in Chinese).

Geological Composite Research Brigade of Bohai Gulf Petroliferous Basin and Structural Group of Wuhan college of Geology, 1976, Some structural traps related to the synsedimentary faults of the Bohai Gulf basin: Petroleum Exploration and Development, 2, 3, 4 (in Chinese).

Geological Institute of Qinghai-Xizang Plateau, Chinese Academy of Sicences, 1980, Geology of Qinghai-Xizang (Tibet) Plateau: Scale 1:1,500,000 (eight sheets), (in Chinese; abstract of summary in English).

Geomechanics Institute, 1976, On tectonic systems: Scientia Sinica, v. 16, no. 4, p. 1–17.

Gervasio, F. C., 1967, Age and nature of orogenesis in the Philippines: Tectonophysics, v. 4, no. 4–6, p. 379–402.

Girardeau, J., Marcoux, J., Allegre, C. J., Bassoullet, J. P. Tang Youking, Xiao Xuchang, Zao Younong and Wang Xibin, 1984a, Tectonic environment and geodynamic significance of the NeoCimmerian donqiao ophiolite, Bangong-Nujiang suture zone, Tibet: Nature, v. 307, p. 27–31.

Girardeau, J., Marcoux, J. and Zao Yougong, 1984b, Lithologic and tectonic environment of the Xigaze ophiolite (Yarlung Zangbo suture zone, southern Tibet, China), and kinematics of its emplacement: Eclogae Geologicae Helvetiae, v. 77, p. 153–170.

Girardeau, J., Mercier, J. C. C., and Zao Yougong, 1985a, Strucutre of the Xigaze ophiolite, Yarlung Zangbo suture zone, southern Tibet, China: genetic implications: Tectonics, v. 4, p. 267–288.

Girardeau, J., Marcoux, J., Fourcade, E., Bassoullet, J. P. and Tang Youking, 1985b, Xainxa ultramafic rocks, central Tibet, China: tectonic environment and geodynamic significance: Geology, v. 13, p. 330–333.

Golitsin, M. V., Prokofyeva, L. M., Kolesnik, V. Ya. and Tyurenkova, G. I., 1984, World resources of oil shales, their production and utilization. *In*: Reports of 27th International Geological Congress, 1984, Moscow, Colloquim 02, Moscow: Izdatel'stvo Nauka, Energy Resources of the World, v. 2, p. 25–48.

Gong Zitong, Chen Hongzhao, Wang Zhenquan, Cai Fengqi and Luo Guobao, 1987, The epigenetic geochemical types of loess in China: Soil Research Report, Institute of Soil Science, Academia Sinica, v. 16, p. 13.

Gopel, C., Allegre, C. J. and Xu Ronghua, 1984, Lead isotopic study of the Xigaze ophiolite (Tibet); the problem of the relationship between magmatites (gabbros, dolerites, lavas) and tectonites (harzburgites): Earth and Planet, Sci. Letters, v. 69, p. 301–310.

Gou Yun-xian, Hou You-tang and Wen Shi-xuan, 1987, Mesozoic Tethys in China. *In*: McKenzie, K. G. (ed.) Shallow Tethys 2: Boston, A. A. Balkema, p. 187–197.

Grabau, A. W., 1923–1924, *Stratigraphy of China, Part I Paleozoic and older*: Peking, Geological Survey of China in Ministry of Agriculture and Commerce, 528 p.

Grabau, A. W., 1928, *Stratigraphy of China: Part II Mesozoic*: Peking, Geological Survey of China, 774 p.

Green, S., 1979, Sichuan Journal: The China Business Review, v. 6, no. 6, p. 14–18.

Group of Tectonics, Institute of Geology and Mineral Resources, Academy of Geological Sciences, 1977, *An outline of the tectonic characteristics of China*: Beijing, Geology Press, 24 p.

Gu Daoyuan, Hou Jianguo, Huang Jianxiang, Zhu Hongyuan and Tao Jinbao, 1987, Isotopic age of the Lower Cretaceous boundary of China: Geological Review, v. 33, no. 3, p. 275–282.

Gu Hao-ding, Chen Yun-tai, Gao Xiang-lin and Zhao Yi, 1976, Focal mechanism of Haicheng, Liaoning Province, earthquake of February 4, 1975: Acta Geophysica Sinica, v. 19, no. 4, p. 270–285.

Gu (Ku) Zhiwei, 1962, *Jurassic and Cretaceous of China*: Peking, Science Press, 84 p. (in Chinese).

Gu Zhiwei, 1982, Distribution and development of nonmarine Mesozoic bivalves and formations in China: Scientia Sinica, ser. B, v. 15, no. 4, p. 438–452.

Gu Zhiwei, ch. ed., 1982, Correlation chart of Jurassic in China with explanatory test. *In*: Stratigraphical correlation chart of China, with explanatory test: Beijing, Science Press, p. 223–240 (in Chinese).

Gu Zhiwei, Sha Jingeng, Li Zishun and Yu Jinshan, 1984, Occurrences of marine Jurassic bivalves in eastern Northeast China and its significance on the nonmarine Jurassic-Cretaceous boundary in east and central Asia. *In*: Su Zongwei (ed.) Contribution to 27th International Geological congress, 1984, Moscow: Developments in Geoscience: Beijing, Science Press, p. 111–118.

Guan Baode, Wu Ruitang, Hambrey, M. J. and Geng Wuchen, 1986, Glacial sediments and erosional pavements near the Cambrian-

Precambrian boundary in western Henan Province, China: Jour. Geological Society, London, v. 143, part 2, p. 311-323.

Guan Shizong, 1985, Shacan-2 gusher is a prelude (Tarim basin): Oil and Gas Geology, v. 6, Special Issue, June 1985, p. 8-11.

Guan Shizong, Qiu Dungzhou, Chen Xianqun, Yuan Fungten, Yan Huaiyu, Wange Shoude, Zhou Jingcai and Chen Sioche, 1981, Geologic history of Late Proterozoic to Triassic in China and associated hydrocarbons. *In*: Mason, J. F. (ed.) 1981, Petroleum geology in China: Tulsa, OK, p. 142-153.

Guan Xunfan, Zhou Youngqing, Xiao Jinghua, Liang Shuzhao and Li Jinmao, 1985, Yinyan porphyry tin deposit - a new type of tin deposits in China: Acta Geologica Sinica, v. 59, no. 2, p. 155-161.

Guo Lingzhi, Shi Yangshen, Lu Huafu, Ma Ruishi, Dong Huogen and Yang Shufen, 1989, The pre-Devonian tectonic patterns and evolution of South China: Jour. Southeast Asia Earth Sci., v. 3, nos. 1-4, p. 87-93.

Guo Lingzhi, Yu Jianhua, Shi Yangshen, Ma Ruishi, Lu Huafu, Yun Lingling, Zhu Huijuan, Yang Shufeng and Chen Shengzao, 1984, On the time and space distribtuion of the granitic rocks and their relation to the tectonic configuration and crustal evolution in southeastern China. *In*: Xu Keqin and Tu Guangchi (eds.) 1984, Geology of Granites and their metallogenetic relations: Beijing, Science Press, p. 55-70.

Guo Shuang-xing, 1981, On the elevation changes of the Qinghai-Xizang Plateau based on fossil angiosperms. *In*: Liu Dong-sheng (ed.) Geological and ecological studies of Qinghai-Xizang Plateau, v. 1, Geology, geological history and origin of Qinghai-Xizang Plateau: Beijing, Science Press; and New York, Gordon and Breach, Science Publishers, Inc., p. 201-206.

Gupta, V. J. and Janvier, Ph., 1979, A review of the Devonian vertebrate localities of the Indian Himalayas (Kashmire, Ladakh and Kumaun), with remarks on their stratigraphical and palaeobiogeographical significance. *In*: Gupta, V. J. (ed.) Upper Palaeozoics of the Himalaya, Contributions to Himalayan Geology, v. 1: Delhi, Hindustan Publishing Corporation, p. 78-83.

Hahn, L., 1984, The Indosinian orogeny in Thailand and adjacent areas: Societe Geologique de France, memoire no. 147, p. 71-82.

Han De-xin (Dexin), Tian Bao lin (Baolin), Shao Zhen-jic (Zhenjie), Pan Sui-xian (Suixian) and Zhuang June, 1979, some sedimentary characteristics of coal measures and factors controlling coal accumulation in Carboniferous rocks in China. *In*: Neuvieme Congres International de Stratigraphie et de Geologie du Carbonifere: Washington and Champaign-Urbana, Compte Rendu, v. 4, p. 237-250.

Han Dexin and Yang Qi, ch. eds., 1979, *Coal Geology of China*: Beijing, Coal Industry Press, v. 1 (in Chinese).

Han Dexin and Yang Qi, ch. eds., 1980, *Coal Geology of China*: Beijing, Coal Industry Press, v. 2 (in Chinese), 415 p.

Han Keyou, 1984, The origin of the Longmen Shan overthrust zone and its oil and gas prospects: Natural Gas Industry, v. 4, no. 3, p. 1-8 (in Chinese).

Han Keyou, 1984, Paleogeography and petrolgraphical facies of the Early Permian period in China: Acta Sedimentologica Sinica, v. 2, no. 3, p. 60-74.

Han Shuanmin, 1983, A tentative discussion on the Xinglongtal vortex structure: Oil and Gas Geology, v. 4, no. 4, ;. 403-415.

Hanzawa, S., 1932, *Neoschwagerina* limestone found in Okinawa: Geological Society of Tokyo Journal, v. 39, no. 469, p. 672.

Hanzawa, S., 1933, On a *Neoschwagerina*-limestone from Okinawajima, Riukiu (Loochoo) Islands: Japanese Journal of Geology and Geography, v. 10, no. 3-4, p. 107-110.

Hao Shisheng, 1984, Prospect of primary oil and gas of the Middle-Upper Proterozoic in Hebei-Liaoning depression: Oil and Gas Geology, v. 5, no. 4, p. 342-348.

Hao Shumin, 1986, The shallow-water evidences from sediments in the Middle Ordovician in Fuping, Shaanxi: Oil and Gas Geology, v. 7, no. 2, p. 164-169.

Hao Yichun, Su Deying, Li Yougui, Yu Jingxian, Zhang Wangping, Li Peixian, Qi Hua, Guan Shaozeng and Gue Fuxiang, 1986, Stratigraphical subdivision of non-marine Cretaceous sequences and the Juro-Cretaceous boundary in China. *In*: Huang Jiqing, ch. ed., Proceedings of the Symposium on Mesozoic and Cenozoic Geology, China: Beijing, Geological Publishing House, p. 109-121.

Hao Yichung, 1983, The Lower-Upper Cretaceous and Cretaceous-Tertiary boundaries in China. *In*: Cretaceous stage boundaries symposium: Meddelelser fra Dansk Geologisk Forening, v. 33, no. 1-2, p. 129-138.

Hao Yichung, Su Deying, Li Yougui, P. H. Ruan and F. T. Yuan, 1974, *Fossil ostracods of the Cretaceous and Tertiary from the Sungliao (Songliao) plain*: Beijing, Geological press., 155 p. (in Chinese).

Hao Yichung, Su Deying, Li Yougui, Yu Jingxian, Zhang Wangping, Li Peisian, Qi Hua, Guan Shaozeng and Guo Fuxiang, 1982, Stratigraphical division of non-marine Cretaceous and the Jura-Cretaceous boundary in China: Acta Geologica Sinica, v. 56, no. 3, p. 187-199.

Hao Yichung, Zeng Xuelu and Li Hanmin, 1982, Late Cretaceous and Tertiary strata and foraminifera in western Talimu basin: Wuhan College of Geology Earth Sciences Journal 17, Special Paper 1, p. 1-161.

Hrding, T. P. and Lowell, J. D., 1979, Structural styles, their plate-tectonic habitats, and hydrocarbon traps in petroleum provinces: Am. Assoc. Petrol. Geol. Bull., v. 63, p. 1016-1058.

Hardy, R. W., 1976, Chinese oil; development prospects and potential impact: Georgetown Univ., Center for Strategic and International Studies, Monogr., p. 88.

Harland, W. B., Cox, A. V., Liewellen, P. G., Pickton, C. A. G., Smith, A. G. and Walters, R., 1982, *A geologic time scale*: New York, Cambridge University Press, 131 p.

Harrison, S. S., 1975, China; the next oil giant: Foreign Policy, no. 20 (Fall), p. 3-27.

Harrison, S. S., 1977, *China, oil and Asia*; *conflict ahead*?: Columbia Univ. Press, 317 p.

He (Ho) C., 1975, *An outline of the geology of Taiwan*: Taiwan, Geol. Surv. of Taiwan (in Chinese).

He Guoqi, Wang Shiguang, Cui Guangzhen and Shao Jian, 1983, On the features of reactivation of the Sino-Korean plate: Tectonophysics, v. 100, nos. 1-3, p. 119-130.

He Xinyi and Weng Fa, 1982, Early Permian rugose corals from Ali (Ngari), northern Xizang (Tibet): Wuhan College of Geology, Earth Science Journal, total 18, no. 3, p. 131-142.

He Xinyi and Weng Fa, 1981, New material of Early Permian corals from Ali, northern Xizang (Tibet): Wuhan College of Geology Earth Sciences Journal, total 19, p. 68-79.

He, Y. and Huh, L. Y., 1978, The Cenozoic foraminifera from the coastal region of Bohai: Beijing, Science Press, 47 p.

He Zhiqiang, 1984, Revitalize the economy by relying on science and technology - major achievements and projects in 35 years of scientific and tecnological development in Yunnan: Kunming, Inquiry into Economic Problems, no. 9, p. 1-19.

He Zhonglin and Xu Yufan, 1985, Geological features of Meso-Cenozoic continental basins in Anhui Province and their oil and gas potential: Oil and Gas Geology, v. 6, no. 1, p. 104-111.

He Ziai, Yang Hong and Luo Xiaozhi, 1981, The nature of the Upper Permian bioherms of Guizhou and their traces: Oil and Gas Geology, v. 2, no. 1, p. 1-10.

Hellinger, S. J., Shedlock, K. M., Sclater, J. G. and Yet Hong, 1985, The Cenozoic evolution of the North China basin: Tectonics, v. 4, no. 4, p. 343-358.

Helmcke, D., 1984, The orogenic evolution (Permian-Triassic) of central Thailand, Implications on paleogeographic models for mainland SE-Asia: Societe Geologique de France, memoire no. 147, p. 83-91 (nouvelle serie).

Helmcke, D., 1985, The Permo-Triassic 'Paleotethys' in mainland Southeast Asia and adjacent parts of China: Geologische Rundschau, v. 74, no. 2, p. 215-228.

Hen Shu Peng, ch. ed., 1986, *Atlas of geo-science analyses of landsat imagery in China*: Beijing, Science Press, 228 p.

Herman, B. M., Anderson, R. N. and Truchan, M., 1979, Extensional tectonics in the Okinawa Trough: American Association of Petroleum Geologists, Memoir 29, p. 199-208.

Hilde, T. W. C., Uyeda, S. and Kroenke, L., 1977, Evolution of the Western Pacific and its margin: Tectonophysics, v. 38, nos. 1–2, p. 145–165.

Hirn, A., Lepine, J. -C., Jobert, G., Spin, M., Whittington, G., Xu Zhong-xin, Gao En-yuan, Wang Xiang-jing, Teng Ji-wen, Xiong Shao-bai, Pandey, M. R. and Tater, J. M., 1984a, Crustal structure and variability of the Himalayan border of Tibet: Nature, v. 307, p. 23–25.

Hirn, A., Nercessian, A., Sapin, M., Jobert, G., Xu Zhong-xin, Gao En-yuan, Lu De-Yuan and Teng Ji-wen, 1984b, Lhasa block and bordering sutures – a continuation of a 500-km Moho traverse through Tibet: Nature, v. 307, p. 25–27.

Ho, C. S., 1967a, Structural evolution of Taiwan: Tectonophysics, v. 4, no. 4–6, p. 367–378.

Ho, C. S., 1967b, Structural evolution and major tectonic forms of Taiwan: Geological Society of China Proceedings, no. 10 (for 1966), 324 p.

Ho, C. S., 1982, Tectonic evolution of Taiwan; explanatory text of the tectonic map of Taiwan: Taipei, Taiwan, The Ministry of Economic Affairs, Republic of China, p. 126 (Chinese text, p. 110).

Ho, C. S., 1983, An introduction to the geology of Taiwan: explanatory text of the geologic map of Taiwan: Taipei, Ministry of Economic Affairs, 154 p.

Ho K'o-jen, 1968, The developments in Red China's petroleum industry: Taipei, Fei'ch'ing Yueh-pao, March 1, p. 52–65 (in Chinese).

Holloway, N. H., 1982, North Palawan block, Philippines – its relation to Asian mainland and role in evolution of South China Sea: American Association of Petroleum Geol. Bull., v. 66, no. 9, p. 1355–1383.

Holmes, A., 1965, *Principles of physical geology*, 2nd ed., revised: London, T. H. Nelson Ltd., 1288 p.

Hong Ye, Shedlock, K. M., Hellinger, S. J. and Sclater, J. G, 1985, The North China basin; an example of a Cenozoic rifted intrapolate basin: Tectonics, v. 4, no. 2, p. 153–169.

Hong Youchong, 1979, On Eocene *Philolimnias*, gen. nov. (Ephemeroptera, Insecta), in Amber from Fushun coalfield, Liaoning Province: Scientia Sinica, v. 22, no. 3, p. 331–339.

Hou H., 1959, Devonian brachiopod fossils from Northeast China: Acta Palaeontologica Sinica, v. 7, no. 2, p. 121–160.

Hou Hong-fei (Hongfie) and Wang Shi-tao (Shitao), 1985, Devonian paleogeography of China: Acta Paleont. Sinica, v. 24, no. 2, p. 186–197.

Hou Jinpeng and Wang Xiaofeng, 1982, Chitinozoan biostratigrapy in China: Soc. Geol. et Mineral. de Bretagne, Bull., Ser. C, v. 14, no. 2, p. 79–89.

Hong Youchong, 1982, Discovery of new fossil spiders in amber of Fushun coalfield: Scientia Sinica, ser. B, v. 25, no. 4, p. 431–437.

Hou, U. T., 1958, Jurassic and Cretaceous freshwater fossil ostracods Subfamilly Cyprideinae from Northwest and Northeast China: Acad. Sci., Coll. Papers Inst. Geol. Paleont., v. 1, p. 33–60.

Hou, Y. T., Huang, B. L., Geng, L. Y., Li, Y. P., Dan, B. G., Cai, Z. G. and Shi, Y. P., 1979, Early Tertiary ostracods from the coastal region of Bohai: Beijing, Science press, 205 p.

Hou You-tang (Youtang), 1983, Characteristics and importance of non-marine ostracods of the Upper Cretaceous: Nanjing Institute of Geology and Palaeontology, Academia Sinica, Bull., v. 6, p. 309–319.

Howell, T. J., 1987, Interest revived in offshore China area: Oil and Gas Journal, v. 85, no. 49, p. 79–82.

Hsu Jen (Xu Ren), 1973, On the discovery of some plant fossils from the Mt. Joima Lungma Region, southern Tibet and its significance: Acta Botanica Sinica, v. 15, no. 2, p. 245–258.

Hsu Jen (Xu Ren), 1976, On the discovery of Glossopteris flora in southern Xizang and its significance in geology and palaeogeography: Scientia Geologica Sinica, v. 4, p. 323–331.

Hsu, K., 1983, Geosynclines in plate-tectonic settings: sediments in mountains. *In*: Hsu, K. J. (ed.) Mountain building processes: London, Academic Press, p. 3–12.

Hu Aiqin and Zhang Zhengen, 1982, On thermal history of Tianshan (Tian Shan) geosyncline according to the data on K-Ar dating of eastern Tianshan (Tian Shan): Scientia Sinica, ser. B, v. 15, no. 11, p. 1213–1226.

Hu Bing, Wang Jingbin et al., 1969, Problems of the Paleozoics of Tarim platform: Amer. Geol. Inst., Internat. Geol. Review, v. 11, no. 6, p. 650–665.

Hu Changming, 1984, Discovery of turbidite from Zhanjin Formation and Qudi Formation in the Late Carboniferous to Early Permian in Duoma district, Rutog, Xizang: Wuhan College of Geology Earth Sciences Journal, v. 25, p. 75–78.

Hu Chaoyuan, 1985, Geologic characteristics and oil exploration of small depressions in eastern China: Geology, v. 13, no. 4, p. 303–306.

Hu Chao-Yuan, 1983, The geological characteristics and oil exploration of small depressions in eastern China: Typewcript for Technical Sessions, AAPG Dallas meeting, 1983, 8 p.

Hu Chaoyuan, 1983, Source bed controls hydrocarbon habitat in continental basins, East China: Acta Petrolei Sinica, no. 2, p. 9–13.

Hu Chao-Yuan and Qiao Hanseng, 1983, Characteristics of oil and gas distribution in the North China basin and the adjacent seas: London, Eleventh World Petroleum Congress, Panel Discussion (PD2), p. 1–9.

Hu Chao-Yuan and Qiao Hanseng, 1984, characteristics of oil and gas distribution in the North China basin and the adjacent seas. *In*: Eleventh World Petroleum Congress, London (1983): V. 2, p. 111–119.

Hu Hong-xiang, Lu Han-xing, Wang Chun-yong, He Zheng-qin, Zhu Liang-bao, Yan Qi-zhong, Fan Yue-xin, Zhang Guo-qing and Deng Ying-e, 1986, Explosion investigation of the crustal structure in western Yunnan Province: Acta Geophysica Sinica, v. 29, p. 133–144.

Hu Huaguang, Hu Shiling, Wang Songshan and Zhu Ming, 1986, On the isotopic ages of Jura-Cretaceous volcanic racks in East China. *In*: Huang Jiqing, ch. ed., Proceedings of the Symposium on Mesozoic and Cenozoic Geology: Beijing, Geological Publishing House, p. 99–107.

Hu Jianyi, Tong Xiaoguang, Xu Shubao and Lin Dianzhong, 1982, Tectonic framework and oil possibilities of Mesozoic rocks in the Bohai bay and its neighbourhood: Acta Petrolei Sinica; v. 3, no. 2, p. 1–8.

Hu Jianyi, Xu Shubao, Tong Xiaoguang and Lin Dianzhong, 1984, Distribution and geological bases forming stratigraphic-lithologic pools in Tertiary oil basins in East China: Acta Petrolei Sinica, v. 5, no. 2, p. 1–9.

Hu Jimin and Zeng Demin, 1985, Charophyta assemblages of the Cretaceous to Paleogene in Hunan Province: Oil and Gas Geology, v. 6, no. 4, p. 409–417.

Hua Yongfeng, 1981, Preliminary study on the origin of mercury deposits in Guizhou and its vicinity: Acta Geologica Sinica, v. 55, no. 2, p. 139–148.

Huan Wen-lin, Wang Su-yun, Shi Zhen-liang and Yan Jia-quan, 1980, The distribution of earthquake foci and plate motion in the Qinghai-Xizang Plateau: Acta Geophysica Sinica, v. 23, p. 269–280.

Huan Wen-lin, Wang Su-yun, Shi Zhen-liang and Yan Jia-quan, 1981, The distribution of earthquake foci and plate tectonics in the Qinghai-Xizang Plateau and its vicinity. *In*: Liu Dong-sheng (ed.) Geological and ecological studies of Qinghai-Xizang Plateau, v. 1, Geology, geological history and origin of Qinghai-Xizang Plateau: Beijing, Science Press; and New York, Gordon and Breach, Science Publishers, Inc., p. 651–659.

Huang Chi'ch'ing (Jiqing), 1961, Basic features of the tectonic structure of China (preliminary conclusions): Internat. Geol. Review, v. 5, no. 3, p. 289–320.

Huang Chi-ching (Jiqing) (T. K. Huang), 1978, An outline of the tectonic characteristics of China: Eclogae Gol. Helvetiae, v. 71, no. 3, p. 611–635.

Huang Ci-xuan and Liang Yu-lian, 1981, Based upon palynological study to discuss the natural enviornment of the central and southern Qinghai-Xizang Plateau of Holocene. *In*: Liu Dong-sheng (ed.) Geological and ecological studies of Qinghai-Xizang Plateau, v. 1, Geology, geological history and origin of Qinghai-Xizang

Plateau: Beijing, Science Press; and New York, Gordon and Breach, Science Publishers, Inc., p. 215–220.

Huang Difan, Li Jinchao and Zhou Zhuhong, 1981, Evolutionary characteristics of soluble matter in Quaternary to Pliocene of the Tsaidam basin: Acta Petrolei Sinica, v. 2, no. 4, p. 100–112.

Huang Difan, Wang Jie, Fan Chenglung, Shang Huiyun and Cheng Keming, 1980, Genesis of Oil and Gas in the Mesozoic and Tertiary Continential sedimentary basins, China: Acta Petrolei Sinica, v. 1, no. 1, p. 31–42.

Huang Huaizeng, 1984, The Meso-Cenozoic tectonic evolution of the continental margins in the central part of eastern Asia: Chinese Acad. Geol. Sciences Bull., ser. II, v. 9, p. 71–85.

Huang Jiqing (T. K. Huiang), ch. ed. and comp., 1979, Tectonic map of China, scale 1:4,000,000 one sheet (in Chinese): Beijing, Cartographic Publishing House.

Huang Jiqing, 1980, On the polycyclic evolution of geosynclinal foldbelts: Scientia Sinica, v. 23, no. 4, p. 475–491.

Huang Jiqing, 1984a, New researchers on the tectonic characteristics of China. *In*: Reports of the 27th International Geological Congress, 1984: Moscow, Colloquium 05, Tectonics of Asia, v. 5, p. 13–28.

Huang Jiqing, 1984b, New researches on the tectonic characteristics of China: Chinese Academy of Geological Sciences Bulletin, v. 9, p. 197–209.

Huang Jiqing, 1986, On the main achievements in geological sciences in China over the last 60 years and our tasks ahead. *In*: Huang Jiqing, Ed. in Chief, 1986, Proceedings of the symposium on Mesozoic and Cenozoic Geology; in connection of the 60th anniversay of the Geological Society of China: Beijing, Geological Publishing House, p. 3–19.

Huang Jiqing (ed.) 1986, Preceedings of the symposium on Mesozoic and Cenozoic geology in connection of the 60th anniversary of the Geological Society of China: Beijing, Geological Publishing House, 899 p.

Huang Jiqing and Chen Bingwei, 1987, *The evolution of the Tethys in China and adjacent regions*: Beijing, Geological Publishing House, p. 100.

Huang Jiqing (Huang, T. K.), Young, C. C., Cheng, Y. C., Chow, T. C., Bien, M. N. and Weng, W. P., 1947, Report on geological investigation of some oil-fields in Sinkiang: Geological Survey of China Mem. 21, ser. A, Nanking, 118 p.

Huang Jiqing, Zhang Zhengkun, Zhang Zhimeng and Chen Guo, 1965, On eugeosynclines and miogeosynclines of China and their polycyclic development: Prof. Paper, China Acad. Geol. Sci. Ser. C., 1965/1 (in Chinese).

Huang Jiqing, Ren Jishun, Jiang Chunfa, Zhang Zhimeng and Zhang Zhengkun, 1974, New observations on the geotectonic characteristics of China: Acta Geologica Sinica, no. 1, p. 36–52.

Huang Jiqing, Ren Jishun, Jiang Chunfa, Zhang Zhengkun and Qin Deyu, 1980, *Geotectonic evolution of China*: Beijing, Science Press, 124 p. (in Chinese).

Huang Jiqing, Ren Jishun, Jiang Chunfa, Zhang Zhengkun and Qin Deyu, 1987, Geotectonic evolution of China: Beijing, Science Press; and Berlin, Springer-Verlap, 214 p. (in English).

Huang Jizhong, 1984, On the occurrence of oil and gas in the Yangxin Limesotne series in Sichuan basin: Acta Petrolei Sinica, v. 5, no. 1, p. 9–17.

Huang Jizhong and Jiang Huaicheng, 1986, On the origin of Ordovician natural gas in Sichuan basin: Acta Petrolei Sinica, v. 7, no. 4, p. 11–23.

Huang Naihe and Wang Genfa, 1987, The new criteria of ancient tidal sediments; tidal cycle sequences: Acta Sedimentologica Sinica, v. 5, no. 2, p. 39–44.

Huang Shaoxian, Jin Fengcai and Di Junheng, 1981, On the relationships between the uranium deposits and the regional geological setting in the Meso-Cenozoic continental clastic rocks of China: Acta Geologica Sinica, v. 55, no. 4, p. 290–296.

Huang Shu-de, 1979, Some problems in the seismic exploration of complicated basins: Acta Geophys. Sinica, v. 22, no. .2, p. 109–139.

Huang Su and Wang Daojing, 1984, Tectonic features of north Xinjiang and formation of oil and gas pools. *In*: Developments in geoscience (Academia Sinica): Beijing, Science Press (Contr. to 27th Internat. Geol. Congress, Moscow, 1984), p. 601–610.

Huang, T., 1968, Foraminiferal study of the Tungliang well TL-1 of the Penghu Islands: United Nations, CCOP, Tech. Bull., v. 1, p. 39–47.

Huang Tifan, Wang Zemin and Shih Guoshi, 1981, Paleogeomorphic features of Shaan-Gan-Ning district during the Indo-Chinese stage and its significance to petroleum geology: Acta Petrolei Sinica, v. 2, no. 2, p. 1–10.

Huang, T. -C., 1977, Calcareous nannoplankton stratigraphy of the upper Wulai Group (Oligocene) in northern Taiwan: Petroleum Geology of Taiwan, no. 14, p. 147–179.

Huang, T. -C., 1978, Calcareous nannofossils of the subsurface pre-Miocene rocks from the Peikang basement high and adjacent areas in western central Taiwan (Part I, Cretaceous): Petroleum Geology of Taiwan, no. 15, p. 49–87.

Huang, T. -C., 1979, A supplementary note on the calcareous nannofossils, age and correlation of the Wuchinshan Formation: Petroleum Geology of Taiwan, no. 16, p. 85–93.

Huang, T. -C., 1980, Calcareous nannofossils from the slate terrane west of Yakou, southern cross-island highway: Petroleum Geology of Taiwan, no. 17, p. 59–74.

Huang, T. -C. and Chi, W. -R., 1979, calcareous nannofossils of the subsurface pre-Miocene rocks from the Peikang basement high and adjacent areas in western central Taiwan (Part II: Paleocene): Petroleum Geology of Taiwan, no. 16, p. 95–129.

Huang, T. K., 1960, The main characteristics of the structure of China; preliminay conclusion: Scientia Sinica, v. 9, no. 4, p. 492–544.

Huang, T. K., Jen Chi-shun, Jiang Chun-fa, Chang Chi-meng and Xu Ahi-qin, 1977, An outline of the tectonic characteristics of China: Acta Geologica Sinica, no. 2, p. 117–135.

Huang, T. K., Young, C. C., Cheng, Y. C., Chow, T. C., Bien, M. N. and Weng, W. P., 1947, Report on geological investigation of some oil-fields in Sinkiang: Nanking, Geol. Survey China Mem. 21, ser. A, p. 118.

Huang Wan-po and Ji Hong-xiang, 1981, The climate and uplift of the Qinghai-Xizang (Tibet) Plateau in the Late Pleistocene and Holocene. *In*: Liu Dong-sheng (ed.) Geological and ecological studies of Qinghai-Xizang Plateau, v. 1, Geology, geological history and origin of Qinghai-Xizang Plateau: Beijing, Science Press; and New York, Gordon and Breach, Science Publishers, Inc., p. 225–230.

Huang Youyuan, 1986, Triassic System in the north part of Tarim basin and its oil prospects: Oil and Gas Geology, v. 7, no. 1, p. 32–41.

Huo Shicheng, Cui Zhilin, Wang Xinlu and Zhang Yinzhou, 1989, The geological succession and geographical distribution of Cambrian Bradoriida from China: Jour. Southeast Asian Earth Sci., v. 3, nos. 1–4, p. 107–118.

Hutchinson, C. S., 1983, *Economic deposits and their tectonic setting*: New York, John Wiley, 365 p.

Hutchison, C. S., 1989, *Geological evolution of south-east Asia*: New York, Oxford University Press, 392 p. (Oxford Monographs on Geology and Geophysics, 13).

Ilich, M. and Meyerhoff, A. A., 1980, Status of geosynclinal and geotectonic-cycle theories (abstract): 26th International Geological Congress, Paris 1980, v. 1, Sec. 5 (Tectonics), p. 351.

Institute of Geology, 1956, *Earth strata*: Beijing, Institute of Geology, Academia Sinica, Science Press, 693 p.

Institute of Geology and Academy of Geological Science, 1980, *Mesozoic stratigraphy and paleontology of the Shaanganning basin*: Beijing, Science Press, v. 1, 212 p. (in Chinese).

Institute of Geology and Chinese Academy of Geological Sciences, 1980, *Mesozoic stratigraphy and paleontology of the Shaanganning basin*: Beijing, Science Press, v. 2, 188 p. (in Chinese).

Institute of Geology, Chinese Academy of Geological Sciences and Wuhan college of Geology, 1985, *Atlas of the palaeogeography of China*: Beijing, Cartographic Publishing House, 143 plates, 85 explanatory note (in Chinese), 28 p. summary (in English), and 2 appendices.

Institute of Geomechanics, Chinese Academy of Geological Sciences,

1976, *On tectonic systems*: Peking (Beijing), August, 17 p.

Institute of Geophysics, Academia Sinica, 1981, Explosion seismic study for velocity distribution and structure of the crust and upper mantle from Damxung to Yidong of Xizang Plateau: Acta Geophysica Sinica, v. 24, no. 2, p. 155–170.

Ishibashi, T., 1969, Stratigraphy of the Triassic formation in Okinawajima, Ryukyus: Kyushu University, Memoirs of the Faculty of Science, series D, Geology, v. 19, no. 3, p. 373–385.

Jaeger, J. C., 1962, *Elasticity, fracture and flow with engineering and geological applications*, 2nd ed.: London, Methuen & Co. Ltd., 208 p.

Jaeger, J. J., Adloff, C., Daubinger, J., Pons, D., Vozenin-serra, C. and Wang, N. W., 1982, The contribution of fossils to the paleogeography of the Lhasa block (Tibet): EOS, v. 63, p. 1093.

Jeffreys, Sir H., 1970, *The earth*, 5th ed.: Cambridge University Press, 525 p.

Jhingran, A. G., 1981, Geology of the Himalaya. *In*: Lall, J. S. et al. (eds.) The Himalaya; aspects of change: New Delhi, Oxford Univ. Press, p. 77–98.

Ji Xiang and Coney, P. J., 1985, Acreted terranes of China. *In*: Howell, D. G. (ed.) Tectonostratigraphic terranes of the circum-Pacific regions: Houston, Texas, Circum-Pacific Council for Energy and Mineral Resources, p. 349–361.

Ji Zixiu, 1981, Unravelling the mystery of Mr. Xixabangma. *In*: Yoshito Mano, Change Cang and Wang Yingheng (eds.) High mountain peaks in China newly opened to foreigners: Beijing, Chinese Mountaineering Association and People's Sports Publishing House; Tokyo, Tokyo Shimbun Publishing Bureau, p. 19–21.

Jiang Fuzhi, 1983, Genetic types and metallogenic characteristics of iron ores and/or marine volcanic copper in China: Mineral Deposits, v. 2, no. 4, p. 11–18.

Jiang, Z., 1985, Evolution of shelly fossils and the end of the Late Precambrian: Precambrian Research, v. 29, nos. 1–3, p. 45–52.

Jiang Zhiwen, Brasier, M. D. and Hamdi, B., 1989, Correlation of the Meishucunian Stage in South China: Acta Geologica Sinica (English edition), v. 2, no. 1, p. 1–9.

Jin Chengwei, 1981, The Mesozoic-Cenozoic volcanic rocks in Xizang and their bearing on geology. *In*: Liu Dong-sheng (ed.) Geological and ecological studies of Qinghai-Xizang Plateau, v. 1, Geology, geological history and origin of Qinghai-Xizang Plateau: Beijing, Science Press; and New York, Gordon and Breach, Science Publishers, Inc., p. 433–442.

Jin Dexing, Cheng Zhaodi, Lin Junmin and Liu Shicheng, 1985, *The marine benthic diatoms in China*; v. 1: Berlin, Springer-Veriag, 313 p.

Jin Fuquan, 1989, Carboniferous paleogeography and paleoenvironment between the North and South China blocks in eastern China: Jour. Southeast Asian Earth Sci., v. 3, nos. 1–4, p. 219–222.

Jin Wanlian, Xhe Shuhao, Xiu Yunzhen, Xiao Zuojun and Yan Huo, 1981, Turbidites in the 3rd member of Shahejie Formation of the depression in the western part of Liaohe basin: Acta Petrolei Sinica, v. 2, no. 4, p. 23–30.

Jin Xiaohua, 1983, Metamorphic regularities of coal and the prospects of oil and gas in the Paleozoic sediments of Guizhou: Oil and Gas Geology, v. 4, no. 2, p. 141–150.

Jin Xuezheng and Zhang Zhensheng, 1986, The sedimentary facies and hydrocarbon accumulation in the Third Member of Shahejie Formation in Beijing-Tianjin depression: Acta Petrolei Sinica, v. 7, no. 3, p. 21–27.

Jin Yugan, 1981, On the paleoecological relation between Gondwana and Tethys faunas in the Permian of Xizang. *In*: Liu Dong-sheng (ed.) Geological and ecological studies of Qinghai-Xizang Plateau, v. 1, Geology, geological history and origin of Qinghai-Xizang Plateau: Beijing, Science Press; and New York, Gordon and Breach, Science Publishers, Inc., p. 171–178.

Jin Yugan, Ma Zhengang et al. (eds.) 1987, *Bibliography of Chinese Carboniferous-Permian palaeontology and stratigraphy*: China, Nanjing University Press, 163 p.

Jinghai Yang and Linhuang Lu, 1981, Geographical and stratigraphical distributions of Permian bryozoa in China. *In*: Larwood, G. P. and Nielsen, C. (eds.) Recent and fossil bryozoa: Fredenborg, Olsen and Olsen, p. 305–307.

Jobert, N., Journet, B., Jobert, G., Hirn, A. and Sun Ke-zhong, 1985, Deep structure of southern Tibet inferred from the dispersion of Rayleigh waves through a long-period seismic network: Nature, v. 313, p. 386–388.

Johnson, B. D., Powell, McA. C. and Veevers, J. J., 1976, Spreading history of the eastern Indian Ocean and Greater India's northward flight from antarctica and Australia: Geol. Soc. America Bull., v. 87, no. 11, p. 1560–1566.

Johnson, D. B., Powell, C. M. and Veevers, J. J., 1978, Spreading history of the eastern Indian Ocean and India's northward flight from Antarctica and Australia: reply: Geol. Soc. America Bull., v. 89, no. 4, p. 650.

Johnson, C. J. and Clar, A. L., 1987, China's minerals future, 1985–2005: Natural Resources Forum, v. 11, no. 2, p. 111–126.

Johnson, M. E., Rong Jia-Yu (Jiayu) and Yang Xue-chang (Xuechang), 1985, Intercontinental correlation by sea-level events in the Early Silurian of North America and China (Yangtze Platform): Geol. Soc. America, Bull., v. 96, no. 11, p. 1384–1397.

Juan, V. C., 1946, Mineral Resources of China: Economic Geology, v. 41, no. 4, pt. 2 (supplement), p. 399–474.

Jun Yuan, C. and Teichert, C., 1983, Cambrian Cephalopoda of China: Palaeontographica, Abt. A, Palaozoologie-Stratigraphie, v. 181, no. 1-3, 102 p.

K-Ar Geochronology Group, 1979, Ka-Ar dating and division of the Himalayan movement in southern Xizang: Scientia Geologica Sinica, v. 1, p. 13–21.

Kaila, K. L., 1981, Structure and seismotectonics of the Himalaya-Pamir-Hindukush region and the Indian plate boundary. *In*: Gupta, H. K. and Delany, F. M. (eds.) Zagros-Hindu Kush-Himalaya geodynamic evolution: American Geophysical Union and Geological Society of America, Geodynamics Series 3, p. 272–293.

Kalandadze, N. N. and Rautuan, A. S., 1983, Mesto Tsentral'noy Azii v zoogeograficheskoy istorii mezozoya, p. 6–44. *In*: iskopayemyye reptilii Mongolii, Sovmestnaya Sovetsko-Mongol'skaya Paleontologicheskaya Ekspeditsiya, Trudy, v. 24: Moscow, Izd. Nauka, p. 136.

Kamen-Kaye, M., 1972, Permian Tethys and Indian Ocean: Amer. Assoc. Petrol. Geol. Bull., v. 56, no. 10, p. 1984–1999.

Kamen-Kaye, M. and Meyerhoff, A. A., 1979, Flood basalts: implications for global-tectonic hypotheses and petroleum exploration: Jour. Petrol. Geol., v. 1, no. 4, p. 29–37.

Kamen-Kaye, M., Meyerhoff, A. A. and Taner, I., 1988, Junggar basin; a Permian to Cenozoic intermontane complex in northwestern China: Senckenbergiana Lethaea, v. 69, nos. 3/4, p. 289–313.

Kan Rongju, Hu Hongxiang, Zeng Rongsheng, Mooney, W. D. and McEvilly, T. V., 1986, Crustal structure of Yunnan Province, People's Republic of China, from seismic refraction profiles: Science, v. 234, no. 4775, p. 433–437.

Kan Rongju, Zhang Sichang et al., 1977, Present tectonic stress field and its relation to the characteristics of recent tectonic activity in southwestern China: Acta Geophysica Sinica, v. 20-, no. 2, p. 98–109.

Kang Yuzhu, 1981, Geological characteristics of petroleum in Talimu basin: Oil and Gas Geology, v. 2, no. 4, p. 329–340.

Kang Yuzhu, Gu Runxu, Jiang Bingnan and Huang Youyuan, 1985, Characteristics of the geological structures of the Shayar uplift in northern Tarim and prospects for petroleum exploration: Oil and Gas Geology, v. 6, no. 1, p. 14–23.

Kao, C. S., Hsiung, Y. H. and Kao, P., 1934, Preliminary notes on Sinian stratigraphy of North China: Bull. Geol. Soc. China, v. 13, p. 24–38.

Kapoor, H. M. and Tokuoka, T., 1985, Sedimentary facies of the Permian and Triassic of the Himalayas. *In*: Nakazawa, K. and Dickins, J. M. (eds.) 1985, The Tethys, her paleogeography and paleobiogeography from Paleozoic to Mesozoic: Tokyo, Takai University Press, p. 23–58.

Kaufman, G. F., 1958, Petroleum developments in Far East in 1957: Am. Assoc. Petrol. Geol. Bull., v. 42, no. 7, p. 1709–1726.

Kaula, W. M., 1979, Thermal evolution of earth and moon growing by planetesimal impacts: Journal of Geophysical Research, v. 84, no. 83, p. 999–1008.

Ke Ru and Piggott, J. D., 1986, Episodic rifting and subsidence in the South China Sea: Amer. Assoc. Petrol. Geol. Bull., v. 70, no. 9, p. 1136–1155.

Khauan, M., 1984, Proposed classification and definitions of heavy crude oils and tar sands, Chapter 2. *In*: Meyer, R. F. et al. (eds.) The future of heavy crude and tar sands: New York, McGraw Hill, p. 7–11.

Khudoley, K. M. and Prosorovskaya, E. L., 1985, Seaways between the western part of the Pacific and adjacent oceans. *In*: Westermann, G. E. G. (ed.) Jurassic biogeography and stratigraphy of east USSR: IG.C.P. Project no. 171, Circum-Pacific Jurassic, Special Paper 10, p. 1–8.

Kidd, W. S. F., 1975, Widespread Late Neogene and Quarternary calc-alkaline volcanism on the Tibetan Plateau (abstract): *Eos* 56, p. 453.

Kim, O. J., 1974, Geology and tectonics of South Korea: United Nations ESCAP, CCOP, Tech. Bull., V. 8, p. 17–37.

Kimura, M., 1985, Back-arc rifting in the Okinawa Trough: Marine and Petroleum Geology, v. 2, no. 3, p. 222–240.

King Wu, 1982, Energy sources in China – present and future: United Nations, Natural Resources Forum, v. 6, p. 53–62.

Kiselev, G. N., 1987, Novyye dannyye o kompleksakh siluriyskikh tsefalopod Kitaya (New data on silurian cephalopod complexes of China): Moskovskogo Obschestva Ispytatley, Prirody, Otdel Geologicheskiy, Byulleten, v. 62, no. 5, p. 82–85 (in Russian, English abstract).

Kizaki, K., 1986, Geology and tectonics of the Ryukyu Islands: Tectonophysics, v. 125, p. 193–207.

Klimetz, M. P., 1983, Speculations on the Mesozoic plate tectonic evolution of eastern China: Tectonics, v. 2, no. 2, p. 139–166.

Komroff, M. (ed.) 1926, *The travels of Marco Polo* (revised from Marsden's translations and edited with an introduction by Manuel Komroff): New York, The Modern Library (Random House), 351 p.

Kong Qingyun, 1981, Organic carbon isotopes in crude oils and sedimentary rocks of Songliao basin: Acta Petrolei Sinica, v. 2, no. 4, p. 61–68.

Kong Zhao-chen and Du Nai-qiu, 1981, Preliminary study on the vegetation of the Qinghai-Xizang Plateau during Neogene and Quaternary periods. *In*: Liu Dong-sheng (ed.) Geological and ecological studies of Qinghai-Xizang Plateau, v. 1, Geology, geological history and origin of Qinghai-Xizang Plateau: Beijing, Science Press; and New York, Gordon and Breach, Science Publishers, Inc., p. 239–246.

Kon'no, Enzo, 1968, The Upper Permian flora from the eastern border of northwest China: Tohoku Univ. Sci. Repts., 2nd ser. (Geol.), v. 39, no. 3, p. 159–211.

Koslovsky, Ye. A., 1987, *The superdeep well of the Kola Peninsula*: Springer-Veriag, 558 p.

Kosygin, Yu. A. and Parfenov, L. M., 1975, Structural evolution of eastern Siberia and adjacent areas: Am. Jour. Science, v. 275A, p. 187–208.

Kou Caixiu, 1983, Existent evidence of Late Cretaecous to early Tertiary of Zhujiangkou basin and its transgression: Oil and Gas Geology, no. 3, p. 304–309.

Kouda, R. and Ni Ruoshui, 1986, Wubu deposit, a representative lead-zinc mineralization associated with Mesozoic volcanic rocks in China: Japan, Geological Survey, v. 3, no. 5, p. 271–293.

Kozur, H. and Gupta, V. J., 1983, The limits of Indian Plate drift. *In*: Contr. Himalayan Geol., v. 2, p. 167–172: Delhi, Hindustan Publishing Corp.

Kravchenko, K. N., 1965, Sec. 31, Gruppa Zapadnokitayskikh basseynov (West China group of basins); sec. 32, Prednan'shan'skiy basseyn (Pre-Nanshan basins); sec. 33, Gruppa basseynov Vostoka i Yugovostoka kitaskoy Narodnoy Respubliki (Eastern and southeastern group of basins of the People's Republic of China). *In*: Neftgazonosyye basseyny zemnogo shara (Petroleum-bearing basins of the globe, I. O. Brod, ed.): p. 286–326.

Kravchenko, K. N., 1968, Kitay (China). *In*: I. V. Vysotskiy (ed.) Geologia nefti, spravochnik, tom 2, kniga 2, Nefteyanyye mestorozhdeniya zarubezhnik stran (Oil fields of foreign countries).

Kravchenko, K. N., 1979, Tectonic evolution of the Tien Shan, Pamir and Karakorum. *In*: Farah, A. and DeJong, K. A. (eds.) Geodynamics of Pakistan: Quetta, Geological Survey of Pakistan, p. 25–40.

Kremp, G. O., 1982, On the continental drift of China, India and other Southeast Asian plates: Paleo Data Banks, v. 17, pt. 2, p. 62–113.

Krishman, M. S., 1960, *Geolôgy of India and Burma*: Madras, Higginbothams (Private) Ltd., 604 p.

Ku, C. W., 1962, *Jurassic and Cretaceous of China*: Beijing, Science Press, 84 p.

Kudo, T., 1966, Chugoku tairiku no Seyiku-shigen (Petroleum resources of continental China): tokyo, Ajia Keizai Kenkjujo (Inst. Asian Economic Affairs), Kaigaitoshisankoshiryo (Research materials on overseas investments), v. 9, p. 61 (in Japanese).

Kumar, S., Singh, I. B. and Singh, S. K., 1977, Lithostratigraphy, structure, depositional environment, palaeocurrent and trace fossils of the Tethyan sediments of Malla-Johar area, Pithoragarh Chamoli district, Uttar Pradesh, India: Jour. Palaeont. Soc. India, v. 20, p. 396–435.

Kunmar, S. P., 1983, Krishna-Godavari-Cauvery. *In*: Petroliferous basins of India: Dehra Dun, Petroleum Asia Journal, v. 6, no. 4, p. 57–65.

Lai Caigen, 1989, Biogeography of the Ordovician cepholopods from China: Jour. Southeast Asian Earth Sci., v. 3, no. 104, p. 125–130.

Lai Cai-gen, 1984, The Tremadoc series of China: Australian Journal of Earth Sciences, v. 31, no. 1, p. 1–6.

Lai Xingrong, 1984, The discovery of Ordovician conodonts in Daxing, Beijing, and its significance: Petroleum Exploration and Development, v. 11, no. 5, p. 23–26.

Lambert, I. B., Walter, M. R., Zang Wenlong, Lu Songnian and Ma Guogan, 1987, Palaeoenvironment and carbon isotope stratigraphy of upper Proterozoic carbonates of the Yangtze platform: Nature, v. 325, p. 140–142.

Lao Qiuyuan, Suo Shutian and Tan Yinjia, 1983, Proterozoic tectonics of China: Geol. Review, v. 29, no. 2, p. 111–120.

Lebauer, L. R., 1982, The refractory raw materials of China: Industrial Minerals, v. 180, September, p. 39–43.

Le , J. S., 1939, *Geology of China*: London, Thomas Murby, 528 p.

Lee, J. S., 1973, Crustal structure and crustal movement: Peking (Beijing), Scientia Sinica, v. 16, no. 4, p. 519–559.

Lee, J. S., 1973, *Outline of geomechanics*: Peking (Beijing), Scientific Publisher, 136 p. (in Chinese).

Lee, J. S., 1984, *Introduction to geomechanics*: Beijing, Science Press; U.S.A., Gordon and Breach, 234 p.

Lee, J. S. and Chao, Y. T., 1924, Geology of the Gorge district of the Yangtze (from Ichang to Tzekuei), with special reference to the development of the Gorges: Bull. Geol. Soc. China, v. 3, p. 351–391.

Lee, K. Y., 1970, Some rare-element mineral deposits in mainland China: U. S. Geol. Survey, Bull., 1312-N, p. 34.

Lee, K. Y., 1984a, Geology of the Qaidamu basin, Qinghai Province, northwest China: United States Geological Survey Open-File Report 84-413, 39 p.

Lee, K. Y., 1984b, Tertiary system and its petroleum potential in the Lunpola basin, Xizang (Tibet): United States Geological Survey Open-File Report 84-420, 5 p., 1 plate.

Lee, K. Y., 1984c, Geology of the Dian-Qian-Gui foldbelt, southwest China: U. S. Geological Survey Open-file report 84-357, 52 p.

Lee, K. Y., 1985a, Geology of the petroleum and coal deposits in the Junggar (Zhungaer) basin, Xinjiang Uygur Zizhiqu, northwest China: U. S. Geol. Survey, Open-File Report 84-230, 53 p.

Lee, K. Y., 1985b, Geology of the Tarim basin, with special emphasis on petroleum deposits, Xinjiang Uygur Zizhiqu, northwest China: U. S. Geol. Survey, Open-file Rept., 85-616, 55 p.

Lee, K. Y., 1986a, Geology of the petroleum and coal deposits in the North China basin, eastern China: U. S. Geological Survey, Open-file report 86-154, 57 p.

Lee, K. Y., 1986b, Geology of the coal and petroleum deposits in the Ordos basin, China: U. S. Geological Survey Open-file report 86-278, 63 p.

Lee, K. Y., 1986c, Petroleum geology of the Songliao basin, Northeast China: U. S. Geological Survey, Open-file report 86-502, 19 p.

Le Fort, P., 1986, Metamorphism and magmatism during the Himalayan collision. *In*: Coward, M. P. and Ries, A. C. (eds.) *Collision tectonics*, Geological Society of London, Special Publication 18, p. 159–172.

Letouzey, J. and Kimura, M., 1985, Okinawa Trough genesis: structure and evolution of a backarc basin developed in a continent: Marine and Petroleum Geology, v. 2, no. 2, p. 111–130.

Letouzey, J. and Kimura, M., 1986, The Okinawa Trough: genesis of a back-arc basin developing along a continental margin: Tectonophysics, v. 125, p. 209–230.

Leyden, R., Ewing, M. and Murauchi, S., 1973, Sonobuoy refraction measurements in East China Sea: Amer. Assoc. Petrol. Geol. Bull., v. 57, no. 12, p. 2396–2403.

Li Baoshi, 1983, Structural traps in northwest Sichuan and a preliminary analysis on the accumulation of hydrocarbons in them: Petroleum Exploration and Development, v. 10, no. 6, p. 13–16, 22.

Li Bing-yuan, Jing Ke, Zhang Qing-song, Yang Yi-chou, Yin Ze-sheng and Wang Fu-bao, 1981, Formation and evolution of the drainage systems in the Xizang area. *In*: Liu Dong-sheng (ed.) Geological and ecological studies of Qinghai-Xizang Plateau, v. 2, Environment and ecology of Qinghai-Xizang Plateau: Beijing, Science Press, and New York' Gordon and Breach, Science Publishers, Inc., p. 1759–1767.

Li Ching-chung, Yu Shou-peng, Liu Wen-liu and Liu Cheng-zheng, 1979, A case history of seismic survey on Dong-xing oil field – a practical example of 3-D seismic interpretation on a complex fault block structure: Acta Geophys. Sinica, v. 22, p. 140–155.

Li Chuan-juei (Chuankui) and Ting Su-yin (Ding Suyin), 1983, The Paleogene mannals of China: Carnegie Mus. Nat. History, Bull., 21, p. 93.

Li Chunyu (C. Y. Lee), Chuan Wang et al., 1980, A preliminary study of plate tectonics of China: Chinese Acad. Geol. Sciences, Bull., ser. I, v. 2, no. 1., p. 22.

Li Chunyu, Wang Quan and Liu Xueya, 1981, The metallogeny and plate-tectonics of China: Acta Geologica Sinica, v. 55, no. 3, p. 195–204.

Li Chunyu (C. Y. Lee), Wang Quan et al., 1982, Explanatory notes to the tectonic map of Asia (in English): Beijing, Research Institute of Geology of the Chinese Academy of Geological Sciences, 49 p.

Li Desheng, 1979, The tectonic framework of Bohai Gulf petroliferous basin: Petroleum Exploration and Development, v. 6, no. 2, p. 1–101.

Li Desheng, 1980, Geology and structural characteristics of Bohai Bay, China: Acta Petrolei Sinica, v. 1, no. 1, p. 6–20.

Li Desheng, 1981, Geological structure and hydrocarbon occurrence of the Bohai Gulf oil and gas basin (China). *In*: Mason, J. F. (ed.) Petroleum geology in China: Tulsa, OK, PennWell Publ. Co., p. 180–192.

Li Desheng, 1982, Tectonic types of oil and gas basins in China identified: Acta Petrolei Sinica, v. 3, no. 3, p. 1–11.

Li Desheng, 1983, Geological evolution history of petroliferous basins on the continental shelf of China: Amer. Assoc. Petrol. Geologists, Technical Session, Dallas Convention, 1983, 18 p.

Li Desheng, 1984, Geologic evolution of petroliferous basins on continental shelf of China: Amer. Assoc. Petrol. Geol. Bull., v. 68, no. 8, p. 993–1003.

Li Desheng, 1985, Tilted fault block-bured hill traps – a new type of oil/gas traps in rift-related tensional basins: Oil and Gas Geology, v. 6, no. 4, p. 386–394.

Li Desheng, 1985, China certain to have several substantial increases in oil and gas reserves and output before the year 2000: Oil and Gas Geology, v. 6, Special Issue, June 1986, p. 7–8.

Li Desheng, 1986, Prospect of the composite megastructural oil and gas field in Bohai Gulf of China: Acta Petrolei Sinica, v. 7, no. 1, p. 1–21.

Li Desheng, Hu Jianyi, Tong Xiaoguang, Xu Shubao and Hu Guonong, 1987, Intracratonic basins of China and their hydrocarbon accumulations. *In*: Twelfth World Petroleum Congress, Proceedings, v. 2, Exploration: New York, John Wiley and Sons Limited, p. 59–71.

Li Desheng and Xue Shuhao, 1983, The eastern China Mesozoic-Cenozoic basins and hydrocarbon occurrence: Acta Geologica Sinica, v. 57, no. 3, p. 224–234.

Li Gansheng, 1986, China's significant advances in oil and gas prospecting during the Sixth Five-Year Plan: Oil and Gas Geology, v. 7, no. 2, p. 123–124.

Li Ji-jun, Li Bing-yuan, Wang Fu-bao, Zhang Qing-song, Wen Shi-xuan and Zheng Ben-xing, 1981, The process of the uplift of the Qinghai-Xizang Plateau. *In*: Liu Dong-sheng (ed.) Geological and ecological studies of Qinghai-Xizang Plateau, v. 1, Geology, geological history and origin of Qinghai-Xizang Plateau, Beijing: Science Press; and New York, Gordon and Breach, Science Publishers, Inc., p. 111–118.

Li Jijun, Wen Shixuan, Zhang Qingson, Wang Fubao, Zheng Benxing and Lie Bingyuan, 1979, A discussion of the period, amplitude and type of the uplift of the Qinghai-Xizang Plateau: Scientia Sinica, v. 23, p. 1314–1328.

Li Qin-zu, 1980, General features of the stress field in the crust of North China: Acta Geophysica Sinica, v. 23, no. 4, p. 376–388.

Li Renwei, Lin Daxing, Huang Wanping and Gan Yinu, 1987, Organic geochemical analysis of environments of Subei basin: Scientia Sinica, series B, v. 30, no. 6, p. 651–662.

Li Rongchuan, 1986, Current tensile stress field in the west Hubei and Jianghan: Acta Geophysica Sinica, v. 29, no. 3, p. 302–306.

Li Siguang (J. S. Lee), 1939, *The geology of China*: London: Thomas Murby & Co., 528 p.

Li Sitian, Huang Jiafu, Yang Shigong, Zhang Xinming, Cheng Shoutian, Zhao Genrong, Li Dianan, Li Guiliang, Ding Jinlin, 1982, Depositional and structural history of the Late Mesozoic Huolinhe basin and its characteristics of coal accumulation: Acta Geologica Sinica, v. 56, no. 3, p. 244–254.

Li Sze-kuang (Li Siguang) (J. S. Lee), 1954, Vortex structure and other problems relating to the compounding of geotectonic systems of Northwest China: Acta Geologica Sinica, v. 34, no. 4, p. 399–410.

Li Sze-kuang (Li Siguang) (J. S. Lee), 1973, Crustal tectonics and crustal movements: Scientia Sinica, v. 16, no. 4, p. 519–559 (in Chinese).

Li Sze-kuang (Li Siguang) (J. S. Lee), 1976, *Methods of geomechanics*: Peking (Beijing), Science Press, 260 p. (in Chinese) (14 papers by author from 1926 to 1965).

Li Tingdong, 1982, The development of geological structures in China: Bull. Chinese Acad. Geol. Sciences, v. 4, p. 1–15.

Li Tingdong, Xiao Xichang, Li Guangcen, Gao Yanlin and Zhou Weiqin, 1986, The crustal evolution and uplift mechanism of the Qinghai-Tibet Plateau: Tectonophysics, v. 127, p. 279–289.

Li Xibin, 1984, Commercial oil flow obtained in eastern Junggar basin: Oil and Gas Geology, v. 5, no. 4, p. 384.

Li Xibin and Jiang Jianbeng, 1987, Survey of petroleum geology and the controlling factors for hydrocarbon distribution in the east part of the Junggar basin: Oil and Gas Geology, v. 8, no. 1, p. 99–107.

Li Xingguo, 1984, Correlation of single sand members in fluvial deltaic phase reservoir rocks in Shengtuo and Gudao oil fields: Petroleum Exploration and Development, v. 11, no. 5, p. 30–38.

Li Xingxue (H. H. Lee), 1963, *Continental strata of Late Paleozoic in China*: Beijing, Science Press, 168 p. (in Chinese).

Li Xingsue and Wu Xiuyuan, 1989, The seccession of late Palaeozoic and Triassic plant assemblages of eastern China: Jour. Southeast Asian Earth Sci., v. 3, nos. 1–4, p. 187–200.

Li Xingxue and Yao Zhaoqi, 1981, Discovery of Cathaysia flora in the Qinghai-Xizang Plateau with special reference to its Permian phytogeographical provinces. *In*: Liu Dong-sheng (ed.) Geological and ecological studies of Qinghai-Xizang Plateau, v. 1, Geology, geological history and origin of Qinghai-Xizang Plateau: Beijing, Science Press; and New York, Gordon and Breach, Science Publishers, Inc., p. 145–148.

Li Xing-xue (Singxue) and Yao Zhao-qi (Zhaoqi), 1985, Carboniferous and Permian floral provinces in East Asia. *In*: Washington, D.

C., and Urbana-Champaign, Illinois, International Congress on Carboniferous stratigraphy and geology, 9th, 1979: Compte rendu, v. 5, paleont., paleoecol., paleogeog., p. 95–101.

Li Xingxue and Zhang Linxin, 1983, The Upper Carboniferous of China. *In*: Martinez-Diaz, C., gen. ed., 1983, The Carboniferous of the World, v. 1, China, Korea, Japan and S. E. Asia: IUGS Publication, no. 16, p. 87–121.

Li Yinhuai, 1986, Geotectonic regionalization and evolution of East China Sea and northern South China Sea (abstr.): Amer. Assoc. Petrol. Bull., v. 70, no. 7, p. 929.

Li Yin-huai, Ma Fu-chen and Zhong Jia-you, 1981, Formation and development of block structures in Himalayas and adjacent region. *In*: Liu Dong-sheng (ed.) Geological and ecological studies of Qinghai-Xizang Plateau, v. 1, Geology, geological history and origin of Qinghai-Xizang Plateau, Beijing: Science Press; and New York, Gordon and Breach, Science Publishers, Inc., p. 85–102.

Li Yinxi (Deputy Chief Geologist, Petroleum Exploration Administrationi of Yunnan, Guizhou and Guangxi), unpubl. ms., The geological characteristics and petroleum geological conditions of two oil-and-gas-basins in Yunnan, Guizhou and Guangxi areas of South China: 35 p.

Li Yongkang, Zhang Minghui, Gao Renqin and Wang Hengjian, 1981, Special features of generation of oil of continental origin in the Songliao basin: Acta Petrolei Sinica, v. 2, no. 1, p. 31–40.

Li Yuehyuen, 1983, IGCP looks at China's phosphorites: Episodes, v. 3, p. 41–42.

Li Yugeng, 1989, Twenty years of oil field development in Daqing - our knowledge and practice: Acta Petrolei Sinica, special issue (October), p. 10–20.

Li Yuwen, 1984, Some new Late Jurassic to Early Cretaceous nonmarine ostracodes from Sichuan basin of China: Jour. Paleontology, v. 58, no. 1, p. 217–233.

Li Zhankui and Li Jianguo, 1987, Aeromagnetic anomaly and Paleozoic hydrocarbon potential of the Lower Yangtze reaches: Oil and Gas Geology, v. 8, no. 3, p. 325–332.

Li Zhiyi, Cai Wenbow et al., 1974, A note on the seismo-geological features of China: Scientia Geologica Sinica, v. 41, no. 1.

Li Zhongwen, 1988, Characteristics of China's W-Sn-bering granites and their relation with mineralization: Geology and Prospecting, v. 24, no. 9, p. 8–13 (in Chinese, English abstract).

Li Ziqiang, Cao Xinling, Chen Jiageng, Liu Jianzhong and Zhang Zhi, 1985, Some considerations on the recent tectonic stress field of China: Tectonophysics, v. 117, p. 161–176.

Li Zishun, Wang Sien, Yu Jingshan, Huang Huaizeng, Zheng Shaolin and Yu Xihan, 1982, On the classification of the Upper Jurassic in North China and its bearing on the Juro-Cretaceous boundary: Acta Geologica Sinica, v. 56, no. 4, p. 347–363.

Liang Dingyi, Nie Zetong, Guo Tieying, Xu Baowen, Zhang Yizhi and Wang Weipin, 1983, Permo-Carboniferous Gondwana-Tethys facies in southern Karakorum, Ali, Xizang (Tibet): Earth Science-Jour. Wuhan College of Geology, no. 1, p. 9–27.

Liang Fuhua, Liu Tianzhong and Cao Yan, 1981, A comparative study of the characteristics of source rocks and oil sources in central Huanghua depression: Acta Petrolei Sinica, v. 2, no. 4, p. 75–82.

Liang, Y., Zhu, S., Zhang, L., Gao, r., Gao, Z. and Bu, D., 1985, Stromatolite assemblages of the Late Precambrian in China: Precambrian Research, v. 29, nos. 1–3, p. 15–32.

Liao Zhi-jie, 1981, The setting of geothermal activities in Xizang and discussion on the heat source problem. *In*: Liu Dong-sheng (ed.) Geological and ecological studies of Qinghai-Xizang Plateau, v. 1, Geology, geological history and origin of Qinghai-Xizang Plateau: Beijing, Science Press; and New York, Gordon and Breach, Science Publishers, Inc., p. 925–930.

Lin Bao-Yu, 1985, Stratigraphical distribution and zoogeographical provinces of the Carboniferous tabulate corals of China: Professional Papers of Stratigraphy and Palaeontology, v. 12, p. 27–46.

Lin Dianzhong, 1982, The shear structure features in Bohai Bay oil and gas-bearing basin: Oil and Gas Geology, v. 3, no. 1, p. 16–24.

Lin Dianzhong, Tong Xiaoguang and Xu Shubao, 1983, Characteristics of seismic reflection in relation to litho-stratigraphic traps in the Cenozoic fault basins in East China: Oil and Gas Geology, v. 4, no. 3, p. 294–303.

Lin Longdong, 1984, The discovery of nappe-trapped oil pools and the prospects of Karamay oilfield: Oil and Gas Geology, v. 5, no. 1, p. 1–10.

Lin Xue-nong, 1981, Tectonic compression characteristics of the Dag Chuka ultrabasic rocks. *In*: Liu Dong-sheng (ed.) Geological and ecological studies of Qinghai-Xizang Plateau, v. 1, Geology, geological history and origin of Qinghai-Xizang Plateau: Beijing, Science Press; and New York, Gordon and Breach, Science Publishers, Inc., p. 587–592.

Ling, H. C., 1975, The petroleum industry of the People's Republic of China: Hoover Inst. Press, Stanford Univ., Pub. no. 142, p. 264.

Liou, J. G., Graham, S. A., Maruyama, S., Wang, X., Xiao, X., Carroll, A. R., Chu, J., Feng, Y., Hendrix, M. S., Liang, Y. H., McKnight, C. L., Tankg, Y., Wang, Z. X., Zhao, M. and Zhu, B., 1989, Proterozoic blueschist belt in western China; best documented Precambrian blueschists in the world: Geology, v. 17, no. 12, p. 1127–1131.

Liou, J. G., Xiaomin Wang, Coleman, R. G., Zhang, Zh. M. and Maruyoma, S., 1989, Blueschists in major suture zones of China: Tectonics, v. 8, no. 3, p. 609–619.

Liu Anlin, 1986, The depositional environment and its relation to the accumulation of oil and gas in the Lower Tertiary Qianjiang Formation in Jianghan salt lake basin: Petroleum Exploration and Development, v. 13, no. 3, p. 10–19.

Liu Baoquan, 1987, Research on oil source of Renqiu oil field buried hill miarolitic oil and inside hill reservoir oil: Oil and Gas Geology, v. 8, no. 3, p. 253–261.

Liu Baoquan, Liang Digang, Fang Jie, Jia Rengfen and Fu Jiamo, 1986, Organic matter maturity and oil/gas prospects in Middle-Upper Proterozoic and Lower Paleozoic carbonate rocks in Northern China: Geochemistry (Beijing), v. 5, no. 1, p. 55–70.

Liu Benpei and Cui Xinsheng, 1983, Discovery of *Eurydesma*-fauna from Rutog, northwest Xizang (Tibet), and its biogeographic significance: Wuhan College of Geology Earth Sciences Journal, v. 19, p. 69–78.

Liu Ben-pei, Zhou Zhang-gua, Xiao Jin-dong, Chen Bei-yue, Zhao Xi-wen, Xin Jian-rang, Li Xiang, Du Yuan-sheng, Xin Wen-jie and Li Guan-cheng, 1989, Characteristics of Devonian sedimentary facies in the Qinling Mountains and their tectono-palaeogeographyic significance: Jour. Southeast Asian Earth Sct., v. 3, no. 1–4, p. 211–217.

Liu Changling, 1988, The genetic types of bauxite deposits in China: Scientia Sinica, Series B, v. 31, no. 8, p. 1010–1024.

Liu Changshi, Zhu Jinchu, Xu Xisheng, Cai Dekun and Yang Pin, 1989, The Hercynian-Indosinian collision type granites of west Yunnan and their tectonic significance: Jour. Southeast Asian Earth Sci., v. 3, nos. 1–4, p. 263–270.

Liu Dongsheng, ch. ed., 1981, Geological and ecological studies of Qinghai-Xizang Plateau: Beijing, Science Press; and New York, Gordon and Breach, Science Publishers, v. 1, 974 p.

Liu Dongsheng, ch. ed., 1981, Geological and ecological studies of Qinghai-Xizang Plateau: Beijing, Science Press; and New York, Gordon and Breach, Science Publishers, v. 2, p. 975–2138.

Liu Dongsheng (Liu Tungsheng), Yuan Baoyin and Lu Yancha, 1986, Recent progress of Quaternary research in China, 331–349 p.. *In*: Tu Guangzhi (ed.) Advances in Science of China-Earth Sciences, v. 1: Beijing, Science Press; New York, John Wiley and Sons, p. 331–349.

Liu Guang-Ding (Guangding), 1988, Geology and exploration of petroleum in the East China Sea: Acta Geophysica Sinica, v. 31, no. 2, p. 184–197.

Liu Guodong, 1987, The Cenozoic rift system of the North China plain and the deep.internal process: Tectonophysics, v. 133, no. 3–4, p. 277–285.

Liu Guodong and Liu Changquan, 1983, Structures of crust and upper mantle and their relation to Cenozoic tectonism in northern part of North China: Scientia Sinica, Series B, v. 26, no. 5, p. 550–560.

Liu Hefu, 1984, Structural styles of mesozoic-Cenozoic petroliferous basins of China. *In*: S. Watson (ed.) Transactions of the Third

Circum-Pacific Energy and mineral Resource Conference: Am. Assoc. Petrol. Geol., p. 173–179.

Liu Hefu, 1986, Geodynamic scenario and structural styles of Mesozoic and Cenozoic basins in China: Amer. Assoc. Petrol. Geol. Bull., v. 70, no. 4, p. 377–395.

Liu Huaizhi, Zhong Ziyun and Yao Ming, 1989, On the tectono-palaeogeography and terrane evolution of southwest China (Guangxi, Guizhou, Yunnan, Sichuan) from later Palaeozoic to Triassic: Jour. Southeast Asian Earth Sci., v. 3, nos. 1–4, p. 223–229.

Liu Hugn-yun, Sha Ching-an and Hu Shih-ling, 1973, The Sinian System in southern China: Scientia Sinica, v. 16, no. 4, p. 266–278.

Liu Kaishi and Sun Diangui, 1983, The hydrocarbon prospecting in Middle Triassic residium, central Sichuan: Oil Geophysical Prospecting, no. 3, p. 231–236.

Liu Kuang-yeh, 1964, Neotectonic movement of the North China platform: Acta Geologica Sinica, v. 44, no. 1, p. 24–36.

Liu Shaolong, 1985, On coal gas prospect in Carboniferous-Permian systems in North China: Acta Petrolei Sinica, v. 6, no. 3, p. 11–18.

Liu Shaolong, 1986, The existence of a large-scale Triassic sedimentary basin in North China: Acta Geologica Sinica, v. 60, no. 2, p. 128–138.

Liu Shifan, 1983, Biogeography of Silurian and Devonian vertebrates in China: Vertebrata PalAsiatica, v. 21, no. 4, p. 292–300.

Liu Songwei, 1981, On the seismic survey in western Haubei basin (western North China basin): Oil Geophysical Prospecting, no. 1, p. 1017.

Liu Tungsheng, 1988, *Loess in China*, 2nd ed.: New York, Springer Series in Physical Environment, no. 5, 224 p.

Liu Tungsheng, An Zhisheng, Yuan Baoyin and Han Jiamao, 1985, The loess-paleosol sequence in China and climatic history: Episodes, v. 8, no. 1, p. 21–28.

Liu Tungsheng (Liu Dongsheng) and Ding Menglin, 1983, The characteristics and evolution of paleoenvironment of China since Late Tertiary: Wuhan College of Geology Earth Sciences Journal, v. 22, p. 15–28.

Liu Wanxiang, 1985, Characteristics of Hetian fault belt and its oil potential: Oil and Gas Geology, v. 6, no. 3, p. 326–334.

Liu Xingli, 1987, Charactersitics of geological structures and hydrocarbon distribution in the Bohai Bay region: Oil and Gas Geology, v. 8, no. 1, p. 45–54.

Liu Xun, 1988, Some recognitions on the sedimentary-tectonic development of the Meso-Cenozoic basins in eastern China: Acta Geologica Sinica, v. 1, no. 4, p. 365–379.

Liu Youmin and Kong Zhiping, 1984, A prospect of oil and gas in the overthrust belt on western border of Orduosi (Ordos) basin: Petroleum Exploration and Development, v. 11, no. 1, p. 33–40.

Liu Yuanlong, Wang Qianshen and Zhao Jianhua, 1978, A preliminary study of the gravity data of the crustal structure of the Peking-Tientsin (Beijing-Tianjin) area and its neighbouring regions: Acta Geophysica Sinica, v. 21, no. 1.

Liu Yuanlong, Zeng Weilu, Wang Qianshen, Wu chuanzheng and Liu Hongcheng, 1984, On forming mechanism of the rifts and the Mesozoic-Cenozoic basins from the deep structures and the stress distributions. *In*: Developments in geoscience; contribution to 27th International Geological Congress, 1984, Moscow: Beijing, Science Press, p. 573–583.

Liu Yuying, 1985, Prospects for oil and gas resources in Mesozoic basins in China: Petroleum Exploration and Development, v. 12, no. 1, p. 22–25, 46.

Liu Zhengrong, Lei Suhua and Hu Suhua, 1977, Yongshan-Dguan Earthquake (M=7.1) on May 11, 1974: Acta Geophysica Sinica, v. 20, no. 2, p. 110–114.

Lo Zhili, 1981, The influence of taphrogeny upon the formation of petroleum and other mineral deposits since late Paleozoic time in southwestern China: Geol. Geo. Sichuan, v. 11, no. 1, pagination N.A. (in Chinese).

Loczy, L., 1893–1899, Die wissenschaftlichen Ergebnisse der Reise des Grafen Bela Szechenyi in Ostasien 1877–1880...1890–1897, erschienenen ungarischen Original: Vienna (Wien), E. Holzel, v. 3 (pagination N.A.); K. K. Miltar-geografischen Institut; atlas, 30 double maps.

Long Yongwen and Zhang Jiquang, 1986, Characteristics of petroleum geology and petroleum potential of Hailar basin: Oil and Gas Geology, v. 7, no. 1, p. 59–67.

Lu Bingquan and Quan Songqing, 1982, A discovery of Paleogene turbidites in the Beibu Gulf: Scientia Geologica Sinica, no. 3, p. 333–335.

Lu Defu, Zhang Wenhua and Wang Weiping, 1984, The major fracture systems in China and their characteristics: Landsat mosaic image and analysis. *In*: Gabrielson, R. H., Ramberg, I. B., Roberts, D. and Steinlein, O. A. (eds.) Proceedings of the Fourth International conference on Basement Tectonics, Oslo, Norway, August 10–14, 1981, Salt Lake City (UT): International Basement Tectonics Association, Publication no. 4, p. 187–192.

Lu Huafu, Yu Hongnian, Ding Youwen and Zhang Qinglong, 1983, Changing stress field in the middle segment of the Tan-Lu fault zone, eastern China: Tectonophysics, v. 98, p. 253–270.

Lu Liangzhao, 1989, The metamorphic series and crustal evolution of the basement of the Yangtze Platform: Jour. Southeast Asian Earth Scie., v. 3, nos. 1–4, p. 293–301.

Lu Songnian and Gao Zhenjia, 1984, Proterozoic diamictites in China: Scientific Papers on Geology for International Exchange – International Geological Congress, 1, p. 205–214.

Lu, S., Ma, G., Gao, Z. and Lin, W., 1985, sinian ice ages and glacial sedimentary facies-areas in China: Precambrian Research, v. 29, nos. 1–3, p. 53–64.

Lu Yanhao, 1962, *Cambrian of China*: Peking (Beijing), Science Press, 133 p. (in Chinese).

Lu Yanhao, 1975, Ordovician trilobite faunas of central and southwest China: Nanking (Nanjing), Paleontologica Sinica, no. 152, new series B, Institute of Geology and Paleontology, Academia Sinica, 351 p. (in Chinese, English abstract).

Lu Yan-hao (Yanhao), 1981, Provincialism, dispersal, development, and phylogeny of trilobites. *In*: Teichert, C., Liu Lu and Chen Peiji (eds.) Paleontology in China, 1979: Geol. Soc. America, Spec. Paper 187, p. 143–151.

Lu Yanhao, 1984, Genesis and distribution of the Cambrian metallic and non-metallic ores of China, and their relation to the bio-environment control hypothesis. *In*: Su Zongwei (ed.) Developments in geoscience; contribution to the 27th International Geological Congress: Beijing, Science Press, p. 135–147.

Lu, Y. H., Chu, C. L. and Chien, Y. Y., 1965, Cambrian paleogeography and lithofacies of China: Acta Geologica Sinica, v. 45, no. 4, p. 349–357.

Lu Yan-hao (Yanhao), Lin Huan-ling (Huanling), Zhou Zhi-yi (Zhiyi) and Peng Shan-chi (Shanchi), 1985, The Cambrian-Ordovician boundary of China and its interval zonation: Acta Palaeontologica Sinica, v. 24, no. 1, p. 5–17.

Lu Yanhao, Li Xingxue and Zhang Wentang (eds.) 1989, *Palaeontologia Cathayana*, 4: Beijing, Science Press, 473 p.

Lu Yousheng, 1987, petroleum geology of the deep-subsidence zones within Tertiary half-graben systems in eastern China: Acta, Petrolei Sinica, v. 8, no. 1, p. 21–26.

Lu, Y. S. and An, Z. S., 1979, The quest for series of natural environment changes in the Loess Plateau during the Brunhes Epoch: Science Report, v. 24, no. 5, p. 221–224 (in Chinese).

Luo Huaizhang, 1983, A regressive correlation between volatility and reflectance of vitrinite in the coal beds in Upper Permian formations in Dian-Qian-Gui region and a discussion on their degree of metamorphism: Petroleum Exploration and Development, v. 10, no. 6, p. 43–50.

Luo Ping, Deng Xunkang and Luo Zhetan, 1986, Diagensis of the Lower Wuerhe (Urho) Conglomerate in Karamay oilfield and its effect on reservoir properties: Oil and Gas Geology, v. 7, no. 1, p. 42–50.

Luo Zhetan, Zhang Ruiniang and He Lianchang, 1980, Tectonics and deposits of the Cenozoic Era in the South China Sea: paper presented at the workshop on the geology and hydrocarbon potential of the South China Sea and possibilities of joint research and development, Univ. Hawaii East-West Center, 5–12 August 1980, 14 p.

Lydekker, R., 1881, Observations on the ossiferous beds of Hundes in Tibet: Geological Survey of India Records, v. 31, p. 179-204.

Lyon-Caen, H. and Molnar, P., 1983, Constraints on the structure of the Himalaya from an analysis of gravity anomalies and a flexural model of the lithosphere: Journal of Geophysical Research, v. 88, p. 8171-8190.

Ma Kaiyi and Jiang Mei, 1984, The basic characteristics of magnetic lineament tectonics in China: Inst. Miner. Deposits, Bull., Chinese Acad. Geol. Sciences, Er. III, no. 1, p. 126-138 (in Chinese, English abstract).

Ma Li, 1985, Subtle oil pools in Xingshugang delta, Songliao basin: Am. Assoc. Petrol. Geol. Bull., v. 69, no. 7, p. 1123-1132.

Ma Li, Ge Taisheng, Zhao Xueping, Zie Taijun, Ge Rong and Dang Zhenrong, 1982, Oil basins and subtle traps in the eastern part of China: Am. Assoc. Petrol. Geol. Mem. 32, p. 287-315.

Ma Wenpu, 1989, Tectonics of the Tangbai-Dabie foldbelt: Jour. Southeast Asian Earth Sci., v. 3, nos. 1-4, p. 77-85.

Ma Xingyuan, 1988, Lithospheric dynamics of China: Episodes, v. 11, no. 2, p. 84-90.

Ma Xingyuan and He Guogi, 1989, Precambrian crustal evolution of eastern Aisa: Jour. Southeast Asian Earth Sci., v. 3, nos. 1-4, p. 9-15.

Ma Xingyuan and Wu Daning, 1987, Cenozoic extensional tectonics in China: Tectonophysics, v. 133, no. 3-4, p. 243-255.

Mainguy, M., 1970, Regional geology and petroleum prospects of the marine shelves of eastern Asia: United Nations ECAFE, CCOP, Tech. Bull., v. 3, p. 91-107.

Maluski, H., Proust, F. and Xiao, X. -C., 1982, $^{39}Ar/^{40}Ar$ dating of the Trans-Himalayan calc-alkaline magmatism of southern Tibet: Nature, v. 298, p. 152-154.

Mao Shaozhi and Norris, G., 1984, Dinoflaggellate and acritarch stratigraphy (Late Cretaceous and Early Tertiary), western Talimu basin, Xinjiang Autonomous Region, northwest China: Wuhan College of Geology Earth Sciences Journal, v. 25, p. 7-22.

Marco Polo (see Yule, Sir H., also Cordier, H., and Komroff, M.)

Marinov, N. A., 1981, *Rol'rysskikh i Sovetskikh geologiv v izuchenii geologii i polezhnykh iskopayemykh Mongolii*: Moscow, Nedra, 198 p.

Martinez-Diaz, C., gen. ed., 1983, The Carboniferous of the World, v. 1, China, Korea, Japan and S. E. Asia: IUGS Publication no. 16, Published by Inst. Geol. y Minero de Espana and Empresa nacional adaro de investigactiones mineras, s. a., p. 247.

Mascle, G. H., 1985, L'Himalaya resulte-t-il du telescopage de trois chaines?' Societe Geologique de France Bulletin (ser. 8) 1, p. 289-304.

Mason, J. F. (ed.) 1981, *Petroleum geology in China*: Tulsa, OK, PennWell Publ. Co., 263 p.

Masters, C. D., Girard, O. W., Jr. and Terman, M. W., 1980, A perspective on Chinese petroleum geology: Southwestern Legal Foundation, Proceedings, Exploration and Economics of the Petroleum Industry, v. 18, p. 199-225.

Mateer, N. J. and Chen Pei-ji (Peiji), 1986, Symposium report; the Cretaceous of South china and its petroleum potential: Cretaceous Research, v. 7, no. 3, p. 301-302.

Matsumoto, T., Yamaguchi, M., Yanagi, T., Matsushita, S., Hayase, I., Ishizaka, K., Kawano, Y. and Ueda, Y., 1968, The Precambrian problem in younger orogenic zones: an example from Japan: Canadian Journal of Earth Sciences, v. 5, no. 3, pt. 2, p. 643-648.

Mattauer, M., 1986, Intracontinental subduction, crust-mantle decollement and crustal-stacking wedge in the Himalayas and other collision belts. *In*: M. P. Coward and A. C. Ries (eds.) Collision tectonics, Geological Society of London, Special Publication 19, p. 37-50.

Mattauer, N., Matte, Ph., Malavieille, J., Tapponier, P., Maluski, H., Xu Zhi, Qin, Lu Yi Lun and Tang Yao qin, 1985, Tectonics of the Qinling belt; build-up and evolution of eastern Aisa: Nature, v. 317, no. 6037, p. 496-500.

Mattauer, M., Tapponier, P. and Proust, F., 1981, Some analogies between the tectonic histories of Afghanistan and Tibet. *In*: Liu Dong-sheng (ed.) Geological and ecological studies of Qinghai-Xizang Plateau, v. 1, Geology, geological history and origin of Qinghai-Xizang Plateau: Beijing, Science Press; and New York, Gordon and Breach, Science Publishers, Inc., p. 315-324.

Matveev, A. K., 1976, Distribution and resources of world coal. *In*: Muir, W. L. G. (ed.) Coal exploration: London, First International Coal Exploration Symposium, Proc., p. 77-78.

McKenzie, K. G. (ed.) 1987, *Shallow Tethys 2*: Boston, A. A. Balkema, 544 p.

Mehta, P. K., 1977, Rb-Sr geochronology of the Kuli-Mandi belt: its implications for the Himalayan tectogenesis: Geologische Rundschau, v. 66, p. 156-175.

Mehta, P. K., 1979, Rb-Sr geochronology of the Kuli-Mandi belt: its implications for the Himalayan tectogenesis - a reply: Geologische Rundschau, v. 68, p. 383-392.

Mehta, P. K., 1980, Tectonic significance of the young mineral dates and the rates of cooling and uplift of the Himalaya: Tectonophysics, v. 62, p. 205-217.

Meng, C. -V., 1970, A conception of the island of Taiwan and its bearing on the development of the Neogene sedimentary basins on its western side: United Nations, ECAFE, CCOP, Tech. Bull., v. 3, p. 109-126.

Meng, C. Y. and Chou, J. T., 1976, The petroliferous Taiwan basins in the framework of the Pacific area: Petrol. Geol. Taiwan, no. 13, p. 15-24.

Meyerhoff, A. A., 1970, Developments in Mainland China, 1949-1968: Am. Assoc. Petrol. Geol. Bull., v. 54, no. 8, p. 1667-1680.

Meyerhoff, A. A., 1970, Continental drift: implications of paleomagnetic studies, meteorology, physical oceanography, and climatology: Jour. Geology, v. 78, no. 1, p. 1-51.

Meyerhoff, A. A., 1970, Continental drift, II: High-latitude evaporite deposits and geologic history of Arctic and North Atlantic Oceans: Jour. Geology, v. 78, no. 4, p. 406-444.

Meyerhoff, A. A., 1973, Mass biotal extinctions: world climate changes and galactic motions: possible interrelations: Canadian Soc. Petroleum Geologists Memoir no. 2, p. 745-758.

Meyerhoff, A. A., 1974, Ocean magnetic anomalies and their relations to continents: Am. Assoc. Petrol. Geol. Memoir 23, p. 411-422.

Meyerhoff, A. A., 1975, China's petroleum potential: World Petroleum Report 1975, v. 21, p. 18-21.

Meyerhoff, A. A., 1978, Petroleum in Tibet and the India-Asia suture (?) Zone: Jour. Petrol. Geol., v. 1, 2, p. 107-112.

Meyerhoff, A. A., 1981, *The oil and gas potential of the Soviet Far East*: Beaconsfield (U.K.), Scientific Press Ltd., 176 p.

Meyerhoff, A. A., 1982, *Petroleum basins of the Union of Socialist Soviet Republics and the People's Republic of China*: Adelaide, Petroleum Exploration Society of Australia, Distinguished Lecture Series, 464 p. (including 141 figs.).

Meyerhoff, A. A., 1984, Carboniferous oil and gas production in the Eastern Hemisphere - 1: Jour. Petroleum Geology, v. 7, no. 2, p. 125-146; 11, Jour. Petroleum Geology, v. 7, no. 3, p. 313-328.

Meyerhoff, A. A. and Meyer, R. F., 1987, Geology of Heavy Crude Oil and Natural Bitumen in the USSR, Mongolia, and China. *In*: Meyer, R. F. (ed.) Exploration for Heavy Crude Oil and Natural Bitumen: American Association of Petroleum Geologists, Studies in Geology, no. 25, p. 31-101.

Meyerhoff, A. A. and Meyerhoff, H. A., 1972, Continental drift, IV: the Caribbean 'plate': Jour. of Geology, v. 80, no. 1, p. 34-60.

Meyerhoff, A. A. and Meyerhoff, H. A., 1972, 'The new global tectonics': Major inconsistencies: Am. Assoc. Petrol. Geol. Bull., v. 56,no. 2, p. 269-336.

Meyerhoff, A. A. and Meyerhoff, H. A., 1972, 'The new global tectonics': Age of linear magnetic anomalies of ocean basins: Am. Assoc. Petrol. Geol. Bull., v. 56, no. 2, p. 337-356.

Meyerhoff, A. A. and Meyerhoff, H. A., 1974, Tests of plate tectonics: Am. Assoc. petrol. Geol. Memoir 23, p. 43-145.

Meyerhoff, A. A. and Meyerhoff, H. A., 1974, Ocean magnetic anomalies and their relations to continents: Am. Assoc. Petrol. Geol. Memoir 23, p. 411-422.

Meyerhoff, A. A. and Meyerhoff, H. A., 1978, Spreading history of the eastern Indian Ocean and India's northward flight from Antarctica and Australia: discussion; Geol. Soc. America Bull., v. 89, no. 4, p. 637-640.

Meyerhoff, A. A. and Taner, I., Unpubl. ms., Qinghai-Xizang (Tibet) Plateau, China: Timing and possible mechanism of uplift, 97 p., 26 figures.

Meyerhoff, A. A. and Teichert, C., 1971, Continental drift, III: Late Paleozoic glacial centers, and Devonian-Eocene coal distribtuion: Jour. Geology, v. 79, no. 3, p. 285–319.

Meyerhoff, A. A. and Willums, J. -O., 1976 (1977), Petroleum geology and industry of the People's Republic of China: United Nations ESCAP (Bangkok), CCOP Technical Bull., no. 10, p. 103–212.

Meyerhoff, A. A. and Willums, J. -O., 1980, China's petroleum industry-geology reserves, technology, and policies: Hong Kong, Asian Profile, v. 8, no. 2, p. 101–121.

Meyerhoff, A. A. and Willums, J. -O., 1981, Petroleum in the People's Republic of China, 1949–1979: Am. Assoc. Petrol. Geol. Studies in Geol., no. 12, p. 195–214.

Meyerhoff, A. A., Chen, C., Kamen-Kaye, M., Wang, K. P. and Willums, J. -O., 1984, Energy in China Summary *in* Watson, S. ed., 1984, Third Circum-Pacific Energy and Mineral Resource Conference: Am. Assoc. petrol. Geologists, p. 199–219.

Meyerhoff, A. A., Meyerhoff, H. A. and Briggs, R. S. Jr., 1972, Continental drift, V: Proposed hypothesis of earth tectonics: Jour. Geology, v. 80, no. 5, p. 663–692.

Meyerhoff, A. A., Taner, I. and Kamen-Kaye, M., 1986, The Indus-Yarlung fault zone: A confusion of terms: Episodes, v. 9, no. 2, p. 103.

Meyerhoff, A. A., Taner, I., Morris, A. E. L., Martin, B. D., Agocs, W. B., and Meyerhoff, H. A., 1991, Surge Tectonics: a new hypothesis of Earth dynamics; *in* S. Chatterjee and N. Horton III (eds.) New concepts of global tectonics: Lubbock, TX, USA, Texas Tech. Univ. Press.

Meyerhoff, H. A. and Meyerhoff, A. A., 1977, Genesis of island arcs. *In*: Geodynamics in south-west Pacific, Noumea-Nouvelle-Caledonie, 27 aout-2, septembre 1976: paris, Editions Technip, p. 357–370.

Mi Hangjun and Jiang Jiayu, 1988, A discussion on the correlation of various blocks in the Western fault-folded belt (northern part) in the western part of Shaanxi-Gansu-Ningxia basin and the location of its east border: Petroleum Exploration and Development, v. 15, no. 1, p. 10–19.

Milanovskiy, Ye. Ye, 1975, Riftogenes i geosinklinal'nyy protsess (Origins of rifts and the geosynclinal process): Mosk. Univ., Vestn. Ser. 4, Geol., no. 4, p. 28–40.

Min Yu, 1981, Hydrocarbon accumulations in sedimentary basins of nonmarine facies in China. *In*: Mason, J. F. (ed.) 1981, Petroleum geology in China: Tulsa, OK, PennWell Publ. Co., p. 1–4.

Min Yu, Shi Baoheng and Cheng Xueru, 1980, Discovery of Daqing oil field and the development of the Chinese petroleum industry: Acta Petrolei Sinica, Special (Daqing) Issue, October, p. 1–9 .

Misch, P., 1945a, Young dynamometamorphism and other alpinotype structures in western Yunnan: Beijing, Academia Sinica, Science Record, v. 1, nos. 3–4, p. 534–540.

Misch, P., 1945b, 'Sinotype' structures in eastern Yunnan: Beijing, Academia Sinica, Science Record, v. 1, nos. 3–4, p. 541–548.

Misch, P., 1945c, Remarks on the tectonic history of Yunnan, with special reference to its relations to the type of the young orogenic deformation: Geological Society of China Bulletin, v. 25, p. 47–153.

Misch, P., 1946, On the discovery of Upper Permian (Lopingian) in western Yunnan: Peking (Beijing), Geol. Soc. China, Bull., v. 26, p. 64–82.

Misch, P., 1960, Regional structural reconnaissance in central-northeast Nevada and some adjacent areas: observations and interpretations. *In*: Boettcher, J. W. and Sloan, W. W. (eds.) Geology of east-central Nevada, Salt Lake City: Intermountain Association of Petroleum Geologists, Eastern Nevada Geological Society, and Utah Geological and Mineralogical Survey, p. 17–42.

Mitchell, A. H. G., 1981, Phanerozoic plate boundaries in mainland SE Asia, the Himalayas, and Tibet: Geol. Soc. London, Jour., v. 138, p. 109–122.

Miyashiro, A., 1973, *Metamorphism and metamorphic belts*: New York, Halsted Press, John Wiley & Sons, 492 p.

Miyashiro, A., Aki, K. and Sengor, A. M. C., 1979, *Orogeny*: New York, John Wiley & Sons, 242 p.

Molnar, P. and Chen Wang-ping, 1983a, Focal depths and fault plane solutions of earthquakes under the Tibetan Plateau: Journal of Geophysical Research, v. 88, p. 1180–1196.

Molnar, P. and Chen Wang-ping, 1983b, Seismicity and mountain building. *In*: Hsu, K. J. (ed.) Mountain building processes: London, Academic Press, p. 41–57.

Molnar, P. and Tapponier, P., 1975, Cenozoic tectonics of Asia: effects of a continental collision: Science, v. 189, p. 419–426.

Molnar, P. and Tapponier, P., 1978, Active tectonics of Tibet: Journal of Geophysical Research, v. 83, p. 5361–5375.

Molnar, P., Chen Wang-ping and Tapponier, P., 1981, Constraints of the amount of north-south shortening in Tibet during the Cenozoic. *In*: Liu Dong-sheng (ed.) Geological and ecological studies of Qinghai-Xizang Plateau, v. 1, Geology, geological history and origin of Qinghai-Xizang Plateau: Beijing, Science Press; and New York, Gordon and Breach, Science Publishers, Inc., p. 757–761.

Moody, J. D., 1973, Petroleum exploration aspects of wrench-fault tectonics: American Association of Petrol. Geol. Bulletin, v. 57, no. 3, p. 449–476.

Moody, J. D. and Hill, M. H., 1956, Wrench-fault tectonics: Geological Society of America Bulletin, v. 67, p. 1207–1246.

Moore, P. S., Hobday, D. K., Mai, H. and Sun, Y. C., 1986, Comparison of selected non-marine petroleum-bearing basins in Australia and China: Australian Petroleum Exploration Association Journal, v. 26, part 1, p. 285–309.

Mu Antze (Enzhi), Wen Shih-hsuan, Wang Yi-kang and Chang Ping-kao, 1973, Stratigraphy of the Mount Jolmo Lungma region in southern Tibet, China: Scientia Sinica, v. 16, no. 1, p. 96–111.

Mu Enzhi, 1962, *Silurian in China*: Peking, Science Press, 95 p. (in Chinese).

Mu Enzhi and Ni Yunan, 1975, Silurian and Devonian graptolites of Qomolangma Feng (Everest) region. *In*: A Scientific Investigation Report 1966–1968, Paleontology, v. 1, Academia Sinica: Peking, Science Press, p. 5–38 (in Chinese).

Mu Enzhi, Boucot, A. J., Chen Xu and Fong Jiayu, 1986, *Correlation of the Silurian rocks of China*: Geol. Soc. America Spec. Paper 202, 81. p.

Mu En-zhi (Enzhi), Li Ji-jin (Jijin), Ge Mei-yu (Meiyu), Chen Xu, Ni Yu-nan (Yunan) and lin Yao-kun (Yaokun), 1981, Paleogeographic maps of the Upper Ordovician of central China and their explanation: Acta Stratigraphica Sinica, v. 5, no. 3, p. 165–170 (in Chinese).

Mu En-zhi (Enzhi) and Rong Jia-yu (Jiayu), 1984, On the international Ordovician-Silurian boundary: Geol. Survey, Canada, Rept. 798696, p. 31 (transl. from Journal of Stratigraphy, v. 7, no. 2, p. 81–91 (in Chinese).

Mueller, S., 1983, Deep structure and recent dynamics in the Alps. *In*: Hsu, K. J. (ed.) Mountain building processes: New York, Academic Press, p. 181–199.

Museum of Geology, Ministry of Geology, 1980, *Minerals in China*: Shanghai, Shanghai Scientific and Technical Publishers, 164 p.

Nakazawa, K. and Dickins, J. M. (eds.) 1985, *The Tethys: her paleobiogeography from Paleozoic to Mesozoic*: Tokyo, Tokai University Press, 317 p.

Nanjing Institute of Geology and Paleontology, Academic Sinica, 1980, Stratigraphy and paleontology of the Upper Permian coal-bearing formations in western Guizhou and eastern Yunnan, China: Beijing, Science Press, 277 p. (in Chinese); (five reports).

Nanjing Department of Geology, 1980, Investigation on the time and spatial distribution of the granitic rocks of southeastern China, their petrographic evolution, petrogenetic types, and the metallogenic relation: Journal of Nanjing University, Speical Issues on Geological Science, Institute of Granitic, Volcanic Petrography and metallogenesis, 56 p.

Nanjing Institute of Geology and Palaeontology, Academia Sinica (comp.), 1984, Stratigraphy and palaeontology of systemic boundaries in China; Cambrian-Ordovician boundary: Anhui Science and Technology Publishing House, v. 1, p. 405; v. 2, 412 p.

Nanking (Nanjing) Institute of Geology and Paleontology, 1974, A handbook to the stratigraphy and paleontology of Southwest China: Peking (Beijing), Science Press, 454 p. (in Chinese).

Nanking (Nanjing) University, Department of Geology, 1974, Granitic rocks of different geological periods of southeastern China and their genetic relations to certain metallic mineral deposits: Scientia Sinica, v. 17, no. 1, p. 55–72.

Nature, 1985, Science in China: Nature, v. 318, no. 6043, p. 205–228.

Needham, J., with Wang Ling, 1959, *Science and civilization in China*, v. 3, *Mathematics and the sciences of the heavens and the earth*: United Kingdom, Cambridge, at the University Press, 877 p.

Needham, J., with Wang Ling, 1961, *Science and civilization in China*; v. 1, introductory origins: Cambridge, Cambridge University Press, 318 p.

Needham, J., 1964, Science and China's infleunce on the world. *In*: Dawson, R. (ed.) The legacy of China: London, Oxford Univ. Press, p. 292.

Ni, J. and York, J. E., 1978, Late Cenozoic tectonics of the Tibetan Plateau: Journal of Geophysical Research, v. 83, p. 5377–5384.

Nur, A. and Ben-Avraham, Z., 1982, Oceanic plateaus, the fragmentation of continents, and mountain building: Journal of Geophysical Research, v. 87, no. 895, p. 3644–3661.

Ogura, T., Sawatari, M. and Murayama, K., 1938, Chi-hsing Shan volcano, Feng-tien and south Hsing-an Province, Manchuria: Ryojun College of Engineering, Survey Reports, Volcanoes in Manchuria, no. 3, p. 1–39 (in Japanese).

Ondrick, R. S. and Woodward, K., 1985, A survey of China's land oil fields, The China Business Review, v. 12, no. 1, p. 14–24.

Owen, E. W., 1975, *Trek of the oil finders; a history of exploration for petroleum*: Amer. Assoc. Petrol. Geologists Mem. 6, 1647 p.

P'an Chung-Hsiang (Pan Zhongxiang), 1982, Petroleum in basement rocks: Am. Assoc. Petrol. Geol. Bull., v. 66, no. 10, p. 1630–1643.

P'an Chung-hsiang (Pan Zhongxiang) and Ryabukhin, G. E., 1961, The geologic structure of the intermontane basins of Central Asia and the presence of oil and gas therein: Am. Geol. Inst., 1963, Internat. Gol. Review, v. 5, no. 8, p. 985–998.

Pan Guoqiang, 1984, The Late Precambrian and Early Palaeozoic marginal basin of South China: Geol. Soc. London special Publ. No. 16, p. 279–284.

Pan, Y. S., 1968, The regional gravity of the Penghu Islands, Taiwan, China: UN ESCAFE, CCOP Tech. Bull., v. 1, p. 87–93.

Pan Zhongxiang, 1983, Significance of unconformity to oil and gas migration and accumulation: Acta Petrolei Sinica, v. 4, no. 4, p. 1–10.

Pan Zhongxiang and Ryabukhin, G. E., 1961, Geologicheskoye stroyeniye vnutrennikh vpadin Tsentral'noy Azii i ikh neftegazonosnost' (Geologic structure of the intermontane basins of central Asia and their petroleum content): Izvestiya Vysshikh Uchebnyk Zavadeniy, Geologyia i razvedka, no. 7, p. 3–21.

Parfenov, L. M. and Natal'in, B. A., 1985, Mesozoic accretion and collision tectonics of northeastern Asia. *In*: D. G. Howell (ed.) Tectonostratigraphic terrances of the circum-Pacific regions: Houston, Texas, Circum-Pacific Council for Energy and Mineral Resources, p. 363–373.

Park, C. S. Jr. and MacDiarmid, R. A., 1964, *Ore deposits*: San francisco, W. H. Freeman and Company, 475 p.

Pei Qi and Wang Xiepei, 1984, Significant role of structural fractures in Renqiu buried-hill oil field in eastern China: Am. Assoc. Petrol. Geol. Bull., v. 68, no. 8, p. 971–982.

Pei, W. C., Chou, M. C. and Zheng, J. J., 1963, *Cenozoic of China*: Peking (Beijing), Science Press, 31 p. (in Chinese).

Peltzer, G., Tapponier, P., Zhang Zhitao and Xu Zhi Qin, 1985, Neogene and Quaternary faulting in and along the Qinling Shan: Natura, v. 317, no. 60–37, p. 500–505.

Peng Guozong, 1986, A preliminary discussion on the origin of lead-zinc ore deposits in the Yutan region of Huayuan County, Hunan Province: Scientia Geologica Sinica, no. 2, p. 179–186.

Peng Zuolin, Chen Xinrong, Zhou Qingjie and Yan Yugui, 1984, Genetic mechanism and accumulation of oil and gas of the Karamay-Urho fault zone in the northwest margin of the Junggar basin. *In*: Su Zongwei (ed.) contribution to 27th International Geological Congress, 1984, Moscow, Developments in Geoscience: Beijing, Science Press, p. 611–620.

Perrodon, A., 1966, *Geologie du petrole*: Paris, Presses Universitaires de France, 440 p.

Petrov, M. P., 1967, *Pustyni Tsentrai'noy Azii, tom2*: Leningrad, Nauka, 288 p.

Petrushevsky, B. A., 1971, On the problem of the horizontal heterogeneity of the earth's crust and uppermost mantle in southern Eurasia: Tectonophysics, v. 11, no. 1, p. 29–60.

Pham Van Ngoc, Boyer, D., Therme, P., Yuan, X. -C., Li, L. and Jin, G. -Y., 1986, Partial melting zones in the crust in southern Tibet from magnetotelluric results: Nature, v. 319, p. 310–314.

Piccoli, G. and McKenzie, K. G. (eds.) 1982, 'Shallow Tethys' International symposium, Padova, 7th-8th June 1982 (first part): Bollettino della Societa Paleontologica Italian, v. 21, no. 2, 3, p. 145–325.

Piccoli, G. and McKenzie, K. G. (eds.) 1983,'Shallow Tethys' International Symposium, Padova, 7th-8th June 1982 (second part): Bollettino della Societa Paleontologica Italiana, v. 22, no. 1, 2, p. 5–188.

Pines, I., Teng, T. -L., Rosenthal, R. and Alexander, S., 1980, A surface wave dispersion study of the crustal and upper mantle of China: Journal of Geophysical Research, v. 85, p. 3829–3844.

Plafker, G., 1976, Tectonic implications of oriented lakes and lineaments in north-eastern Bolivia. *In*: R. A. Hodgson, S. P. Gay Jr. and J. Y. Benjamin (eds.) Proceedings of the First International Conference on basement Tectonics, Salt Lake city, Utah, June 3–7, 1974: Utah Geological Association, p. 519–527.

Polo, Marco (see Yule, Sir H., also Cordier, H. and Komroff, M.).

Powell, C. M. and Conaghan, P. J., 1973, Plate tectonics and the Himalayas: Earth and Planetary Science Letters, v. 20, p. 1–12.

Powell, C. M. and Conaghan, P. J., 1975, Tectonic models of the Tibetan Plateau: Geology, v. 3, p. 727–731.

Powell, C. M. and Conaghan, P. J., 1979, Rb-Sr geochronology of the Kuli-Mandi belt: its implications for the Himalayan tectogenesis-discussion: Geologische Rundschau, v. 68, p. 380–383.

Pratt, W. E. and Good, D. (eds.) 1950, *World geography of petroleum*: Am. Geogr. Soc. - Princeton Univ. Press, 464 p.

Project Manager, 1974, The offshore hydrocarbon potential of East Asia; a review of investigations, 1966–1973: United Nations UNDP/CCOP, Supplementary to the Report of the Tenth Session of CCOP, 67 p.

Qi Shoumei and Tang Xingnin, 1983, A general survey of the vanadium-bearing titanomagnetite deposits in China: Bulletin of the Institute of Mineral Deposits, Chinese Academy of Geological Sciences, Series III, no. 3, p. 73–77 (in Chinese).

Qi Yannian, Wang Sumin, Zhu Haihong, Yao Bingheng and Su Shoude, 1983, Relationship between paleogeographic environment of Early-Cretaceous and oil-gas distribution in Songliao basin: Scientia Sinica, ser. B, v. 16, no. 4, p. 424–437.

Qian Yi, 1984, Early Cambrian-Late Precambrian small shelly faunal assemblage with discussion on Cambrian-Precambrian boundary in China. *In*: Su Zongwei (ed.) Developments in Geoscience, Contribution to 27th International Geological Congress, 1984, Moscow: Beijing, Science Press, p. 9–20.

Qiang Zitong, Guo Yihua, Zhang Fan, Yan Chuantai and Zheng Jiafeng, 1985, The Upper Permian reef and its diagenesis in Sichuan basin: Oil and Gas Geology, v. 6, no. 1, p. 82–90.

Qiao Xiufu, Ma Lifang and Zhang Huimin, 1988, The terminal precambrian paleogeographic framework in China: Acta Geologica Sinica, v. 62, no. 4, p. 290–300.

Qin Bao-yan, Kuo Zeng-jian and Liu Guang-yuan, 1981, Discussion on forecasting great earthquakes (M=8) in the Qinghai-Xizang Plateau and its vicinity. *In*: Liu Dong-sheng (ed.) Geological and ecological studies of Qinghai-xiang Plateau, v. 1, Geology, geological history and origin of Qinghai-Xizang Plateau: Beijing, Science Press; and New York, Gordon and Breach, Science Publishers, Inc., p. 807–814.

Qiu Dongzhou, 1984, Turbidite in the 4-5th Members of Pakabulak Formation in Kekeya oil field (Tarim basin): Oil and Gas Geology, v. 5, no. 1, p. 55–59.

Qiu Qun, 1976, On the background and seismic activity of the M = 7,8 Tangshan earthquake, Hopei Province of July 28, 1976: Acta Geophysica Sinica, v. 19, no. 4, p. 259–269.

Qu Jie, 1985, The distribution and development characteristics of sedimental system in paralic lacustrine basins in eastern China: Oil and Gas Geology, v. 6, no. 2, p. 197–208.

Radchenko, O. A., 1968, *Geochemical regularities in the distribution of the oil-bearing regions of the world*: Israel Program for Scientific Translations, Jerusalem, and National Science Foundation, Washington, 312 p.

Raiverman, V., Kunte, S. V. and Mukherjea, A., 1983, Basin geometry, Cenozoic sedimentation and hydrocarbon prospects in north western Himalaya and Indo-Gangetic plains. *In*: L. L. Bhandari, B. S. Venkatachala, R. Kumar, S. N. Swamy, P. Garga and D. C. Srivastava (eds.) Petroliferous basins of India: Dehra dun, Petroleum Asia Journal, v. 6, p. 67–92.

Raju, D. S. N., 1968, Preliminary study on the Cretaceous planktonic foraminifera from the sub-surface section of Cauvery basin, south India (abst.): Geol. Bundesanstalt Verh. (Vienna, Austria), v. 1, no. 3, p. A91–A92.

Ramanathan, S., 1968, Stratigraphy of Cauvery basin with respect to its oil prospects: Geol. Soc. India Mem. 2, p. 152–167.

Rangarao, A., 1983, Geology and hydrocarbon potential of a part of Assam-Arakan basin and its adjacent region. *In*: Bhandari, L. L., Venkatachala, B. S., Kumar, R., Swamy, S. N., Garga, P. and Srivastava, D. C. (eds.) Petroliferous basins of India: Dehra Dun, Petroleum Asia Journal, v. 6, p. 127–158.

Ranneft, T. S. M., 1979, Segmentation of island arcs and application to petroleum geology: Journal of Petroleum Geology, v. 1, no. 3, p. 35–53.

Rastogi, B. K., 1974, Earthquake mechanisms and plate tectonics in the Himalayan region: Tectonophysics, v. 21, p. 47–56.

Ray, D. K., general convener, 1982, *Commentary on the tectonic map of south and east Asia*: Paris, Commission for the Geological Map of the world, Subcommission for South and East Asia, 26 p.

Ren Jishun, 1984, The Indosinian orogeny and its significance in the tectonic evolution of China: Bull. Chinese Acad. Geol. Sci., no. 9, p. 31–44.

Ren Jishun and Chen Tingyu, 1989, Tectonic evolution of the continental lithosphere in eastern China and adjacent area: Jour. Southeast Asian Earth Sci., v. 3, nos. 104, p. 17–27.

Ren Jishun, Jiang Chunfa, Zhang Zhengkun and Qin Deyu, 1980, *The geotectonic evolution of China*: Beijing, Science Press, 124 p. (in Chinese).

Research Group or Center of Precambrian and Metamorphic Geology, Institute of Geology, Academy of Geological Sciences, Ministry of Geology, 1962, *Precambrian of China*: Peking, Science Press, 80 p. (in Chinese).

Ren Jishun, Jiang Chunfa, Zhang Zhengkun and Qin Deyu (under the direction of Prof. Huang Jiqing), 1987, Geotectonic evolution of China: Beijing, Science Press; and Berlin, Springer-Verlag, 224 p.

Richard, L., 1908, Comprehensive geography of the Chinese Empire: Shanghai, Tu-Se-Wei Press, (English transl. by M. Kennelly).

Richthofen, F. P. W., Freiherr von, 1877–1911, China, Ergebnisse eigener Reisen und darauf gegrundeter Studien, von Ferdinand Freiherr von Richthofen: Berlin, D. Reimer, v. 5 (v. 1, 1877; v. 2, 1882; v. 3, 1882; v. 4, 1883; v. 5, 1911).

Richthofen, F. P. W., Freiherr von, 1885 and 1912, Atlas von China, Orographische und geologische Karten von Ferdinand Freiherr von Richthofen, zu des Verfassers Werk; China, Ergebnisse Reisen und darauf gegruendeter Studien: Berlin, D. Reimer, v. 2 (v. 1, 1885; v. 2, 1912).

Robertson Research International Ltd., 1979, The People's Republic of China; its petroleum geology and resources: Great Britain and France, The Robertson Research Group; v. 1, text; v. 2, plates (forty-seven).

Roeder, D. H., 1973, Subduction and orogeny: Journal of Geophysical Research, v. 78, p. 5005–5024.

Rong Jiayu, 1979, The *Hirnantia* fauna of China, with comments on the Ordovician-Silurian boundary: Acta Stratigr. Sinica, v. 3, no. 1, p. 1–29.

Rong Jiayu and Yang Xuechang, 1981, Middle and Late Silurian brachiopod faunas in Southwest China: Nanjing Inst. Geol. Palaeont., Academia Sinica, Mem 13, p. 163–270.

Rong Jia-yu, 1985, The state of current Silurian stages in China: Journal of Stratigraphy, v. 9, no. 2, p. 96–107 (in Chinese, English abstract).

Roybarman, A., 1983, Geology and hydrocarbon prospects of West Bengal. *In*: Bhandari, L. L., Venkatachala, B. S., Kumar, R., Swamy. S. N., Garga, P. and Srivastava, D. C. (eds.) Petroliferous basins of India: Dehra Dun, Petroleum Asia Journal, v. 6, p. 51–56.

Rui Lin, 1981, The Permian of the Yangzi stratigraphic province: Acta Stratigraphica Sinica, v. 5, no. 4, p. 263–275.

Rui Lin, Jiang Nayan and Chen Chuzhen, 1983, Latest Permian and earliest Triassic sedimentary provinces and their facies types in South China: Scientia Sinica, Series B, v. 26, no. 10, p. 1099–1108.

Rui Liu and Shen Jinzhang, 1981, On the genus Palaeofusulina: Geol. Soc. America, Spec. Paper 187, p. 33–37.

Rui Zongyao, Huang Chongke, Qi Guoming, Xu Jue and Zhang Hongtao, 1987, Chapter 2, Spatial distribtuion of China's major porphyry copper (molybdenum) deposits; a summary of the 21 major metallogenic belts: Report no. 2643026, p. 33 (from Geol. Survey, Canada, Ottawa: transl. from Chinese. *In*: Porphyry copper (molybdenum) deposits of China; Beijing, Geological Publishing House, p. 5–12, 1979).

Russell, D. E. and Zhai Ren-jie (Renjie), 1987, The Paleogene of Asia; mammals and stratigraphy: Memoires du Museum National d'Histoire Naturelle, Serie C, Sciences de la Terre, Tome 52, 488 p.

Saidov, N. M., 1956, Mezo-kaynozoyskiye kontinental'yye otlozheniya Dzhungarskoy vpadiny (Mesozoic and Cenozoic continental deposits of the Dzungaria (Junggar) basin: Akad. Nauk SSSR, Izvestiya, ser. geol., no. 10, p. 85–97.

Samylina, V. A. and Yefimova, A. F., 1968, Pervyye nakhodkhi ranneyurskoy flory v basseyne r. Kolymy: Doklady Adademii Nauk SSSR, v. 179, no. 1, p. 116–168.

Santo, T., 1971, Seismicity around the Himalaya: Science (Tokyo), v. 41, p. 231–244.

Sastri, V. V. and Raiverman, V., 1968, On the basin study programme of the Cretaceous-Tertiary sediments of Cauvery basin: Geol. Soc. India Mem. 2, p. 143–151.

Sastri, V. V., Raju, A. T. R., Sinha, R. N., Venkatchala, B. S. and Banerji, R. K., 1977, Biostratigraphy and evolution of the Cauvery basin, India: Geological Society of India, Journal, v. 18, no. 8, p. 355–377.

Saxena, M. N., 1971, The crystalline axis of the Himalaya, India Shield and continental drift: Tectnophysics, v. 12, p. 433–447.

Saxena, M. N., 1971, Geological classification and tectonic history of the Himalaya: Proc. Indian Nat. Sci. Acad., v. 37(A), p. 28–54.

Saxena, M. N., 1975, Synchronous orogenic and epeirogenic episodes in the Himalaya and Indian shield, and the origin of the Himalaya: Chayancia Geologica (Hindustan), v. 1, no. 2, p. 123–142.

Saxena, M. N., 1978, The Himalaya – a modified arc system and gravity tectonics. *In*: Sakiani, P. S. (ed.) Current trends in geology, v. 1, Tectonic geology of the Himalaya: New Delhi, Today and Tomorrow's Printers & Publishers, p. 313–340.

Saxena, M. N., 1981, Chronology of tectonic events of the Himalaya. *In*: Sakiani, P. S. (ed.) Metamorphic tectonics of the Himalaya: New Delhi, Today and tomorrow's Printers & Publishers, p. 303–347.

Saxena, M. N., Gupta, V. J., Meyerhoff, A. A. and Archbold, N. W., 1986, Tectonic and spatial relations between India and Asia since Proterozoic time. *In*: Gupta, V. J. (ed.) 1986, Contributions to Himalayan Geology, v. 3, Geology of Western Himalayas: Delhi (India), Hindustan Publishing Corp., p. 187–207.

Schreiber, A., Weippert, D., Wittekind, H. -P. and Wolfart, R., 1972, Geology and petroleum potentials of central and south Afghanistan: Am. Assoc. Petrol. Geol. Bull., v. 56, p. 1494–1519.

Schwab, F. L., 1982, Editor's comments on papers 35 through 42, p. 308–313. *In*: Schwab, F. L. (ed.) Geosynclines: concept and place within plate tectonics: Stroudsbury (PA), Hutchinson Ross Publishing Co., 411 p.

Seeber, L., Armbruster, J. G. and Quittmeyer, R. C., 1981, Seismicity and continental subduction in the Himalayan arc. *In*: Gupta, H. K. and Delany, F. M. (eds.) Zagros-Hindu Kush-Himalaya geodynamic evolution: American Geophysical Union and Geological Society of America, Geodynamics Series 3, p. 215–242.

Sengor, A. M. C., 1984, *The Cimmeride orogenic system and the tectonics of Eurasia*: Geol. Soc. America, Spec. Paper, 195, 82 p.

Sengor, A. M. C., 1985, East Asian tectonic collage: Nature, v. 318, no. 6041, p. 16–17.

Sengor, A. M. C., 1985, Die Alpiden und die Kimmeriden; die verdoppelte Geschichte der Tethys: Geologische Rundschau, v. 74, no. 2, p. 181–213.

Sengor, A. M. C. (ed.) 1989, *Tectonic evolution of the Tethyan Region*: London, Kluwer Academic Publishers, 698 p.

Shah, S. C. and Sastry, M. V. A., 1975, Significance of Early Permian marine faunas of peninsular India. *In*: Campbell, K. S. W. (ed.) Gondwanan geology - papers presented at the Third Gondwana Symposium, Canberra, Australia, 1973: Australian National University Press, p. 391–395.

Shao Ji'an, 1989, Continental crust accretion and tectono-magmatic activity of the northern margin of the Sina-Korean plate: Jour. Southeast Asian Earth Sci., v. 3, nos. 1–4, p. 57–62.

Shareq, A., 1981, Geological observations and geophysical investigations carried out in Afghanistan over the period of 1972–1979. *In*: Gupta, H. K. and Delany, F. M. (eds.) Zagros-Hindu Kush-Himalaya geodynamic evolution: American Geophysical Union and Geological Society of America, Geodynamics Series 3, p. 87–110.

Sharma, K. K., 1983, Ladakh-Deosai batholith and its surrounding rocks. *In*: Gupta, V. J. (ed.) Stratigraphy and structure of the Kashmir and Ladakh Himalaya: Delhi, Hindustan Publishing corporation, p. 180–187.

Shen Xianjie, 1986, Mechanism of thermo-tectonic evolution of the uplift of the Qinghai-Xizang Plateau: Scientia Geologica Sinica, no. 2, p. 101–113.

Shen Xiuying and Duan Shufu, 1986, The seismic stratigraphic analysis of salt lake facies in north Dongpu depression: Oil Geophysical Prospecting, v. 21, no. 4, p. 380–390, 404.

Sheng Jinzhang, 1962, *Permian of China*: Beijing, Science Press, 95 p. (in Chinese).

Sheng Jinzhang, Jin Yugan, Rui Lin, Zhang Linxin, Zhang Zhuoguan, Wang Yujing, Liao Zhuoting and Zhao Jiamin, 1979, correlation charts and explanatory notes of the Permian system in China: Nanjing Inst. Geology and Paleontology, Academia Sinica, p. 111–127 (in Chinese).

Sheng Jinzhang, Rui Lin and Chen Chuzhen, 1985, Permian and Triassic sedimentary facies and paleogeography of South China. *In*: Nakazawa, K. and Dickins, J. M. (eds.) 1985, The Tethys: Tokyo, Tokai University Press, p. 59–81.

Sheng Jin-zhang, Chen Chu-zhen, Wang Yigang, Rui Lin, Liao Zhuoting, Bando Yuji, Ishiii Ken-ichi, Nakazawa Keiji and Nakamura Koji, 1984, Permian-Triassic boundary in middle and eastern Tethys: Jour. Faculty of Science, Hokkaido Univ., Ser. 4, Geol. Miner, v. 21, no. 1, p. 133–181.

Sheng Shen-Fu, 1980, The Ordovician system in China; correlation chart and explanatory notes: Internat. Union Geol. Sciences, Publ. No. 1, p. 7.

Sheng Xinfu, 1974, Subdivision and correlation of Ordovician system in China: Peking, Geological Publishing House (in Chinese).

Shibata, K., Nozawa, T. and Waniess, R. K., 1970, Rb-Sr geochronology of the Hida metamorphic belt, Japan: Canadian Journal of Earth Sciences, v. 7, no. 6, p. 1383–1401.

Sinha-Roy, S., 1982, Interactions of Tethyan blocks and evolution of Asian fold belts: Tectonophysics, v. 82, p. 277–297.

Skorduli, V. D., Khudyk, M. V., Le Van Ky, Nguen Ngok Ky and K. M. Sevost'yanov, 1983, Geologicheskoye stroyeniye i neftegazonosnost' Khanoyskogo progiba (Geologic structure and oil and gas possibilities in Hanoi basin): Geologiya Neftl i Gaza, v. 5, p. 55–60.

Smith, A. G., 1976, Plate tectonics and orogeny: a review: Tectonophysics, v. 33, nos. 1–2, p. 215–285.

Smith, B. J. (ed.) 1982, *Dinosaurs from China*: Natl. Mus. Victoria, Melbourne, Australia, 52 p.

Societe Geologique de France, N. S., 1984, Paleogeographie de l'Inde, du Tibet et du sud-est Asiatique; confrontation des données paleontologiques avec les modeles géogynamiques: Mem., 147, 194 p.

Song Jianguo and Liao Jian, 1982, Structural characteristics and petroliferous regions in the Chaidamu (Tsaidam) basin: Acta Petrolei Sinica, Special Issue 1982, p. 14–23.

Song, J. S., He, C. C., Qian, Z. S., Pan, C. L., Zheng, G. G. and Zhen, Y., 1978a, *On the Paleogene dinoflagellates and acritarchs from the coastal region of Bohai*: Beijing, Science Press, 190 p. (in Chinese, English abstract).

Song, J. S., Gao, L., Zhan, H. Y., Guan, X. I. and Wang, K. D., 1978b, *Early Tertiary spore and pollen grains from the coastal region of Bohai*: Beijing, Science Press, 117 p. (in Chinese, English abstract).

Song Zhi-chen and Liu Geng-wu, 1981, Tertiary palynological assemblages from Xizang with reference to their paleogeographical significance. *In*: Liu Dong-sheng (ed.) Geological and ecological studies of Qinghai-Xizang Plateau, v. 1, Geology, geological history and origin of Qinghai-Xizang Plateau: Beijing, Science Press; and New York, Gordon and Breach, Science Publishers, Inc., p. 207–214.

Song Zhi-chen (Zhichen), Li Hao-min (Haomin), Zheng Ya-hui (Yahui) and Liu Geng-wu (Gengwu), 1981, Miocene floristic regions of China. *In*: Teichert, C. et al. (eds.) Paleontology in China, 1979: Geol. Soc. America, Spec. Paper, 187, p. 249–254.

Song Zhichen, Li Wenben and He Chengquan, 1983, Cretaceous and Palaeogene (Paleogene) palynofloras and distribution of organic rocks in China: Scientia Sinica, ser. b, v. 26, no. 5, p. 538–549.

Srikantia, S. V. and Bhargava, O. N., 1978, The Indus tectonic belt of Ladakh Himalaya: its geology, significance and evolution. *In*: Saklani, P. S. (ed.) current trends in geology, v. 1, Tectonic geology of the Himalaya: New Delhi, Today and tomorrow's Printers & Publishers, p. 43–62.

Stanford Research Institute, 1963, The economic potential of communist China: v. 2, June.

Staritskiy, Yu. G., Maymin, Yu. S. and Trofimov, V. A., 1973, Tectonic development of North Vietnam: International Geology Review, v. 15, p. 1381–1390.

Staub, R., 1928, *Der Bewegungsmechanismus der Erde*: Berlin, Borntraeger, 270 p.

Stocklin, J., 1980, Geology of Nepae and its regional frame: Jour. Geol. Soc. London, v. 137, p. 1–34.

Stocklin, J., 1981, A brief report of geodynamics in Iran: American Geophy. Union and Geol. Soc. America, Geodynamic Series, v. 3, p. 70–74.

Stocklin, J., 1984, Orogeny and Tethys evolution in the Middle East: an appraisal of current concepts. *In*: Reports of the 27th International Geological Congress, 1984, Moscow, Colloqium 05, Tectonics of Asia, v. 5, p. 65–84: Mowcow, Izdatel'stvo Nauka.

Stocklin, J., 1989, Tethys evolution in the Afghanistan-Pamir-Pakistan region. *In*: Sengor, A. M. C. (ed.) Tectonic Evolution of the Tethyan Region: London, Kluwer Academic Publishers, p. 241–264.

Stoneley, R., 1975, On the origin of ophiolite complexes in the southern Tethys region: Tectonophysics, v. 25, p. 303–322.

Strachey, R., 1851, On the geology of part of the Himalaya Mountains and Tibet: Geological Society of London Quarterly Journal, v. 7, p. 292–310.

Strubell, W., 1968, Uber die Gewinnung und Verwendung von Erdol im alten China: Erdol und Kohle-Erdgas-Petrochemie, Bd. 21, Hft. 7, p. 435–436.

Su Yingxue and Zhang Bingzheng, 1987, The discovery of bured hill reversed fault in the Gubei area in Jiyang depression and its geological significance: Petroleum Exploration and Development, v. 14, no. 4, p. 42–47.

Sugisaki, R., Mizutani, S., Hattori, H., Adachi, M. and Tanaka, T., 1972, Late Paleozoic geosynclinal basalt and tectonism in the Japanese Islands: Tectonophysics, v. 14, no. 1, p. 35–56.

Sun Ai-lin (Ailin), 1973, Permo-Triassic reptiles of Sinkiang (Xinjiang): Scientia Sinica, v. 16, no. 1, p. 152–156.

Sun Dapeng, 1984, Distribution and formation of salts and their relationship with petroleum generation in Qaidam basin: Oil and Gas Geology, v. 5, p. 132–139.

Sun Dazhong and Lu Songnian, 1985, A subdivision of the Precambrian of China: Precambrian Research, v. 28, no. 2, p. 137–162.

Sun Dianqing, Gao Qinghua and Deng Naigong, 1982, Crustal movements and tectonic systems in China since Mesozoic: Acta Geologica Sinica, v. 56, no. 3, p. 200–211.

Sun Dong-li (Dongli), 1983, Aspects of the marine Cretaceous of China: Cretaceous Research, v. 4, no. 2, p. 145–158.

Sun Guofan, Liu Jingping and Miao Youngwang, 1983, Preliminary study on petroleum ptoential of Upper Paleozoic in northern Ordos basin: Oil and Gas Geology, v. 4, no. 1, p. 109–115.

Sun Guofan, Xie Qiuyuan, Liu Jingping, Xie Shiying, Liu Keqi, Yuan Weiguo, 1986, Evolution stacking and hydrocarbon potential of the Ordos basin – prototype analysis of a huge basin in Chinese continental plate: Oil and Gas Geology, v. 7, no. 4, p. 356–367.

Sun Ronggul, Cui Guangzhen and Li Maosong, 1983, Tanchen-Lujiang (Tan-Lu) fault and Indosinian movement in southeast Anhui Province: Scientia Geologica Sinica, no. 1, p. 1–9.

Sun, T. C. and Sun, T. Y., 1959, On the prospects of petroleum in Tsaidam (Qaidam) basin in view of geotectonic types: coll. Papers and Notes on Geomechanics, no. 1, p. 46–59 (in Chinese).

Sun Xiangjun, 1979, Palynofloristical investigation of the Late Cretaceous and Paleocene of China: Acta Phytotax. Sinica, v. 17, no. 3, p. 8–23 (in Chinese).

Sun Zhaocai, 1980, The tectonic framework and petroleum prospectivity of pre-Mesozoic in the northern part of the Eerduos (Ordos) basin: Acta Petrolei Sinica, v. 1, no. 3, p. 7–17.

Suneja, I. J., 1978, Ordovician palaeogeography of northwestern Kashmir (India): Delhi, Recent Researches in Geology, v. 4, p. 323–328.

Sung Kwun Chough, 1983, *Marine geology of Korean seas*: Boston, International Human Resources Development Corporation, 157 p.

Sung Ying-hsing (Song Yingxing), 1637, Tien Kung K'ai Wu (tian Gond Kai Wu) (Exploitation of the works of nature).

Sungshan Wang and Ian McDougall, 1980, K-Ar and $^{40}Ar/^{34}Ar$ ages on Mesozoic volcanic rocks from the lower Yangtze (Tangzi) volcanic zone, southeastern China: Jour. Geol. Soc. Australia, v. 27, nos. 1, 3, p. 121–128.

Sweeting, M. M., 1986, Limestone landscapes of South China: Geology Today, v. 2, no. 1, p. 11–16.

Sze, H. J. and Chou, C. Y., 1962, *Continental strata of Mesozoic of China*: Peking (Beijing), Science Press, 180 p. (in Chinese).

Talent, J. A. and Mawson, R., 1979, Palaeozoic-Mesozoic biostratigraphy of Pakistan in relation to biogeography and the coalescence of Asia. *In*: Farah, A. and DeJong, K. A. (eds.) Geodynamics of Pakistan: Quetta, Geological Survey of Pakistan, p. 81–102.

Tamaki, K., 1988, Geological structure of the Japan Sea and its tectonic implications: Bulletin of the Geological Survey of Japan, v. 39, no. 5, p. 269–365.

Tan Shidian, 1985, Overthrust belts in northwest China and their petroleum trap type: Oil and Gas Geology, v. 6, no. 2, p. 179–186.

Tan Tjongkie (Chen Zongji), 1987, Geodynamics and tectonic evolution of Panxi rift: Tectonophysics, v. 133, nos. 3–4, p. 287–206.

Taner, I., Kamen-Kaye, M. and Meyerhoff, A. A., 1988, Petroleum in the Junggar basin, northwestern China: Journal of Southeastern Asian Earth Sciences, v. 2, nos. 3/4, p. 163–174.

Taner, I., and Meyerhoff, A. A., 1990, Petroleum at the roof of the World – The geological evolution of the Tibet (Qinghai-Xizang) Plateau: Jour. Petroleum Geology, v. 13, no. 2, p. 157–178 (Part I), no. 3, p. 289–314 (Part II).

Junggar basin, northwestern China: Journal of Southeastern Asian Earth Sciences, v. 2, nos. 3/4, p. 163–174.

Tang Chongguang, 1980, The petroleum-geological characteristics of the rifting in eastern China: Acta Petrolei Sinica, v. 1, no. 4, p. 19–26.

Tang Tianfu, Yang Hengren, Lan Xiu, Hu Lanying, Yu Congliu, Zhong Shilan, Zhang Yiyong and Wei Jingming, 1984, Upper Cretaceous-lower Tertiary transgression and sedimentation of western Tarim basin, China. *In*: Su Zongwei (ed.) Developments in geoscience. Contribution to 27th International Geological congress, 1984, Moscow: Beijing, Science Press, p. 203–214.

Tang Tienfu, Xue Yaosong and Yu Congliu, 1981, Distribution and depositional environments of the Sinian carbonate rocks in South China: Acta Petrolei Sinica, v. 2, no. 2, p. 11–19.

Tang Xin and Zhou Mingzhen, 1964, The vertebrate-bearing early Tertiary of South China: a review: Beijing, Vertebrate PalAsiatica, v. 8, no. 2, p. 119–133.

Tang Xiyuan and Luo Zhujin, 1986, Features of petroleum geology for the block-faulted zone in the northern margin of Qaidam basin: Oil and Gas Geology, v. 7, no. 2, p. 182–191.

Tang Xiyuang, Guo Zhongming and Wang Dingyi, 1988, The characteristics and evolution of the thrust nappe tectonic belt and its petroleum potential in the west Ordos basin: Oil and Gas Geology, v. 9, no. 1, p. 1–10.

Tang Zheyao and Kong Kinxiang, 1984, Structural characteristics of Sinian reservoirs in Weiyuan Gas field, Sichuan: Acta Petrolei Sinica, v. 5, no. 4, p. 43–54.

Tang Zhi, 1982, Tectonic features of oil and gas basins in eastern part of China: Am. Assoc. Petrol. Geol. Bull., v. 66, no. 5, p. 509–521.

Tang Zhi and Chang Chenyong, 1978, Discussion of the original hydrocarbon occurrence in the Paleozoic and Sinian formations of North China: Petroleum Exploration and Development, no. 5.

Tang Zhongyu, 1984, Relationship of oil and gas with volcanism in Sanshui basin: Oil and Gas Geology, v. 5, no. 2, p. 89–100.

Tao Weiping, 1986, The Triassic marine gypsum deposits in China. *In*: Huang Jiqing (ed.) Proceedings of the symposium on Mesozoic and Conozoic geology in connection of the 60th anniversary of the Geological Society of China: Beijing, Geological Publishing House, p. 593–607.

Tapponier, P. and Molnar, P., 1977, Active faulting and tectonics in China: Jour. Geophys. Research, v. 82, no. 10, p. 2905–2930.

Tapponier, P., Mercier, J. L., Armijo, R., Han Tonglin and Zhou Ji, 1981, Field evidence for active normal faulting in Tibet: Nature, v. 294, p. 410–414.

Tapponier, P., Pelzer, P., Le Dain, A. Y. and Armijo, R., 1982, Propagating extrusion tectonics in Asia: new insights from simple experiments with plasticine: Geology, v. 10, p. 611–616.

Tapponier, P., Peltzer, G. and Armijo, R., 1986, On the mechanics of the collision between India and Asia. *In*: Coward, M. P. and Ries, A. C. (eds.) Collision tectonics: Geological Society of London, Special Publication 19, p. 115–157.

Taraz, H., 1973, Correlation of uppermost Permian in Iran, Central Asia and South China: Amer. Assoc. Petrol. Geol. Bull., v. 57, no. 6, p. 1127–1133.

Tectonic Map Compiling Group, Institute of Geology, Academia Sinica, 1974, A preliminary note on the basic tectonic features and their development in China: Scientia Geologica Sinica, 1.

Teichert, C. and Meyerhoff, A. A., 1972, Continental drift and the marine environment: Montreal, 24th Internat. Geol. Congress, sec.7, Paleontology, p. 3339–3349.

Teichert, C., Liu Lu and Chen Peiji (Beiji) (eds.) 1981, *Paleontology in China, 1979*: Geol. Soc. America, Spec. Paper 187, 264 p.

Teng Chi-tung (Deng Qidong), Chang Yu-ming (Zhang Yuming), Hsu Kwei-lin (Xu Qweilin) and Fan Fu-tian (futian), 1979, On the tectonic stress field in China and its relation to plate movement: Physics of the Earth and Planetary Interiors, v. 18, no. 4, p. 257–273.

Teng Chi-tung (Deng Qidong), Xu Guilin et al., 1979, The seismotectonic map of people's Republic of China: Beijing, Cartographic Publishing House, 37 p. (Chinese, English summary, one sheet).

Teng Ji-wen, 1981, Characteristics of geophysical fields and plate tectonics of the Qinghai-Xizang Plateau and its neighbouring regions. *In*: Liu Dong-sheng (ed.) Geological and ecological history and origin of Qinghai-Xizang Plateau: Beijing, Science Press; and New York, Gordon and Breach, Science Publishers, Inc., p. 633–649.

Teng Ji-wen, Wang Qian-shen, Liu Yuan-long and Wei Si-yu, 1983, The characteristics of geophysical fields and the formation of oil- and gas-bearing basins in eastern China: Acta Geophysica Sinica, v. 26, no. 4, p. 319–330.

Terman, M. J., 1974, Tectonic map of China and Mongolia: Geol. Soc. America, 2 sheets, map and cross section, scale 1:5,000,000.

Thein, M., 1973, A preliminary synthesis of the geological evolution of Burma with reference to the tectonic development of southeast Asia: Geological Society of Malaysia Bulletin, v. 6, p. 87–116.

Thompson, M. L. and Foster, C. L., 1937, Middle Permian fusulinids from Szechuan (Sichuan), China: Jour. Paleontology, v. 11, no. 2, p. 126–144.

Tian Chonglu, Li Detong, Liu Tiequan, Jiang Hongtao, Wu Zhengming and Lu Songnian, 1987, characteristics of tectonic-sedimentary facies in the lower Tertiary of Jizhong depression: Oil and Gas Geology, v. 8, no. 1, p. 90–98.

Tian Zaiyi, 1983, Characteristics of petro-geological structures in China and evaluation of their prospects of oil and gas: Acta Petrolei Sinica, v. 4, no. 1, p. 1–10.

Tian Zaiyi and Du Yonglin, 1987, Formation and evolution of the Yilan-Yitong graben: Tectonophysics, v. 133, no. 3–4, p. 165–173.

Tian Zaiyi and Wang Shanshu, 1985, Geological structures and oil and gas sedimentary basins in the China seas: Acta Petrolei Sinica, v. 6, no. 3, p. 1–10.

Tian Zaiyi, Chai Guilin and Lin Liang, 1985, Tectonic evolution of Tarim basin and its oil and gas potential: Oil and Gas Geology, v. 6, no. 3, p. 250–259.

Tian Zaiyi, Chang Chengyong, Huang Difan and Wu Chongyun, 1983, Sedimentary facies, oil generation in Meso-Cenozoic continental basins in China: Oil and Gas Journal, v. 81, no. 20, p. 120–127.

Ting, J. -S., 1979, Late Paleocene planktonic foramifera from WG-1 well, western central Taiwan: Petroleum Geology of Taiwan, no. 16, p. 168–185.

Tong Chonguan, 1980, Some characteristics of petroleum geology of the rift system in eastern China: Acta Petrolei Sinica, v. 1, no. 4, p. 19–26.

Tong Wei and Zhang Ming-tao, 1981, Characteristics of geothermal activities in Xizang Plateau and their controlling influence on plateau's tectonic model. *In*: Liu Dong-sheng (ed.) Geological and ecological studies of Qinghai-Xizang Plateau, v. 1, Geology,m geological history and origin of Qinghai-Xizang Plateau: Beijing, Science Press; and New York, Gordon and Breach, Science Publishers, Inc., p. 841–846.

Tong Xiaoguang, 1984, Petroleum geological characteristics of the Liaohe depression, Liaoning Province: Acta Petrolei Sinica, v. 5, no. 1, p. 19–27.

Tong Xiaoguang, 1984, A forecast of prospect and petroleum geology in Liaodongwan depression - geollogical analogy between Liaodongwan depression and Liaohe depression: Petroleum Exploration and Development, v. 11, no. 5, p. 1–8.

Trewartha, G. T., 1961, *The earth's problem climates*: Madison (WI), University of Wisconsin Press, 334 p.

Trommsdorff, V., Dietrich, V. and Honegger, K., 1983, The Indus suture zone: paleotectonic and igneous evolution. *In*: Hsu, K. J. (ed.) Mountain building processes: New York, Academic Press, p. 213–219.

Tu Guang-chi, Zhang Yu-quan, Zhao Zhen-hua and Wang Zhonggang, 1981, Characteristics and evolution of granitoids of southern Xizang. *In*: Liu Dong-sheng (ed.) Geological and ecological studies of Qinghai-Xizang Plateau, v. 1, Geology, geological history and origin of Qinghai-Xizang Plateau: Beijing, Science Press; and New York, Gordon and Breach, Science Publishers, Inc., p. 353–361.

Tung, J. P., 1974, *The surface wave study of crustal and upper mantle structure of mainland China*: University of Southern California, unpublished Ph. D. dissertation, 248 p.

Ulmishek, G., 1984, *Geology and petroleum resources of basins in western China*: Argonne National Laboratory, ANL/ES-146, 131 p.

Valdiya, K. S., 1976, Himalayan transverse faults and faults and their parallelism with subsurface structures of north Indian plains: Tectonophysics, v. 32, p. 353–386.

Valdiya, K. S., 1981, Tectonics of the central sector of the Himalaya. *In*: Gupta, H. K. and Delany, F. M. (eds.) Zagros-Hindu Kush-Himalaya geodynamic evolution: American Geophysical Union, Geodynamic Series 3, p. 87–111.

Varentsov, M. I. (ed.) 1967, *Tektonika neftegazonosnykh vpadiny Sredney Azil i Kazakhstana*: Moscow, Nauka, 124 p.

Virdi, N. S., 1981, Geotectonic evolution of the Indus suture zone. *In*: Sinha, A, K. (ed.) Contemporary geoscientific researches in Himalaya, v. 1, Tectonics-regional geology and biostratigraphy: Dehra Dun, Bishen Singh Mahendra Pal Singh, p. 131–136.

Voskresenskiy, A. A., Kravchenko, K. N., Movshovitch, E. B. and Sokolov, B. A., 1971, *Ocherk geologii Pakistana* (Short description of geology of Pakistan): Moscow, Nedra, 168 p.

Vysotsky, I. V. (ed.) 1968, *Geologiya nefti, spravochnik, Tom 2, kniga 2, Neftyaniye mestorozhdeniya zarubezhnykh stran*: Moscow, Nedra, 804 p.

Wadia, D. N., 1953, *Geology of India*, 3rd ed: London, MacMillan & Co. ltd., 531 p.

Wageman, J. M., Hilde, T. W. C. and Emery, K. O., 1970, Structural framework of East China Sea and Yellow Sea: Am. Assoc. Petrol. Geol. Bull., v. 54, no. 9, p. 1430–1439.

Wagner, H. C., Wagner, L. C., Wang, F. F. H. and Wong, F. L. (eds.) 1988, *Petroleum resources of China and related subjects*: Circum-Pacific Council for Energy and Mineral Resources, Earth Science Series, v. 10, 909 p.

Wales, National Wales, 1987, Dinosaurs from China: Natl. Mus. Wales, Cardiff, United Kingdom, 6 p.

Wan Tianfeng, 1989, Tectonic evolution of Proterozoic-Jurassic in six provinces of southeastern China: Jour. Southeast Asian Earth Sci., v. 3, nos. 1–4, p. 95–103.

Wan Ziyi, 1986, A brief introduction to the major types of iron deposits in Tibet and their geological setting: Mineral Deposits, v. 5, no. 4 (Serial No. 18), p. 32–48 (in Chinese, English abstract).

Wang, C., Shih, Y. and Zhou, W., 1982, Dynamic uplift of the Himalaya: Nature, v. 298, p. 533–556.

Wang Changlie, Xu Youzhi, Xie Ciguo and Xu Wenguang, 1982, The geological characteristics of Shizhuyuan W-Sn-Mo-Bi deposit. *In*: Hepworth, J. V. and Zhang Yuhong (eds.) Tungsten geology, Jiangxi, China: Bandung (Indonesia), ESCAP/RMROC, and Beijing, Geological Publishing House, p. 413–425.

Wang Chaodong, 1986, Application of gravity exploration results to discuss the basal structure of the Chuxiong basin, Yunnan Province: Oil and Gas Geology, v. 7, no. 4, p. 419–426.

Wang Cheng-yuan (Chengyuan), 1985, A progress report of Carboniferous conodonts in China: International Congress on carboniferous Stratigraphy and Geology, v. 9, no. 5, p. 301–302 (meeting of May 1976).

Wang Chengyuan (ed.) 1987, *Carboniferous boundaries in China*: China, Beijing, Science Press, Contribution to the 11th International Congress of Carboniferous Stratigraphy and Geology, 180 p.

Wang Chengyuan and Wang Zhihao, 1981, Permian conodont biostratigraphy of China. *In*: Teichert, C. et al. (eds.) Paleontology in China, 1979: Geol. Soc. America, Spec. Paper 187, p. 227–236.

Wang, C. S., 1972, Geosynclines in new global tectonics: Geological Society of America Bulletin, v. 83, no. 7, p. 2105–2110.

Wang, C. S., 1979, Concept of geosyncline: Geologische Rundschau, v. 58, no. 2, p. 696–706.

Wang Defa, Sun Yongchuan and Zheng Junmao, 1983, Depositional features and oil and gas distribution in some of the fault-depression lake basins in North China: Acta Petrolei Sinica, v. 4, no. 3, p. 13–21.

Wang Dewen and Li Guiqun, 1985, New viewpoints on polycyclic orogenies in China: Earth Science, v. 39, no. 3 (198), p. 220–228 (in Japanese, English abstract).

Wang Dezi, Zhou Xinmin and Xu Xisheng, 1989, Types and genetic model of Precambrian granitoids of South China: Jour. Southeast Asian Earth Sci., v. 3, nos. 1–4, p. 255–261.

Wang Dingyi, Che Zireng, Zhang Shutian, Lin Zuowen, Dun Tiejun and Liu Laimin, 1987, Characteristics of tectonic evolution and formation mechanism in the Nanxiang basin: Oil and Gas Geology, v. 8, no. 4, p. 363–372.

Wang Dongfan and Quan Heng, 1984, Mesozoic tectonic-magmatism in Daxinganling, China: Wuhan College of Geology, Earth Science Journal, no. 3, total 26, p. 81–90.

Wang Erchie and Chu, J. J., 1988, Collision tectonics in the Cenozoic orogenic zone bordering China, India and Burma: Tectonophysics, v. 147, no. 1–2, p. 71–84.

Wang Fu-bao and Li Bing-yuan, 1985, The lower boundary of the Quaternary in the Himalayan region in China. *In*: Liu Tung-sheng (Dong-sheng) (ed.) Quarternary geology and environment of China: Beijing, China Ocean Press; and Berlin, Springer-Verlag, p. 18–22.

Wang, H. C., 1978, On the stratigraphic provinces of China: Jour. Stratigraphy, v. 2, no. 2, p. 81–104 (in Chinese).

Wang Hongchen (Wang, H. C.), 1980, Megastages in the tectonic development of Asia: Scientia Sinica, v. 23, no. 3, p. 331–345.

Wang Hongzhen, 1981, Geotectonic units of China according to the theory of mobilism: Jour. Wuhan Coll. Geol. Earth Science, v. 15, no. 1, p. 42–66.

Wang Hongzhen, 1981, Silurian rugose coral assemblages and paleobiogeography in China. *In*: Teichert, C. et al. (eds.) Paleontology in China, 1979: Geol. Soc. America, Spec. Paper 187, p. 55–63.

Wang Hongzhen, 1983, On the geotectonic units of Xizang (Tibet) region: Wuhan college of Geology Earth Sciences Journal, total 19, p. 1–8.

Wang Hongzhen, ch. compiler, 1985, Atlas of the palaeogeography of China: Beijing, Cartographic Publishing House; 143 p., of maps and cross sections; 85 p., of summary (Chinese), 28 p. of explanation (English).

Wang Hongzhen and Liu Benpei, 1979, On the stratigraphic provinces and palaeogeographical development of China: 2nd All-China Stratigraphic Congress, Proceedings, Beijing, p. 35 (in Chinese with English abstr.).

Wang Hongzhen and Liu Benpei, 1980, *Historical Geology*: Beijing, Science Press, 352 p. (in Chinese).

Wang Hongzhen and Qiao Xiufu, 1984, Proterozoic stratigraphy and tectonic framework of China: Geological Magazine, v. 121, no. 6, p. 599–614.

Wang Jian and Wang Yongling, 1986, Geothermal and paleogeothermal characteristics of Liaohe graben district and their relationship to oil-gas resources: Oil and Gas Geology, v. 7, no. 2, p. 21–30.

Wang Ji'an, Wang Jiyang, Wang Yongling and Zhang Zhongyi, 1985, Geothermal characteristics of Liaohe graben: Oil and Gas Geology, v. 6, no. 4, p. 347–358.

Wang Ji-yang, Wang Ji-an, Xiong Liang-ping and Zhang Ju-ming, 1985, Analysis of factors affecting heat flow density determination in the Liaohe basin North China: Tectonophysics, v. 121, p. 63–78.

Wang Jiyang and Huang Shaopeng, 1988, Compilation of heat flow data for continental China: Scientia Geologica Sinica, v. 2, p. 196–204.

Wang Jun-wen, Chen Zhong-li, Gui Xun-tang, Xu Rong-hua and Zhang Yu-quan, 1981, Rb-Sr isotopic studies on some intermediate-acid plutons in southern Xizang. *In*: Liu Dong-sheng (ed.) Geological and ecological studies of Qinghai-Xizang Plateau, v. 1, Geology, geological history and origin of Qinghai-Xizang Plateau: Beijing, Science Press; and New York, Gordon and Breach, Science Publishers, Inc., p. 512–520.

Wang Jutian, 1984, A seismic stratigraphic study in Biyang depression and further exploration of oil and gas: Petroleum Exploration and Development, v. 11, no. 2, p. 17–25.

Wang Liang-cheng and Wang Dong-an, 1981, Character of the sedimentary facies-belts and the sedimentary model of the subsidence belt in the Yarlung Zangbo-Xiangquan River. *In*: Liu Dong-sheng (ed.) Geological and ecological studies of Qinghai-Xizang Plateau, v. 1, Geology, geological history and origin of Qinghai-Xizang Plateau: Beijing, Science Press; and New York, Gordon and Breach, Science Publishers, Inc., p. 599–609.

Wang Naiwen, 1989, Micropaleontological study of lower Palaeozoic siliceous sequences of the Yangtze Platform and Eastern Qinling Range: Jour. Southeast Asian Earth Sci., v. 3, nos. 1–4, p. 141–161.

Wang Nai-wen (Naiwen), 1983, The Tethyan Jurassic stratigraphy of China: Contribution to the Geology of the Qinghai-Xizang (Tibet), v. 3, p. 82–86 (in Chinese).

Wang Ningguo, 1982, The Aerjin arcuate structural zone and its relationship with oil and gas: Acta Petrolei Sinica, v. 3, no. 2, p. 21–27.

Wang Ping, 1981, Occurrence of oil in the complex fault blocks in the Jiyang Sag: Acta Petrolei Sinica, v. 2, no. 3, p. 37–44.

Wang Pinxian, 1985, *Marine micropaleontology of China*: New York, Springer-Verlag, 370 p.

Wang Pinxian, Min Qiubao, Bian Yunhua and Cheng Xinrong, 1981, Strata of Quaternary transgressions in East China; a preliminary study: Acta Geologica Sinica, v. 55, no. 1, p. 1–13.

Wang Qijun, Zhu Shuian, Gao Pinwen and Li Chunju, 1983, Sedimentary environment and characteristics of parent oil-forming matter and resrvoir rock of Biyang depression: Earth Science - Journal of Wuhan College of Geology, no. 2 (tol. 20), p. 135–146.

Wang Quan and Liu Xueya, 1976, Paleo-oceanic crust of the Chilienshan region, western China and its tectonic significance: Scientia Geologica Sinica, No. 1, p. 42–55.

Wang Quan and Liu Xueya, 1986, Paleoplate tectonics between Cathaysia and Angarland in Inner Mongolia of China: Tectonics, v. 5, no. 7, p. 1073–1088.

Wang, S., Huang, L. J. and Li, H. W., 1978, Early Tertiary charophytes from the coastal region of Bohai: Beijing, Science Press, 49 p. (in Chinese with English abstract).

Wang Shangwen (ed.) 1983, *The petroleum geology of China*: Beijing, Petroleum press, 348 p. (in Chinese).

Wang Shangwen, Hu Wenhai and Tan Shidian, 1982, Habitat of oil and gas fields in China: Oil and Gas Jour., v. 80, no. 24, p. 119–121, 124–128.

Wang Shizhao, 1982, On The geologic type and genesis of the circular images of Nanpanjiang area and its relationship to petroleum prospecting: Oil and Gas Geology, v. 3, no. 1, p. 81–87.

Wang Tonghe, 1986, Tectonical flow of basins in the west of Daxinganling: Acta Petrolei Sinica, v. 7, no. 3, p. 29–37.

Wang Tonghe, 1987, Tectonic evolution and oil distribution of pull-apart basins in the Hexi Corridor region: Oil and Gas Geology, v. 8, no. 3, p. 271–280.

Wang Wuhe and Lu Xiliang, 1985, Brief introduction of the overtrust belt in the south of Jiuxi basin (Jiuxi subbasin of Juiquan basin): Oil and Gas Geology, v. 6, no. 3, p. 341–343.

Wang Xie-pei, Fei Qi and Zhang Jia-hua, 1981, Mechanism of the formation of diapiric structural traps in Dongying depression, Shandong Province: Acta Petrolei Sinica, v. 2, no. 3, p. 13–22.

Wang Xie-pei, Fei Qi and Zhang Jia-hua, 1985, Cenozoic diapiric traps in eastern China: Am. Assoc. Petrol. Geol. Bull., v. 69, no. 12, p. 2098–2109.

Wang Xie-pei, Fei Qi, Zhang Jia-hua, Wang Shu-zhen and Yang Xiang-min, 1981, The tectonic framework and mechanism of formation of local structural traps in the oil-gas-bearing basins in northern Jiangsu: Petroleum Exploration and Development, v. 8, no. 3, p. 1–15.

Wang Yichang and Tan Shidian, 1981, The evolution and hydrocarbon-bearing prospects of Tarim plate structure: Petroleum Exploration and Development, v. 8, no. 3, p. 1–9, 59.

Wang Yichang and Wan JingLian, 1984, Results of a detailed gravity survey in a mature exploration area: Oil Geophysical Prospecting, v. 19, no. 2, p. 148–155.

Wang Yi-gang, 1981, Special remarks on geological history of the Himalayas. *In*: Liu Dong-sheng (ed.) Geological and ecological studies of Qinghai-Xizang Plateau, v. 1, Geology, geological history and origin of Qinghai-Xizang Plateau: Beijing, Science Press; and New York, Gordon and Breach, Science Publishers, Inc., p. 297–306.

Wang Yigang and Sun Dongli, 1983, A survey of the Jurassic system of China: Canadian Jour. Earth Science, v. 20, no. 11, p. 1646–1656.

Wang Yigang and Sun Dongli, 1985, The Triassic and Jurassic paleogeography and evolution of the Qinghai-Xizang (Tibet) Plateau: Canadian Jour. Earth Science, v. 22, p. 195–204.

Wang Yigang, Chen Chuchen, He Guoxiong and Chen Jinhua, 1981, An outline of the marine Triassic in China: International Union of Geol. Sciences, Publication no. 7, 21 p.

Wang, Y. and Yu, C. M., 1962, *Devonian of China*: Peking, Science Press, 140 p. (in Chinese).

Wang Yu, Boucot, A. J., Rong Jiayu and Yang Xuechang, 1984, Silurian and Devonian biogeography of China: Geol. Soc. America Bull., v. 95, no. 3, p. 265–279.

Wang Yu, Jin Yu-gan (Yugan), Liu Di-yong (diyong), Su Han-kui (Hankui), Rong Jia-yu (Jiayu), Liao Zhuo-ting (Zhuoting), Sun Dong-li (Dongli) and Yang Xue-chang (Huechang), 1981, Stratigraphic distribution of Brachiopoda in China. *In*: Teichert, C. et al. (eds.) Paleontology in China, 1979: Geol. Soc. America, Spec. Paper 187, p. 97–105.

Wang Yu-jing and Mu Xi-nan, 1983, Upper Carboniferous and Lower Permian strata in the Gondwana-Tethys province of Xizang (Tibet): *Palaeontologica Cathayana* 1, p. 411–419.

Wang Yunliang, S. S. Hughes, Tong Chunhang, Xiong Shunhua, Li Juchu, Zhou Rongsheng and Li Jianlin, 1989, Geochemistry and petrology of Emeishan basalts and subcontinental mantle evolution in southwestern China: Chinese Journal of Geochemistry, v. 8, no. 1, p. 37–53.

Wang Zengji, 1989, Carboniferous coral fauna provinces of China: Jour. Southeast Asia Earth Sci., v. 3, nos. 1–4, p. 163–169.

Wang Zhiping and Yang Fengqing, 1988, Carboniferous paleobiogeography and paleoclimatology of China: Earth Science, Journal of Wuhan College of Geology, v. 13, no. 5, p. 495–502.

Wang Zhitian and Qin Kezhang, 1988, Types, metallogenic environments and characteristics of temporal and spatial distribution of copper deposits in China: Acta Geologica Sinica, v. 62, no. 3, p. 257–267.

Wang Zuzhi, 1984, The distribution of the northeast structural axes in Qaidam basin and their action in oil and gas exploration: Petroleum Expl. and Devlp., v. 11, no. 5, p. 9–15.

Waterhouse, J. B., 1983, How wide was the Tethyan 'ocean'? *in* Gupta, V. J. (ed.) Contributions to Himalayan geology, v. 2, Stratigraphy and structure of Kashmir and Ladakh Himalaya: Delhi, Hindustan publishing Corporation, p. 246–249.

Wei Siyu and Deng Xiaoyue, 1989, Geothermal activity, geophysical anomalies and the geothermal state of the crust and upper mantle in the Yarlung Zangbo river zone: Tectonophysics, v. 159, p. 247–254.

Wei Si-yu, Teng Ji-wen, Yang Bing-ping and Hu Zhong-yi, 1981, Distribution of geothermal activity and characteristics of geophysical fields on the Xizang Plateau. *In*: Liu Dong-sheng (ed.) Geological and ecological studies of Qinghai-Xizang Plateau, v. 1, Geology, geological history and origin of Qinghai-Xizang Plateau: Beijing, Science Press; and New York, Gordon and Breach, Science Publishers, Inc., p. 865–874.

Weippert, D., Wittekindt, H. and Wolfart, R., 1970, Zur geologischen Entwicklung von Zentral - und Sudafghanistan: Hannover, Beihefte zum Geologischen Jahrbuch, v. 91, p. 1–99.

Weller, J. M., 1944, Outline of Chinese geology: Am. Assoc. Petrol. Geol. Bull., v. 28, no. 10, p. 1417–1429.

Weller, J. M., 1944, Petroleum possibilities of Red basin of Szechuan (Sichuan) Province, China: Amer. Assoc. Petrol. Geol. Bull., v. 28, no. 10, p. 1430–1439.

Wen Shixuan, 1984, Paleobiogeography of Qinghai-Xizang Plateau, evidence for continental drift. *In*: Liu Dong-sheng (ed.) Geological and ecological studies of Qinghai-Xizang Plateau, v. 1, Geology, geological history and origin of Qinghai-Xizang Plateau: Beijing, Science Press; and New York, Gordon and Breach, Science Publishers, Inc., p. 149–157.

Wen Shi-xuan, Zhang Bing-gao, Wang Yi-gang, Su Dong-li, Wang Yu-jing, Chen Chu-zhen, Dong De-yuan, Liao Wei-hua, Chen Ting-en, He Guo-xiong, Mu Xi-nan, Yin Ji-xiang and Wu Hao-ruo, 1981, Sedimentary development and formation of stratigraphic region in Xizang. *In*: Liu Dong-sheng (ed.) Geological and ecological studies of Qinghai-Xizang Plateau, v. 1, Geology, geological history and origin of Qinghai-Xizang Plateau: Beijing, Science Press; and New York, Gordon and Breach, Science Publishers, Inc., p. 119–130.

Willis, Bailey, 1907–1913, Research in China; Washington D. C., Carnegie Institution of Washington: Atlas; Research in China, 1903–1904, Geographical and geological maps, Bailey Willis, geologist in charge, 1906; v. 1, pt. 1, Descriptive topography and geology, v. 1, pt. 2, Petrography and zoology, Eliot Balckwelder, Syllabary of Chinese sounds, Friedrich Hirth; v. 2, Systematic geology, Bailey Willis; v. 3., The Cambrian faunas of China, by Charles D. Walcott, A report on upper Paleozoic fossils collected in China in 1903–1904, by George H. Girty.

Wintle, A. and Derbyshire, E., 1985, The loess region of China: Nature, v. 318, no. 6043, p. 234.

Wise, D. U., 1969, Regional and sub-continental sized fracture systems detectable by topographic shadow techniques: Geological Survey of Canada Paper 68-52, p. 176–199.

Wise, D. U., 1982, Linesmanship and the practice of linear geo-art: Geological Society of America Bulletin, v. 93, no. 9, p. 886–888.

Wolfart, R., 1967, Zur Entwicklung der palaozoischen Tethys in Vorderasien: Erdol and Kohle-Erdgas-Petrolchemie, v. 20, p. 168–180.

Wolfart, R., 1970, Fauna, Stratigraphie und Palaogeographie des Ordoviziums in Afghanistan: Hannover, Beihefte zum Geologischen Jahrbuch, v. 89, p. 1–169.

Wolfart, R., 1983, The Cambrian System in the Near and Middle East: Correlation chart and Explanatory notes: Ottawa, International Union of Geological Sciences Publication no. 15, p. 72.

Wolfart, R. and Kursten, M., 1974, Stratigraphie und Palaogeographie des Kambriums in mittleren Sud-Asien (Iran bis Nord-Indien): Geologisches Jahrbuch, Reihe B, v. 8, p. 185–234.

Wood, B. G. M., 1985, The mechanics of progressive deformation in crustal plates - a working model for southeast Asia: Geological Society of Malaysia Bulletin, v. 18, p. 55–99.

Wu Chungyu, 1982, Mesozoic and Cenozoic lake deltas and oil-gas distribution in eastern China: Beijing, Scientific Research Inst. for Petroleum Exploration and Development, reprint, p. 25.

Wu Hao-ruo, 1981, Geological outline and geological history of the Yarlung Zangbo suture zone. *In*: Liu Dong-sheng (ed.) Geological and ecological studies of Qinghai-Xizang Plateau, v. 1, Geology, geological history and origin of Qinghai-Xizang Plateau: Beijing, Science Press; and New York, Gordon and Breach, Science Publishers, Inc., p. 567–577.

Wu Jilong and Lu Xueqi, 1986, The Structural framework of eastern Jizhong depression and its evolution: Oil and Gas Geology, v. 7, no. 1, p. 20–31.

Wu Jinmin, 1988, Cenozoic basins of the South China Sea: Episodes, v. 11, no. 2, p. 91–96.

Wu Liren, 1988, Mesozoic granitoids and two great genetic types of porphyry ore deposits in the east of China: Scientia Geologica Sinica, no. 4, p. 313–323.

Wu Liren and Qi Jinying, 1985, Mesozoic volcanic rocks in the lower reaches of the Chanjiang River: Tectonophysics, v. 112, p. 519–532.

Wu Qingfu, 1987, Growth thrust fault belts and their oil and gas exploration: Oil and Gas Geology, v. 8, no. 2, p. 119–125.

Wu Qingfu, 1987, The mechanism of the formation of growth-thrust fault zones and their problems of oil and gas exploration in the compressive basins, western China: Acta Petrolei Sinica, v. 8, no. 4, p. 1–8.

Wu Wangshi and Liao Weihua, 1981, Stratigraphic distribution of fossil corrals and their palaeobiogeographic provinces in Xizang. *in* Liu Dong-sheng (ed.) Geological and ecological studies of Qinghai-Xizang Plateau, v. 1, Geology, geological history and origin of Qinghai-Xizang Plateau: Beijing, Science Press; and New York; Gordon and Breach, Science Publishers, Inc., p. 165–170.

Wu Wang-shi (Wangshi) and Zhao Jia-min (Jiamin), 1985, Carboniferous coral assemblages of China. *In*: Washington, D. C. and Urbana-Champaign, Illinois, International congress on Carboniferous stratigraphy and geology, 9th, 1979: Compte Rendu, v. 5, paleont., paleoecol., paleogeog., p. 200–204.

Wu Zhenline and Lu Jiuhua, 1983, Location of faults at Tanlu in the Liaohe depression and their control over oil accumulations of the region: Acta Petrolei Sinica, v. 4, no. 1, p. 17–22.

Wu Zhenquan, 1983, A highly explored (mature) area, Jiuquan basin and its exploration prospects: Amer. Assoc. Petrol. Geologists, Technical Session, Dallas, 1983, typescript, 24 p.

Wu Zhizhong, 1985, A model of continental crust collision in relatin to oil exploration in the northwestern margin of Junggar basin: Oil and Gas Geology, v. 6, no. 2, p. 211–219.

Xiang Dingpu, 1982, The characteristics of geological structures in the Chillenshan (Qilian Shan) region, China: Scientia Geologica Sinica, no. 2, p. 364–370.

Xiang Liwen, 1987, The Permian Tethys in China. *In*: McKenzie, K. G. (ed.) Shallow Tethys 2: Boston, A. A. Balkema, p. 127–130.

Xiao Xuchang and Wang Fangguo, 1984, An introduction to the ophiolite of China: Bull. Chinese Acad. Geol. Sci., no. 9, p. 19–30.

Xiao Xuchang, Qu Jingchuan, Chen Guoming, Zhu Zhiahi and Gu Qinge, 1980, Ophiolites of the Tethys-Himalayas of China and their tectonic significance. *In*: Auboin, J., co-ord; Debeimas, J. and Latreille, N. (eds.) Geology of the Alpine chains born of the Tethys; Bur. Rech. Geol. Minieres, Mem. 115, p. 149–151.

Xiaoguang Tong and Huang Zuan, 1988, Buried-hill discoveries in Damintun depression of North China basin (abst.): Am. Assoc. Petrol. Geol. Bull., v. 72, no. 2, p. 260.

Xie Gongjian, 1981, Occurrence of the oil pool in the metamorphosed basement in Yaerxia in the western part of the Jiuquan basin: Acta Petrolei Sinica, v. 2, no. 3, p. 23–30.

Xie Gongjian, 1983, Oil prospect in Huahai-Jinta basin: Oil and Gas Geology, v. 4, no. 3, p. 318–323.

Xie Mingqian, 1987, Disintegration of fault block and formation and development of hydrocarbon basin in North China: Oil and Gas Geology, v. 8, no. 4, p. 391–401.

Xie Qingdao and Peng Funan, 1981, On the evolution of the Ryukyu arc and the basin of the East China Sea: Scientia sinica, v. 24, no. 11, p. 1553–1565.

Xie Taijun, Qi Zuoming and Zhu Zhendong, 1983, Salt structures in Jianghan basin and their control to oil and gas accumulation: Petroleum Exploration and Development, v. 10, no. 6, p. 1–8.

Xinan Mu and Riding, R., 1988, Silicification of Permian calcareous algae from Nanjing, China: Geological Magazine, v. 125, no. 2, p. 123–139.

Xing Yusheng, 1984, The Sinian and its position in geological time scale: Moscow, International Geological Congress, 27th, v. 1, Stratigraphy, p. 271–287.

Xing Yusheng and Liu Guizhi, 1982, Late Precambrian microflora of China and its stratigraphical significance: Bull. Chinese Acad. Geol. Sciences, v. 4, p. 55–67.

Xing Yusheng and Luo Huilin, 1984, Precambrian-Cambrian boundary candidate, Meishucun, Jinning, Yunnan, China: Geological Magazine, v. 121, no. 3, p. 143–154.

Xing Yusheng, Ding Qixiu and Luo Huilin, 1982, Biotic characteristics of the Sinian-Cambrian boundary beds in China and the boundary problems. *In*: Cowie, J. W. (ed.) The Precambrian-Cambrian boundary: Precambrian Research, v. 17, no. 2, p. 77–85.

Xing Yusheng, Ding Qixiu, Luo Hullin, He Tinggui and Wang Yangeng, 1984, The Sinian-Cambrian boundary of China and its related problems: Geological Magazine, v. 121, no. 3, p. 155–170.

Xing Yusheng, Duan Chenghua, Liang Yuzuo and Cao Renguan, 1985, *Late Precambrian palaeontology of China*: Beijing, Geological Publishing House, 288 p. (in Chinese, English summary).

Xixi Zhao and Coe, R. S., 1987, Paleomagnetic constraints on the collision and rotation of North and South China: Nature, v. 327, no. 6118, p. 141–144.

Xizang Scientific Expedition, Academia Sinica (Compilers), 1987, Stratigraphy of the Mount Qomolangma region: China, Beijing, 225 p. (translated by Qi Baoji).

Xu Huaida, Lu Weiwen, Wang Shifeng and Wan Jingping, 1983, Paleogene sedimentary systems and direction of searching for oil and gas in Bohaiwan basin: Acta Geologica Sinica, v. 57, no. 3, p. 243–253.

Xu Jiawei, Tong Weixing, Zhu Guang, Lin Shoufe and Ma Gaufeng, 1989, An outline of the pre-Jurassic tectonic framework in east Asia: Jour. Southeast Asian Earth Sci., v. 3, nos. 1–4, p. 29–45.

Xu Jiawei, Zhu Guang, Tong Weixing, Cui Kerei and Liu Qing, 1987, Formation and evolution of the Tancheng-Lujiang wrench fault system: a major shear system to the northwest of the Pacific Ocean: Tectonophysics, v. 134, p. 272–310.

Xu Ren, 1978, On the palaeobotanical evidence for continental drift and the Himalayan uplift: The Palaeobotanist, v. 25, p. 131–145.

Xu Ren, 1981, Vegetational changes in the past and the uplift of Qinghai-Xizang Plateau. *In*: Liu Dong-sheng (ed.) Geological and ecological studies of Qinghai-Xizang Plateau, v. 1, Geology, geological history and origin of Qinghai-Xizang Plateau: Beijing, Science Press, New York; Gordon and Breach, Science Publishers, Inc., p. 139–148.

Xu Shice and Wang Hengjian, 1981, Deltaic deposits of a large lake basin. *In*: Mason, J. F. (ed.) 1981, Petroleum geology in China: Tulsa, OK, PennWell Publ. Co., p. 202–213.

Xu Shirong and Xu Jinhua, 1986, The new results of seismic exploration in Huayingshan fault zone: Acta Petrolei Sinica, v. 7, no. 3, p. 39–48.

Xu Shu-ying, 1981, The evolution of the palaeogeographic environments in the Tanggula Mountains in the Pliocene-Quaternary. *In*: Liu Dong-sheng (ed.) Geological and ecological studies of Qinghai-Xizang Plateau, v. 1, Geology, geological history and origin of Qinghai-Xizang Plateau: Beijing, Science Press, New York; Gordon and Breach, Science Publishers, Inc., p. 247–255.

Xu Weimin and Xia Yan, 1986, Characteristic of petroleum generation in Jurong basin of Jiangsu Province and its oil-gas prospects: Petroleum Exploration and Development, v. 13, no. 4, p. 11–19.

Xu Zhengyu, 1980, The Tertiary and its petroleum potential in the Lunpola basin, Tibet: Oil and Gas Geology, v. 1, no. 2, p. 153–158.

Xu Zhichuan, 1984, A discussion on the potential oil accumulation of organic reefs in the Early Permian age in an area around Nanpanjiang: Petroleum Exploration and Development, v. 11, no. 2, p. 10–16.

Xue Muqia, ch. ed., 1982, *Almanac of China's economy 1981 with economic statistics for 1949–1980*: Beijing, Modern Cultural Co. Ltd., New York, Harper and Row Publ.; and Eurasia Press, 1, 144 p.

Xue Xiaofeng, 1984, Research on the isotopic age of the Sinian-Cambrian boundary at the Meishucun section in Jinning county, Yunnan Province, China: Geol. Magazine, v. 121, no. 3, p. 171–173.

Xuecheng Yuan, Shi Wang, Li Li and Jieshou Zhu, 1986, A geophysical investigation of deep structure in China. *In*: Barazangi, M. and Brown, L. (eds.) Reflection seismology; a global perspective: Washington, American Geophysical Union, p. 151–160.

Yan Binghen, Lin Liang and Liu Qiusheng, 1985, Seismic facies patterns of the fault-depressed lake basins of eastern China: Acta Petrolei Sinica, v. 6, no. 4, p. 1–11.

Yan Dunshi and Zhai Guangming, 1981, Exploration practice and exploration prospects of the buried-hill oil fields in North China. *In*: Mason, J. F. (ed.) 1981, Petroleum geology in China: Tulsa, OK, PennWell Publ. Co., p. 92–100.

Yan Dunshi, Wang Shangwen and Tang Cho, 1980, Block faulting and formation of oil and gas fields associated with buried hills in Bohai Bay basin: Acta Petrolei Sinica, v. 1, no. 2, p. 1–10.

Yn Jia-quan, Shi Zhen-liang, Wang Si-yun and Huan Wen-lin, 1981a, Recent tectonics in the Qinghai-Xizang Plateau: Acta Geophysica Sinica, v. 24, p. 385–393.

Yan Jia-quan, Shi Zhen-liang, Wang Su-yun and Huan Wen-lin, 1981b, Recent tectonics in the Qinghai-Xizang Plateau. *In*: Liu Dong-sheng (ed.) Geological and ecological studies of Qinghai-Xizang Plateau, v. 1, Geology, geological history and origin of Qinghai-Xizang Plateau: Beijing, Science Press; and New York, Gordon and Breach, Science Publishers, Inc., p. 789–795.

Yan Wenbo, 1984, Preliminary study of oil generation in Baiyanhua Group of Erlian basin: Acta Petrolei Sinica, v. 5, no. 1, p. 29–35.

Yan Yuqui, Yang Shizhua, Hu Boliang, Wen Changqing and Jin Huijuan, 1983, Some problems concerning petroleum geology of the Tarim basin: Scientia sinica, Series B, v. 26, no. 11, p. 1201–1215.

Yan Zhuyun, Tang Kedong, Bei Jingwen and Ma Youchen, 1989, High pressure metamorphic rocks and their tectonic environment in northeastern China: Jour. Southeast Asian Earth Sci., v. 3, nos. 1–4, p. 303–313.

Yang Bing-ping, Shu Pei-yi and Zhang Zhi-he, 1981, Some characteristics of small earthquakes in the Zhamo and Damxung districts in Xizang Plateau. *In*: Liu Dong-sheng (ed.) Geological and ecological studies of Qinghai-Xizang Plateau, v. 1, Geology, geological history and origin of Qinghai-Xizang Plateau: Beijing, Science Press, and New York, Gordon and Breach, Science Publishers, Inc., p. 797–805.

Yang Hua, 1983, Magnetic field and the future of petroleum prospecting in the Tarim basin. *In*: Zhu Xia (ed.) Tectonics and evolution of the Mesozoic and Cenozoic basins of China: Beijing, Science Press, p. 214–219.

Yang Huazhou, 1989, The aluminous deposits of China: Journal of

Changchun College of Geology, v. 19, no. 1, p. 31-42 (in Chinese, English abstract).

Yang Jialu, 1989, A survey of Cambrian palaeo-tectonogeography of East Qinlin: Jour. Southeast Asian Earth Sci., v. 3, nos. 1-4, p. 203-210.

Yang Jiliang, 1983, Characteristics of oil pools in Daqing oilfield and the accumulation regularities of oil and gas in Songliao basin: Oil and Gas Geology, v. 4, no. 2, p. 171-180.

Yang Jiliang and Zhang Shuying, 1987, Turbidity deposit of Cretaceous in Songliao basin: Petroleum Exploration and Development, v. 14, no. 2, p. 10-15.

Yang Jingzhi, Sheng Jinzhang, Yang Jingshi, Wu Wangshi, Zhang Linxin, Liao Zhuoting and Ruan Yiping, 1962, *The Carboniferous system of China*: Beijing, Science Press, 112 p. (in Chinese).

Yang Junjie and Zhang Borong, 1986, Torsional break type graben and the associated overthrust belt - Yinchuan graben and Hengshanbu overthrust belt: Petroleum Exploration and Development, v. 13, no. 2, p. 1-8.

Yang Kesheng, 1987, The types and the evolution of structures in Hexi Corridor petroliferous basins: Petroleum Expl. and Devlp., v. 14, no. 5, p. 1-6.

Yang Kesheng and Wang Tonghe, 1985, Contemporaneous thrust fault in Jizhong depression and its significance in petroleum exploration: Petroleum Exploration and Development, v. 12, no. 4, p. 1-7.

Yang Qi, Ren Deyi and Pan Zhigui, 1982, Preliminary investigation of the metamorphism of Chinese coals: International Jour. Coal Geology, v. 2, p. 31-48.

Yang Runlin, 1984, On the Miocene sedimentary environment in western Tarim basin in view of microfossil assemblages: Oil and Gas Geology, v. 5, p. 148-155.

Yang Runlin, 1985, The Paleocene Series in eastern China: Oil and Gas Geology, v. 6, no. 4, p. 419-425.

Yang Shihpu, Hou Hungfei, Gao Lianda, Wang Zengji and Wu Xianghe, 1980, The Carboniferous system of China: Acta Geologica Sinica, v. 54, no. 3, p. 167-175.

Yang Shihpu, Lin Yintang, Tang Guanxiu, Wang Zhiping and Wu Shizhong, 1983, The Lower Carboniferous (Fengninian) of China. *In*: Martinez-Diaz, C., gen. ed., 1983, The Carboniferous of the World, v. 1, China, Korea, Japan, and S. E. Asia: IUGS Publication no. 16, p. 16-56.

Yang Shihpu, P'an Kiang and Hou Hungfei, 1981, The Devonian system in China: Geol. Magazine, v. 118, no. 2, p. 113-138.

Yang Wanli, 1985, Daqing oil field, People's Republic of China, a giant field with oil of nonmarine origin: Am. Assoc. of Petroleum Geol. Bull., v. 69, no. 7, p. 1101-1111.

Yang Wanli, 1986, Basic characteristics of hydrocarbon distribution of Songlia basin and its exploration potential. *In*: Tu Guangzhi (ed.) 1986, Advances in science of China, Earth Sciences, v. 1: Beijing, Science Press, and New York, John Wiley and Sons, p. 575-598.

Yang Weiran, Guo Tieying, Lu Yuanliang, Zheng Jiandong, Su Jian and Ma Xingyuan, 1984, 'Opening' and 'closing' in the tectonic evolution of China: Earth Science Journal, Wuhan College of Geology, no. 3, p. 39-56.

Yang Wenkuan, Chen Zhengfu, Luo Yangchen and Tang Feilong, 1984, Occurrence of hydrocarbons in the Paleozoic carbonate rocks of marine origin in the Xiangzhong depression: Acta Petrolei Sinica, v. 5, no. 4, p. 9-18.

Yang Zhende and Xie Mingqian, 1984, On the tectonic evolution of East Asia in late Precambrian: Scientia Geologica Sinica, no. 4, p. 373-383.

Yang Zhilin, 1984, A discussion of the sedimentary environment in Qaidam basin's Yiliping depression: Petroleum Expl. and Devlp., v. 11, no. 4, p. 31-38.

Yang Zunyi, Cheng Yuqi and Wang Hongzhen, 1986, *The geology of China*: Oxford Monographs on Geology and Geophysics 3: Oxford, Clarendon Press, 303 p.

Yang Zunyi, Feng Shaonan and Lin Jiaxing, 1984, New recognition of the boundary between the Lower and Upper Permian in South China: Earth Science-Jour. Wuhan College of Geology, Total 25, no. 2, p. 1-6.

Yang Zunyi, Li Zishun, Qu Lifan, Lu Chongming, Zhou Huiqin, Zhou Tongshun, Liu Guifang, Liu Benpe and Wu Ruitang, 1982, The Triassic System of China: Acta Geologica Sinica, v. 56, no. 1, p. 1-21.

Yang Zunyi, Nie Zetong, Wu Shunbao and Liang Dingyi, 1986, Cretaceous rudists from Ngari, Tibet Autonomous Region, China and their geologic significance. *In*: Huang Jiqing, ed. in chief, Proceedings of the Symposium on Mesozoic and Cenozoic geology - in connection in 60th anniversary of the geological Society of China: Beijing, Geological Publ. House, p. 231-242.

Yang Zunyi, Wu Shunbao and Yang Fengqing, 1980, Permo-Triassic boundary in the marine regime of South China: Wellington, New Zealand, fifth Int. Gondwana Symposium, p. 71-77.

Yang Zunyi, Wu Shunbao and Yang Fengqing, 1981, Permo-Triassic boundary in the marine regime of South China: Earth Sciences, no. 1 of 1981, p. 4-15.

Yang Zuxu, Long Xinyen, Chen Fengchi and Cui Xuezhou, 1983, Structural zones and distribution of oil and gas fields in the Songliao basin: Acta Petrolei Sinica, v. 4, no. 2, p. 1-8.

Yao Yimin and Xiang Weida, 1986, A discovery of Eocene Siphonoalgoclastic dolomite from Well Zhuang 65, Gubei, Shandong Province and its significance in oil exploration: Petroleum Exploration and Development, v. 13, no. 3, p. 25-29.

Yao Zhenkai, 1984, Geological-geochemical condition of mineralization of stratabound carbonate uranium deposits in China: Acta Sedimentologica Sinica, v. 2, no. 1, p. 65-75.

Ye Dequan and Zhang Ying, 1981, Application of fossil ostracoda to the correlation of the oil-bearing deposits of Daqing: Acta Palaeontologica Sinica, v. 20, no. 1, p. 81-87.

Ye Dequan, Zhao Chuanben and Zhang Ying, 1980, Paleontological characteristics of continental fossils from Cretaceous in Songliao basin: Acta Petrolei Sinica, Daqing Spec. Issue (October), p. 49-57.

Ye Du-zheng and Gao You-xi, 1981, The seasonal variations of the heat source and sink over Qinghai-Xizang Plateau and its role in the general circulations. *In*: Liu Dong-sheng (ed.) Geological and ecological studies of Qinghai-Xizang Plateau, v. 1, Geology, geological history and origin of Qinghai-Xizang Plateau: Beijing, Science Press; and New York, Gordon and Breach, Science Publishers, Inc., p. 1453-1461.

Ye Duzheng, Fu Congbin, Chao Jiping and M. Yoshino (eds.) 1987, *The climate of China and global climate*: Berlin, Springer-Verlag: Beijing, China Ocean Press, 441 p.

Ye Hefei, 1985, The geological application of landsat images in Lunpola basin, northern Tibet: Petroleum Expl. and Devlp., v. 12, no. 1, p. 8-14.

Ye Hong, Shedlock, K. M., Hellinger, S. J. and Sclater, J. G., 1985, The North China basin: an example of a Cenozoic rifted intraplate basin: Tectonics, v. 4, no. 2, p. 153-169.

Ye Hong, Zhang Botao and Mao Fungying, 1987, The Cenozoic tectonic evolution of the Great North China: two types of rifting and crustal necking in the Great North China and their tectonic implications: Tectonophysics, v. 133, no. 3-4, p. 217-227.

Ye Hong, Zhang Wen-yu, Yu Zhi-shui and Xia Qin, 1981, The seismicity and recent movement in the Himalayan region. *In*: Liu Dong-sheng (ed.) Geological and ecological studies of Qinghai-Xizang Plateau, v. 1, Geology, geological history and origin of Qinghai-Xizang Plateau: Beijing, Science Press; and New York, Gordon and Breach, Science Publishers, Inc., p. 65-80.

Ye, T. -D., Ting, L. -S. and Chang, Y., 1976, *Fossil Ostracoda of Cretaceous age from the Sungliao basin*: Peking, Science Press, 102 p. (in Chinese).

Yen, T. P., 1953, On the occurrence of the late Paleozoic fossils in the metamorphic complex of Taiwan: Bull. Geol. Survey of Taiwan, no. 4, p. 23-27.

Yen, T. P., Sheng, C. C. and Keng, W. P., 1951, The discovery of fussuline limestone in the metamorphic complex of Taiwan: Geological Survey of Taiwan Bulletin, no. 3, p. 23-25., 45.

Yi Ronglong and Jiang Shengbang, 1980, The oil and gas potential in the Paleozoic of the Tarim basin: Oil and Gas Geology, v. 1, no. 1, p. 26-36.

Yin, J. -X. and Fang, Z. -J., 1973, Marine Jurassic in western Yunnan: Scientia Geologica Sinica, v. 3, p. 236–243.

Yin Jixiang, Chang Chengfa, Deng Wanming, Zhang Qi, Zheng Xilan, Pan Yusheng, Wu Haoruo, Sun Yiyin, Teng Jiwen, Wen Shixuan, Zhang Yuquan, Wang Fubao, Zhang Zhifel and Zhang Qingson, 1980, Geology and geophysics. *In*: Zhang Rongzu (ed.) A scientific guidebook to south Xizang (Tibet), Symposium on Qinghai-Xizang (Tibet) Plateau, Beijing, China, June 2–June 14, 1980: Beijing, Academia Sinica, p. 17–104.

Yin Ji-xiang, Wu Hao-ruo, Sun Yi-yin, Wen Chuan-fen and Liu Tung-sheng (Dong-sheng), 1983, Study on Himalayan stratigraphy in south Xizang Tibet. *In*: Gupta, V. J. (ed.) Contributions to Himalayan geology, v. 2, Stratigraphy and structure of Kashmir and Ladakh Himalaya: Delhi, Hindustan Publishing Corporation, p. 59–74.

Yin Xiuhua, Shi Zhihong, Liu Zhangpo and Zhang Yumei, 1983, The basic features of the regional gravity field on the Chinese continent: Chinese Geophysics, v. 2, no. 2, p. 365–372 (in Chinese).

Yin Zanxun (Yin, T. H.), 1949, Tentative classification and correlation of Silurian rocks of South China: Geol. Soc. China, Bull., v. 29, p. 1–62.

Yin Zan-xun (Zanxun), 1966, China in the Silurian period: Geol. Soc. Australia, Jour., v. 13, pt. 1, p. 277–297.

Yin Zhenxun (Yin, T. H.), Zhang Xhouxin et al., 1966, *Stratigraphic lexicon of China, VII, Carboniferous system*: Peking (Beijing), Science Press, I-VIII, 232 p.

Yin Ziming, 1988, The texture and structure features of Carboniferous bauxite in China and its emplacement mechanism: Journal of Central-South Institute of Mining and Metallurgy, v. 19, no. 4, p. 357–363 (in Chinese).

Yong Jiafu, 198, Depositional types of sand body and reservoir characteristics in Shengtuo oil field: Acta Petrolei Sinica, v. 9, no. 2, p. 43–56.

Yong Tianshou, 1987, Carboniferous oil pool of basement weathered crust in Junggar basin: Acta Petrolei Sinica, v. 8, no. 2, p. 33–38.

York, J. E., Cardwell, R. and Ni, J., 1976, Seismicity and Quaternary faulting in China: Seis. Soc. America, Bull., v. 66, p. 1983–2001.

Young, G. M., Chen, J. -B., Zhang, H. -M. (eds.) 1985, *Late Precambrian geology*: Precambrian Research, v. 29, nos. 1–3, 329 p.

Yu Changmin, Wang Chengyuan, Ruan Yiping, Yin Bao'an, Li Zhenliang and Wei Weilie, 1987, A desirable section for the Devonian-Carboniferous boundary stratotype in Guilin, Guangxi, South China: Scientia Sinica, Series B, v. 30, no. 7, p. 751–765.

Yu Deyuan, 1959, *Geology of China*: Peking (Beijing), Geological Publishing House (in Chinese).

Yu Jiaren and Tang Jieting, 1988, Preliminary study on the A-Bei andesite reservoir in Erlian basin: Petroleum Expl. and Devlp., v. 15, no. 1, p. 38–44.

Yu Jingshan and Li Zishun, 1989, *Monotis* fauna of China and its palaeobiogeographic significance: Jour. Southeast Asia Earth Sci., v. 3, nos. 1–4, p. 179–186.

Yu Wen, 1983, Sequence and distribution of gastropod associations of the Upper Cretaceous and late Tertiary of China: Nanjing Institute of Geology and Palaeontology, Academia Sinica, Bull., v. 6, p. 321–353 (in Chinese, English summary).

Yu Wen, Mao, X. L., Chen, Z. Q. and Huang, L. S., 1978, Early Tertiary gastropod fossils from coastal region of Bohai: Beijing, Science Press, p. 145 (in Chinese with English abst.).

Yu Zhang, 1984, Sichuan mineral resources; reserves of 80 different industrial-use minerals; Sichuan Ribao, 11 June, p. 2.

Yu Zhihong and Liu Zhongping, 1984, The relationship between linear structures and mineral resources in China: Inst. Mineral Deposits, Bull., Chinese Acad. Geol. Sciences, Ser. III, no. 1, p. 1–13 (in Chinese, English abstract).

Yu Zhihong, Liu Zhongping, Wan Defang and Fu Yijie, 1981, Some tectonic features of the linear structures on the territory of China; and explanation of the 1:6,000,000 scale tectonic map of the linear structures of China: Beijing, Cartographic Publishing House, 19 p. (in Chinese, English abstract) (map 1:6,000,000; in Chinese, legend in English).

Yuan Bingcun, Chen Rongshu, Wang Jingping, Cai Xifen, Liu Anlin and Zhai Chaiyu, 1983, Early Eocene sedimentary environment of Jianghan basin and its oil and gas potential: Oil and Gas Geology, v. 4, no. 4, p. 395–402.

Yuan Jianqi and Gao Jianhua, 1987, Structural control of Meso-Cenozoic salt basins in China: Earth Science, Journal of Wuhan College of Geology, v. 12, no. 4, p. 337–347.

Yuan Ke-xing (Gexing) and Zhang Sen-gui (Sengui), 1981, Lower Cambrian archaeocyathid assemblages of central and southwestern China: Geological Society of America, Special Paper 187, p. 39–52.

Yuan Ke-xing (Gexing) and Zhang Sen-gui (Sengui), 1983, Archaeocyatha biogeographic provinces of the Lower Cambrian of China: Nanjing Institute of Geology and Palaeontology, Academia Sinica, Bull., v. 6, p. 100–116 (in Chinese, English abstract).

Yuan, P. L. and Young, C. C., 1934, On the discovery of a new *Dicynodon* in Sinkiang (Xinjiang): Geological Society of China, Bull., v. 13, no. 4, p. 563–573.

Yuan Xuecheng, Wang Si, Li Li and Zhu Jieshou, 1986 (true publication date, November 1985), A geophysical investigation of deep structure in China. *In*: Barazangi, M. and Brown, L. (eds.) Reflection seismology: a global persepctive: American Geophysical Union, Geodynamics Series 13, p. 151–160.

Yue Wenzhe, Wei Naiyi, Song Wei and Liu Fengme, 1986, Sedimentary facies of the Late Carboniferous Weining age and control of the metallogenesis of massive sulfide deposits: Nanjing Institute of Geology and Mineral LResources, Bull., Chinese Academy of Geological Sciences, v. 7, no. 4, p. 19–41 (in Chinese, English abstract).

Yule, Sr. H., ed. and transl., 1926, *The book of Ser Marco Polo the Venetian concerning the kingdoms and marvels of the East* (third edition, revised by H. Cordier): London, John Murray, v. 1, 462 p., v. 2, 662 p., v. 3, 161 p. (see H. cordier for v. 3).

Yule, Sir H. and H. Cordier (eds.) 1903, *The book of Ser Marco Polo the Venetian concerning the kingdoms and marvels of the East*: London, John Murray, v. 1, 462 p., v. 2, 622 p.

Zakharov, S. A. and Porshnyakov, G. S., 1984, The evolution of geologic structures in Soviet Central Asia and their position in the tectonic setting of south-western Asia. *In*: Reports of the 27th International Geological Congress, 1984, Moscow, Colloqium 05, Tectonics of Asia, v. 5, p. 85–99: Moscow, Izdatel'stvo Nauka.

Zeitler, P. K., Johnson, N. M., Naeser, C. W. and Tahirkheli, R. A. K., 1982, Fission-track evidence for Quarternary uplift of the Nanga Parbat region, Pakistan: Nature, v. 298, p. 255–257.

Zha Quanheng, 1984, Jizhong Depression, China – its Geologic Framework, Evolutionary History, and Distribution of Hydrocarbon: Am. Assoc. Petrol. Geol. Bull., v. 68, no. 8, p. 983–992.

Zhai Guang-Ming (Guangming), 1986, Geology and petroleum potential of northwestern China: Am. Assoc. Petrol. Geol. Memoir 40, p. 503–513.

Zhai Guangming and Zha Quanheng, 1982, Buried-hill oil and gas pools in the North China basin: Am. Assoc. Petrol. Geol. Mem. 32, p. 317–335.

Zhai Guangming, Zhang Wenzhao and Hu Chaoyuan, 1982, Oil, gas accumulations in China's continental basins: Oil and Gas Journal, v. 80, no. 50, p. 129–130, 132–136.

Zhan Xie, 1980, Lithofacies and bioherms of Upper Paleozoic in Yunnan-Guizhou-Guangxi Provinces, China: Am. Assoc. petrol. Geol. Bull., v. 65, no. 5, p. 80–7.

Zhang Bing-gao, 1981, latest marine sediments in Xizang and process of early Tertiary regression. *In*: Liu Dong-sheng (ed.) Geological and ecological studies of Qinghai-Xizang Plateau, v. 1, Geology, geological history and origin of Qinghai-Xizang Plateau: Beijing, Science Press; and New York, Gordon and Breach, Science Publishers, Inc., p. 329–332.

Zhang Buchun, Jia Sanfa, Wang Tonghe and Zheng binghua, 1985, Intraplate seismotectonic features of North China: Tectonophysics, v. 117, p. 177–191.

Zhang Delin, 1985, The seismic structural layers and the geological history of Manite depression: Oil Geophysical Prospecting, v. 20, no. 6, p. 605–614.

Zhang Fakui, 1984, The fossil record of Mesozoic mammals of China:

Vertebrate PalAsiatica, v. 22, no. 1, p. 29-38 (in Chinese, English abstr.).

Zhang Fuming and Jian Zongyu, 1981, The sedimentation history and environmental characteristics of the Early Tertiary in Huanghua basin: Oil and Gas Geology, v. 2, no. 2, p. 141-157.

Zhang Fuxiang, 1986, On the basin-marginal ramp model formed by Maoshan nappe structure: Oil and Gas Geology, v. 7, no. 2, p. 149-155.

Zhang Guoqei, Yu Zaiping, Sung Yang, Chen Shunyou, Li Taohang, Xu Feng and Zhang Cherngli, 1989, The major suture zone of the Qinling orogenic belt: Jour. Southeast Asian Earth Sci., v. 3, nos. 1-4, p. 63-76.

Zhang Jinbo, 1985, Method to predict the high yield fractured zone in mixed granite rocks in a buried hill reservoir: Petroleum Exploration and Development, v. 12, no. 4, p. 8-12.

Zhang Jinshan, 1982, Tectonic evolution of Ordos basin and its oil and gas potential: Oil and Gas Geology, v. 3, no. 4, p. 304-315.

Zhang Kai, Luo Zhili, Zhang Qing, Dai Jinxing and Yao Huijun, 1980, classification of basins and potential hydrocarbon resources in China: Acta Petrolei Sinica, v. 1, no. 4, p. 1-18.

Zhang Kai, Luo Zhili, Zhang Qing and Yao Huijun, 1981, On the evolution of the continental plate and the characteristics of oil- and gas-bearing basin: Petroleum Exploration and Development, no. 1, p. 13-25.

Zhang Kai, Zhang Qing, Yao Huijun and Gao Minyuan, 1983, Meso-Cenozoic tectonics and evolution of riftogenic petroliferous basins along China's sea area and its vicinity: Oil and Gas Geology, v. 4, no. 4, p. 353-364.

Zhang Kang, 1983, The evolution properties of a large Mesozoic sedimentary basin in North China fault-block region: Oil and Gas geology, v. 4, no. 2, p. 202-208.

Zhang Kang and Wu Zidian, 1985, Tectonics of the west edge fault belt of Ordos region and its (prospects) in petroleum exploration: Oil and Gas Geology, v. 6, no. 1, p. 71-81.

Zhang Keng (Ch'ang K'eng), Ch'eng Ch'ing-ta and Zabarinskiy, P. O., 1958, Neftanyye i gazovyye mestorozhdeniya Kitayskoy Naradnoy Respubliki: Moscow, Izd. Gos. Nauchno-Technicheskoyo, Neftyanoye i Gorno-Topliynoy Literaturi, p. 112.

Zhang Linxin, 1987, *Carboniferous stratigraphy in China*: Beijing, Science Press, Contribution to the 11th International congress of Carboniferous Stratigraphy and Geology, 160 p.

Zhang Lujin, 1983a, On the character of Permian microflora in the Junggar basin of Xinjiang: Palaeontologia Cathayana, no. 1, p. 327-365.

Zhang Lujin, 1983b, On the age of the Badaowan Formation in northern Xinjiang: Scientia Sinica, Series B, v. 26, no. 7, p. 774-781.

Zhang Peishan and Tao Kejie, 1984, A study on rare earth and alkaline earth double carbonate minerals in China. *In*: Su Zongwei (ed.) Developments in geoscience; contribution to 27th International Geological Congress, 1984, Moscow: Beijing, Science Press, p. 271-286.

Zhang Qi, Li Dazhou and Zhong Kuiwu, 1989, Preliminary study of Palaeo-Tethyan ophiolites in Hengduan Mountain Region (HMR), China: Jour. Southeast Asian Earth Sci., v. 3, nos. 1-4, p. 249-254.

Zhang Qi, Li Da-zhou and Li Shao-hua, 1981, A pair of metamorphic belts in south Xizang. *In*: Liu Dong-sheng (ed.) Geological and ecological studies of Qinghai-Xizang Plateau, v. 1, Geology, geological history and origin of Qinghai-Xizang Plateau: Beijing, Science Press; and New York, Gordon and Breach, Science Publishers, Inc., p. 443-450.

Zhang Qing-song, Li Bing-yuan, Yang Yi-chou, Yin Ze-sheng and Wang Fu-bao, 1981, Basic characteristics of neotectonic movement of Qinghai-Xizang Plateau. *In*: Liu Dong-sheng (ed.) Geological and ecological studies of Qinghai-Xizang Plateau, v. 1, Geology, geological history and origin of Qinghai-Xizang Plateau: Beijing, Science Press; and New York, Gordon and Breach, Science Publishers, Inc., p. 103-109.

Zhang Qinwen and Huang Huaizeng, 1982, The history of the Mesozoic-Cenozoic structural and magmatic reactivation in eastern China: Acta Geologica Sinica, no. 2, p. 111-122.

Zhang Qinwen, Qu Jingchuan and Chen Bingwei, 1989, Tectonic evolution of the Yangtze tectonic regime. *In*: Sengor, A. M. C. (ed.) Tectonic Evolution of the Tethyan Region: London, Kluwer Academic Publishers, p. 513-549.

Zhang Qiusheng, 1987, Banded iron formations in China. *In*: Appel, P. W. U. and LaBerge, G. L. (eds.) Precambrian iron-formations: Athens, Greece, Theophrastus Publications, S. A., p. 423-448.

Zhang Renjie and Pojeta, J., Jr., 1986, Overview of the Devonian in China. *In*: J. Pojeta, Jr. (ed.) Devonian rocks and Lower and Middle Devonian pelecypods of Guangxi, China, and the Traverse Group of Michigan: U. S. Geol. Survey, Prof. Paper, 1,394 A-G, p. 5-21.

Zhang Shao-quan, Wu Li-jun, Guo Jian-ming, Chen Xue-bo, Zhao Jing-xian, Ding Yun-yu, Huang Chang-lin, Zhang Chang and Chen Zhi-tai, 1985, An interpretation (of) the DS data on Menyuan-Pingling-Weinan profile in west China: Acta Geophysica Sinica, v. 28, p. 460-472.

Zhang Shen, 1988, The Cretaceous in Henan and its relationship to petroleum exploration: Petroleum Exploration and Development, v. 15, no. 1, p. 34-37.

Zhang Shuye, Zhao Gaoshi and Jin Yanren, 1989, Characteristics of the glaucophane schist-eclogite belt in central China and its geological significance: Jour. Southeast Asian Earth Sci., v. 3, nos. 1-4, p. 315-319.

Zhang Wei, 1983, Cambrian system in the southeast margin of Ordos basin: Oil and Gas Geology, v. 4, no. 3, p. 246-253.

Zhang Wen-tang (Wentang) 1986, Correlation of the Cambrian in China: Palaeontologia Cathayana, v. 3, p. 267-285.

Zhang Wentang and Jell, P. A., 1987, Cambrian trilobites of North China; Chinese Cambrian trilobites housed in the Smithsonian Institution: Beijing, Science Press, 459 p.

Zhang Wenyou, 1942, On the incipient-type structure of Duangxi (Guangxi): Geological Review, v. 7, no. 6 (in Chinese).

Zhang Wenyou, 1979, The geological structures and their bearing on the formation and enrichment of iron ores: Lecture to the seminar of the Ministry of Metallurgy (in Chinese).

Zhang Wenyou, chief compiler, 1983, The marine and continental tectonic map of China and its environs: Beijing, China, Science Press, 1:5,000,000, 6 sheets (in Chinese and English editions).

Zhang Wenyou and Zhang Chi, 1984, The characteristic features of faultblock tectonic evolution of China. *In*: Gabrielsen, R. H., Ramberg, I. B., Roberts, D. and Steinlein, O. A. (eds.) Proceedings of the Fourth International conference on Basement Tectonics: internatl. Basement Tectonics Assoc. Publ. 4,p. 193-200.

Zhang Wenyou and Zhong Jiayou, 1977, On the development of fracture systems in China: Scientia Geologica Sinica, 3.

Zhang Wenyou and Zhong Jiayou, 1978, The development of fracture systems in China: Scientific Papers on Geology for International Exchange, 1 (in Chinese).

Zhang (Chang) Wenyou, Li Yinhuai, Zhong Jiayou and Ma Fuchen, 1981, Fracture systems and fault-block tectonics: Scientia Sinica, v. 24, no. 12, p. 1717-1731.

Zhang Wenyou, Xie Mingqian, Yang Zhende and Ma Fuchen, 1984, Evolution of the margin of the continent of China and prospective oil and gas resources: London, World Petroleum Congress, 11th, Proc., 2, p. 127-134.

Zhang Wenyou, He Hong and Zhong Jiayou, 1978, On fault blocks and plates: Scientia Sinica, v. 21, no. 2.

Zhang Wenyou (W. Y. Chang), Ye Hong (Yeh Hung) and Zhong Jiayou (J. Y. Chong), 1979, On 'fault blocks' and 'plates': Scientia Sinica, v. 22, no. 12, p. 1407-1429.

Zhang Wenzhao, 1983, On the formation of oil fields in continental basins in China: Acta Petrolei Sinica, v. 4, no. 3, p. 1-11.

Zhang Xiaolin and Wang Jianmin, 1987, A discussion on the sedimentary facies and possible oil and gas resources in Chuxiong basin and its neighbouring region: Petroleum Exploration and Development, v. 14, no. 6, p. 1-7.

Zhang Xiaoyun and Zhong Jianhua, 1984, On the age of Nenjiang Formation in Songliao basin based on ostracoda: Oil and Gas Geology, v. 5, no. 1, p. 73-76.

Zhang Xingjing, 1983, The Cenozoic stratigraphical sequence of Linhe region in Neiming Autonomous Region: Petroleum Explor. Development, no. 4, p. 1–8.

Zhang Yecheng, Hu Jinjiang and Zhou Ruiliang, 1985, Status of China's geothermal resources and prospects for their development and utilization: China Geology, no. 1, p. 12–22 (in Chinese).

Zhang Yi-gang (Yigang), 1981, Cool shallow origin of petroleum-microbiol. genesis and subsequent degradation: Jour. Petroleum Geology, v. 3, no. 4, p. 427–444.

Zhang Yi-gang and Chen Huan-jiang, 1985, Concepts on the generation and accumulation of biogenic gas: Jour. Petroleum Geology, v. 8, no. 4, p. 405–422.

Zhang Yiwei, 1983, On the formation of dustpan-shaped depressions in the western part of Shandong – a preliminary discussion on wave-like crustal movement: Acta Petrolei Sinica, v. 4, no. 4, p. 19–25.

Zhang Yongxia, 1982, The characteristics of regional geological structure of Tarim basin, China: Acta Geophysica Sinica, v. 25, no. 3, p. 243–251.

Zhang Yongxia and Yang Hua, 1983, Aeromagnetic survey and geological structure of offshore China: Scientia Sinica, ser. B, v. 26, no. 9, p. 996–1008.

Zhang Yongxia, Li Luling, Zhou Fuhong and Wu Qida, 1983, A study on the geological structure of the offshore areas of China: Geological Review, v. 29, no. 2, p. 101–110.

Zhang Yulan, Wang Kaifa, Wang Yongyuan, Wang Jiawen, Wang rong, Wang Congfeng and Qian Shaohua, 1986, Paleogene spore-pollen assemblages of Nanling basin in Anhui Province and their geological significance: Oil and Gas Geology, v. 7, no. 1, p. 75–81.

Zhang Yunping and Tang Kedong, 1989, Pre-Jurassic tectonic evolution of intercontinental region and the suture zone between the North China and Siberian platform: Jour. Southeast Asian Earth Sci., v. 3, nos. 1–4, p. 47–55.

Zhang Yu-quan, Wang Zhong-gang, Zhao Zhen-hua, Xu Yong-hua, Lin Xue-nong and Xie Ying-wen, 1981, On the crustal movement characteristics of the Xizang Plateau with special reference to intermediate-acid magmatic activities. *In*: Liu Dong-sheng (ed.) Geological and ecological studies of Qinghai-Xizang Plateau, v. 1, Geology, geological history and origin of Qinghai-Xizang Plateau: Beijing, Science Press; and New York, Gordon and Breach, Science Publishers, Inc., p. 370–386.

Zhang, Zh. M., Liou, J. G. and Coleman, R. G., 1984, An outline of the plate tectonics of China: Geol. Soc. America Bull., v. 95, no. 3, p. 295–312.

Zhang Zhi-Meng, Jhun G. Liou and Coleman, R. G., 1989, The Mesozoic and Cenozoic tectonism in eastern China. *In*: The evolution of Pacific Ocean margins: New York, Oxford University Press, p. 124–141.

Zhang Zichao, Ma Guogan and Lee Huaqin, 1984, The chronometric age of the Sinian-Cambrian boundary in the Yangtze (Yangzi) platform, China: Geol. Magazine, v. 121, no. 3, p. 175–178.

Zhang Zuqi, 1984, The Permian System in South China: Newsl. Stratigr., v. 13, no. 3, p. 156–174.

Zhang Zuqi, 1987, Carboniferous chronostratigraphy in China: Oil and Gas Geology, v. 8, no. 2, p. 126–137.

Zhao Bai, 1982, The prospects of petroleum exploration of Permo-Carboniferous in Junggar basin: Oil and Gas Geology, v. 3, no. 1, p. 75–80.

Zhao Bai, 1983, Oilfields in Karamay overthrust belt explored: Oil and Gas Geology, v. 4, no. 2, p. 235.

Zhao Bai, 1988, Oil prospect in Tulupan (Turpan or Turfan) basin: Petroleum Expl. and Devlp., v. 15, no. 2, p. 8–11.

Zhao Chenglin, 1988, Preliminary approach on the sedimentary facies of the Wutong formation in Chaoxian county (Chao County), Anhui: Oil and Gas Geology, v. 9, no. 1, p. 40–45.

Zhao Conglun, 1984, On the characteristic types and mechanism of vertical structural variations in Sichuan basin: Acta Petrolei Sinica, v. 5, no. 2, p. 11–21.

Zhao Jungpu, 1984, Ophiolite and continental suture: Scientia Geologica Sinica, no. 4, p. 359–372.

Zhao Songqiao, 1986, *Physical geography of China:* Beijing, Science Press; New York, John Wiley & Sons, Inc. Publ., 209 p.

Zhao Wu-ling and Morgan, W. J., 1985, Uplift of Tibetan Plateau: Tectonics, v. 4, p. 359–369.

Zhao Younian, Lai Xiangfu and Yu Rulong, 1985, A note on Longmenshan nappe: Oil and Gas Geology, v. 6, no. 4, p. 359–368.

Zhao Yushi and Sheng Chunxiang, 1985, The outline of shale diapirs in Damintun depression: Oil Geophysical Prospecting, v. 20, no. 1, p. 58–65.

Zhao Zhen-hua, Wang Yi-xian, Qian Zhi-sin, Zhang Yu-quan and Wang Zhong-gang, 1981, Geochemistry of REE in intermediate-acidic intrusive rocks in southern Xizang. *In*: Liu Dong-sheng (ed.) Geological and ecological studies of Qinghai-Xizang Plateau, v. 1, Geology, geological history and origin of Qinghai-Xizang Plateau: Beijing, Science Press; and New York, Gordon and Breach, Science Publishers, Inc., p. 507–514.

Zhao Zhongyuan, 1984, Structural pattern and evolution of Bohaiwan basin, China: Acta Petrolei Sinica, v. 5, no. 1, p. 1–8.

Zhao Zhongyuan, Guo Zhongming and Hui Binyao, 1984, Hetao arcuate tectonic system and their mechanism of formation and evolution: Oil and Gas Geology, v. 5, no. 4, p. 347–361.

Zhen Wei-Huang, 1983, The geology of industrial minerals and rocks in China: Ontario Geol.Survey Misc. Paper 114, p. 175–179.

Zheng Ben-xing and Li Ji-jun, 1981, Quaternary glaciation of the Qinghai-Xizang Plateau. *In*: Liu Dong-sheng (ed.) Geological and ecological studies of Qinghai-Xizang Plateau, v. 1, Geology, geological history and origin of Qinghai-Xizang Plateau: Beijing, Science Press; and New York, Gordon and Breach, Science Publishers, Inc., p. 1631–1640.

Zheng Jiandong and Zheng Ninghua, 1982, The seismologic and geologic characteristics in Xizang (Tibet) from satellite images: Seismology and Geology, v. 4, no. 4, p. 35–44 (in Chinese, English abstract).

Zheng Xuezheng, 1985, Meso-Cenozoic volcanic rocks in East China and adjacent areas with relation to plate tectonics: Tectonophysics, v. 112, p. 533–550.

Zheng Yan-zhong, 1981, A preliminary investigation on tectonic features in Tanggula-Yushu region, Qinghai. *In*: Liu Dong-sheng (ed.) Geological and ecological studies of Qinghai-Xizang Plateau, v. 1, Geology, geological history and origin of Qinghai-Xizang Plateau: Beijing, Science Press; and New York, Gordon and Breach, Science Publishers, Inc., p. 57–64.

Zheng Yiyi, Xu Kaizhi, Yang Bingzhong, Liu Maoxiu, Yan Lin, Wan Weidong, Liu Wanzhu and Dai Wenzhi, 1984, Geological characteristics of ophiolite-melange in all regions and its relationshi with regional tectonics: Journal of Changchun College of Geology, v. 4, no. 38, p. 29–37 (in Chinese, English abstract).

Zheng Zhi, Lu Daren, Feng Molin, Feng Baohua and Jin Taiquan, 1982, Kaolin deposits of China. *In*: Van Olphen, H. et al. (eds.) Boiogna and Pavia, International clay conference, 7th, 1981: Amsterdam, Elsevier Sci. Publ. Co., Developments in sedimentology, p. 719–731.

Zhou Chaoji and Jiang Jianhang, 1984, The petroleum potential of three major depressions in the Tarim basin: Oil and Gas Geology, v. 5, no. 3, p. 277–284.

Zhou Guoqing, 1989, The discovery and significance of the northeastern Jiangxi Province ophiolite (NEJXO), its metamorphic periododite and associated high temperature-high pressure metamorphic rocks: Jour. Southeast Asian Earth Sci., v. 3, no. 1–4, p. 237–247.

Zhou Jingcai, Yang Hong, Liao Shinan and Wei Xinhua, 1984, Turbidites form the Middle Triassic Jiangdonggou formation in Sanchahe area, southern Guizhou: Oil and Gas Geology, v. 5, no. 4, p. 385–395.

Zhou Jiu, Huang Xinwu and Wang Xuechang, 1983, Gravity tectonics of the Qinghai-Tibetan Plateau: Scientia Geologica Sinica, no. 4, p. 343–354.

Zhou Liangren, Zhao Zhichang and Zhang Jinsheng, 1987, The essential features of geotectonic development and magmatic evolution on the western Junggar area, Sinkiang Uighur, China: Bulletin of the Xi'an Institute of Geology and Mineral Resources, Chinese Academy of Geological Sciences, v. 16, p. 3–55 (in Chinese, English abstract).

Zhou Ming-zhen (Mingzhen), 1981, Vertebrate paleontology in China, 1949–1979: Geological Society of America Special Paper 187, p. 15–20.

Zhou Tongshun and Zhou Huiqin, 1983, Triassic nonmarine strata and flora of China: Bull. Chinese Acad. Geol. Sciences, v. 5, p. 95–110 (in Chinese, English abstract).

Zhou Yongchang, 1985, Some problems concerning correlation of the Pliocene and Miocene series in southwest Tarim basin: Oil and Gas Geology, v. 6, p. 316–325.

Zhou Yun-sheng, Zhang Qi, Jin Cheng-wei and Deng Wan-ming, 1981, The migration and evolution of magmatism and metamorphism in Xizang since Cretaceous and their relation to the Indian plate motion: a possible model for the uplift of Qinghai-Xizang Plateau. *In*: Liu Dong-sheng (ed.) Geological and ecological studies of Qinghai-Xizang Plateau, v. 1, Geology, geological history and origin of Qinghai-Xizang Plateau: Beijing, Science Press; and New York, Gordon and Breach, Science Publishers, Inc., p. 363–378.

Zhou Zhiyi and Dean, W. T., 1989, Trilobite evidence for Gondwanaland in east Asia during the Ordovician: Jour. Southeast Asian Earth Sci., v. 3, nos. 1–4, p. 131–140.

Zhu Ming, 1989, The research on the isotopic ages of Mesozoic porphyries and porphyritic deposits in eastern China: Scientia Geologica Sinica, v. 4, no. 2, p. 190–200.

Zhu Shangqing, 1982, Geologic characteristics of the stratiform phosphate deposits of China: Jour. Wuhan coll. Geol. - Earth Science, v. 16, no. 1, p. 157–166.

Zhu Shixing, 1982, An outline of studies on the Precambrian stromatolites of China: Precambrian Research, v. 18, no. 4, p. 367–396.

Zhu Shuian, Li Chunju, Chen Yongzheng, Gao Yurong, 1983, Sedimentary features of the Shuanghe alluvial fan in the Miyang (Biyang) depression: Acta Petrolei Sinica, v. 4, no. 1, p. 11–16.

Zhu Shuian, Xu Shirong, Zhu Shaobi and Wang Yixian, 1981, Petroleum geological characteristics of the Biyang depression, Henan: Acta Petrolei Sinica, v. 2, no. 2, p. 21–28.

Zhu Songnian, 1984, The vicissitudes of latitudinal tectonic zones of China and their tectonic significance: Wuhan College of Geology, Earth Sciences Journal, no. 3 (total 26), p. 57–70.

Zhu Xia (ed.) 1983, *Tectonics and evolution of the Mesozoic and Cenozoic basins of China*: Beijing, Science Press, 220 p. (in Chinese).

Zhu Xia, 1984, The polycyclic tectonic theory and the evolution of petroliferous basins: Chinese Acad. Geol. Sciences, Bull., no. 9, p. 197–209.

Zhu Xia, Chen Huanjiang, Sun Zhaocai and Zhang Yuchang, 1986, The Mesozoic-Cenozoic tectonics and petroliferous basins of China. *In*: Huang Jiqing (ed.) Proceedings of the symposium on Mesozoic and Cenozoic geology in connection of the 60th anniversity of the Geological Society of China: Beijing, Geological Publishing House, p. 889–894.

Zhu Ying, 1984, Deep rifts and basement seams: rethinking long-range oil and gas forecasts: China Geology, no. 4, p. 14–17.

Zhu Zhicheng, Ji Kecheng and Fan Guangming, 1982, Preliminary research on the thrust wedges in some red basins of south China: Earth Sci. Jour., Wuhan College of Geology, v. 16, no. 3, p. 77–86.

Zhu Zhi-wen, Zhu Xiang-yuan and Zhang Yi-ming, 1981, Palaeomagnetic observation in Xizang and continental drift. *In*: Liu Dong-sheng (ed.) Geological and ecological studies of Qinghai-Xizang Plateau, v. 1, Geology, geological history and origin of Qinghai-Xizang Plateau: Beijing, Science Press; and New York, Gordon and Breach, Science Publishers, Inc., p. 931–939.

Ziegler, A. M., 1985, Paleogeographic evolution of China (abstr.): Amer. Assoc. Petrol. Geol. Bull., v. 69, no. 11, p. 2046–2047.

Zimina, V. G., 1967, O Glossopteris i Gangamopteris iz permskikh otlozheniy Yuzhnogo Primor'ya: Paleontol. Zhurnal, no. 2, p. 113–124.

Zvi Ben-Avraham (ed.) 1989, *The Evolution of the Pacific Ocean margins*: New York, Oxford University Press, 234 p.

Subject Index

MIX
Papier aus verantwortungsvollen Quellen
Paper from responsible sources
FSC® C105338

If you have any concerns about our products,
you can contact us on
ProductSafety@springernature.com

In case Publisher is established outside the EU,
the EU authorized representative is:
Springer Nature Customer Service Center GmbH
Europaplatz 3, 69115 Heidelberg, Germany

Printed by Libri Plureos GmbH
in Hamburg, Germany